Spatial Information Systems

Geospatial data infrastructure

Concepts, cases, and good practice

Edited by

Richard Groot and John McLaughlin

OXFORD
UNIVERSITY PRESS

Great Clarendon Street, Oxford OX2 6DP

Oxford University Press is a department of the University of Oxford.
It furthers the University's objective of excellence in research, scholarship,
and education by publishing worldwide in

Oxford New York
Athens Auckland Bangkok Bogotá Buenos Aires Calcutta
Cape Town Chennai Dar es Salaam Delhi Florence Hong Kong Istanbul
Karachi Kuala Lumpur Madrid Melbourne Mexico City Mumbai
Nairobi Paris São Paulo Singapore Taipei Tokyo Toronto Warsaw

and associated companies in Berlin Ibadan

Oxford is a registered trade mark of Oxford University Press
in the UK and certain other countries

Published in the United States
by Oxford University Press Inc., New York

First published 2000

British Library Cataloguing in Publication Data
Data available

Library of Congress Cataloging in Publication Data
Geospatial data infrastructure : concepts, cases, and good practice / edited by Richard
Groot and John McLaughlin.
p. cm. — (Spatial information systems)
Includes bibliographical references.
1. Digital mapping. 2. Geographic information systems. I. Groot, Richard.
II. McLaughlin, John D. III. Series.
GA139.G46 2000 910'.285—dc21 00-037319

ISBN 0-19-823381-7 (acid-free paper)

1 3 5 7 9 10 8 6 4 2

Typeset by Best-set Typesetter Ltd., Hong Kong
Printed in Great Britain
on acid-free paper by
Biddles Ltd,
www.biddles.co.uk

Foreword

Fostering improved management and protection of scarce living space, introducing effective stewardship for the economic development of our renewable and non-renewable natural resources, improving access to education and health care, and particularly providing facilities for life-long learning, are global issues played out at the level where people live.

Resolution of these issues requires timely and efficient access to reliable geospatial data at affordable cost. Specialists at government policy level and in the private sector must have this access, but so too should the people most directly affected. They need to have access to relevant data and to the capacity to use them in order to participate in finding solutions that work best in their local situation.

To realize the promise of the 'Information Society' to benefit mankind therefore requires an information infrastructure which includes Geospatial Data Infrastructures (GDI). This can be thought of as networked geospatial data service facilities, designed to share available data and make them broadly accessible and available at the lowest possible cost, where and when they are needed. This implies the emergence of an information market-place, *the demands of which will largely determine the content, form and extent of the geospatial data infrastructure*.

The design, development and implementation of such a geospatial data infrastructure is a daunting task. It is not only a matter of technology but also one of designing the institutions, the legislative and regulatory framework, and developing new types of skills. It further calls for a considerable degree of well-disciplined management expertise. In fact the challenge lies in finding the right balance and combination of all of these elements to define and develop the geospatial data infrastructure in a socially and economically justifiable manner. Measuring up to the challenge requires the concerted effort of many different disciplines working in governments, the private sector, and academic institutions.

At the level of the European Community much thought and effort has gone into defining how to organize the conditions for the optimal use of geospatial information in Europe through its GI2000 initiative. Closely in phase with this, and through the co-ordinating efforts of the Dutch Ministry of Housing, Spatial Planning and the Environment (VROM) supported by the Netherlands Council for Geographic Information (RAVI), slowly but surely the institutional environment, foundation data, and data access mechanisms are being put in place.

In this context, I welcome the publication of this book. Written for practitioners, it explains the elements of geospatial data infrastructures and their relationships, and presents recommended practices for their design, development and implementation. I hope this book will enjoy the broadest possible distribution and use.

J. W. Remkes
State Secretary of Housing,
Spatial Planning and the Environment,
The Netherlands

Acknowledgements

The recognition that it is invariably cheaper to use available data than to construct new databases has raised a whole host of issues related to data sharing and data accessibility within the geomatics community. This became an urgent issue when Canadian geomatics agencies at different levels of government were forced in the late 1970s and early 1980s to explore the possibility of exploiting topographical data collected at the provincial levels for mapping programs at the national level. The term 'Geospatial Data Infrastructure' was introduced to promote a better understanding of the complex of technical, organizational, and institutional requirements needed to facilitate effective data sharing. Subsequently the concept of information infrastructure or 'electronic highway' began to receive attention at the highest levels of government, notably in the USA and in the European Community, as something of fundamental importance for realizing the promise of the 'information society'.

Much has been written about this subject since the early 1990s. However, to date there has been no single source for describing all the critical elements in building and sustaining a Geospatial Data Infrastructure. The original idea of publishing this book emerged at the first conference on Global Spatial Data Infrastructure held in the Fall of 1996 in Bonn, Germany. With the encouragement of Peter Burrough we decided to respond to the challenge.

We were especially pleased with the enthusiastic response received from prospective chapter authors. We very much wish to thank them for their contributions, their patience, and their cooperation in the process of putting this book together as coherently as possible. As well, we were fortunate to have the services of a very skilled technical editor, Jackie Senior, who was of enormous assistance in our quest for relevance, balance, and consistency.

We would like to acknowledge the financial and logistical assistance of the International Institute for Aerospace Survey and Earth Sciences (ITC) and the University of New Brunswick throughout this project. Finally, we would like to thank Saskia Tempelman for her reliable secretarial support.

THE EDITORS

Contents

List of Contributors xix
List of Figures xxiv
List of Tables xxvii
List of Abbreviations xxix

1. Introduction
Richard Groot and John McLaughlin

1.1 The challenge 1
1.2 The motivation to write this book 2
1.3 What is geospatial data? 3
1.4 Towards the Geospatial Data Infrastructure concept 3
1.5 Terminology 4
1.6 Direction and contents of the book 6
Bibliography 11

2. Who wants a GDI?
Lance McKee

2.1 Introduction 13
2.1.1 Synthesis of geospatial information with other information 13
2.1.2 Mobile, wireless, and location-aware devices 13
2.1.3 Commerce in geospatial information products and services 14
2.1.4 Geospatial becomes mundane 14
2.1.5 The progress will be chaotic 14
2.2 Two examples 16
2.3 Shaping the inevitable commercial development of the GDI 19
2.4 Effects on society 22
Bibliography 24

Policy framework for GDI

3. GDI from a legal perspective
Jan Kabel

3.1 Introduction 25

3.2 Commercialization of public sector information 27

3.2.1 Pros and cons 27
3.2.2 What guidelines are available? 27
3.2.3 Freedom of information 28
3.2.4 Democratic access rights used for commercial purposes: different views 29
3.2.5 A level playing field: unfair competition and commercialization of public information 30

3.3 Protection of investments in geospatial databases 31

3.3.1 Methods of protection 31
3.3.2 The extraction right 32
3.3.3 The other side: freedom of information versus protection of investments 32

3.4 Privacy and protection of personal data 33

3.4.1 Ratio and scope of protection 33
3.4.2 Principles of protection 35
3.4.3 Specific problems 35
3.4.4 Free flow of personal data 35

3.5 Liability of intermediaries 36

3.6 Conclusion 36

Acknowledgement 37

Bibliography 37

4. Funding an NGDI
David Rhind

4.1 Introduction 39

4.2 The definition of NGDI 39

4.3 The NGDI players 40

4.4 The distribution of expenditures in an NGDI 41

4.4.1 The cost of raw data capture or maintenance 43
4.4.2 The cost of the physical infrastructure 43
4.4.3 The cost of people 43
4.4.4 Other costs 44
4.4.5 So how much *does* NGDI cost currently? 44

4.5 Who pays and how is the money raised? 46
4.6 Different funding models 48
4.6.1 Arguments for cost recovery by public sector bodies 50
4.6.2 Arguments for dissemination of data at zero or copying cost by public bodies 51
4.7 Measuring the benefits of an NGDI 51
4.8 The global GDI 52
4.9 Conclusion 53
Acknowledgements 54
Bibliography 55

5. The role of standards in support of GDI
Peter L. Croswell

5.1 Introduction 57
5.2 Developers of standards 60
5.2.1 Overview of standards organizations 60
National government organizations 60
Independent standards bodies 61
Industry consortia and trade associations 61
Professional organizations 61
5.3 Types of standards impacting GDI 62
5.3.1 Hardware and physical connection standards 62
5.3.2 Communication and network management standards 66
5.3.3 Operating system software standards 68
Core operating system issues 68
System and network management 69
Object management architectures 70
5.3.4 User interface 70
5.3.5 Data format, exchange, and access 71
Geospatial data exchange 71
Attribute data access and exchange 72
5.3.6 Programming and application development standards 73
5.3.7 User design standards 74
Database schemas 74
Geospatial data coding and classification 75
Geospatial metadata 75
Map compilation and map accuracy standards 75
Map presentation standards 77
5.4 A practical context for adoption and implementation of standards 77

5.5 Outline for a geospatial data standards manual 78
I Introduction 79
II Hardware and network standards 79
III System administration standards 79
IV Software and application standards 79
V Data format standards 80
VI Data compilation and update standards 80
VII Product presentation standards 80
VIII System access and data/product distribution standards 80
Bibliography 81

6. Quality management in GDI
Mark Doucette and Chris Paresi

6.1 Introduction 85
6.2 The evolving GDI environment 86
6.3 Integrity of GDI 87
6.3.1 Lineage 88
6.3.2 Consistency 88
6.3.3 Completeness 88
6.3.4 Semantic accuracy 89
6.3.5 Temporal accuracy 89
6.3.6 Positional accuracy 90
6.3.7 Attribute accuracy 90
6.4 Some issues of standards 91
6.5 Users of geospatial data infrastructure 91
6.6 Presentation issues relating to geospatial data infrastructure 93
6.7 The International Organization for Standardization 93
Bibliography 95

7. Anticipating cultural factors of GDI
Willem van den Toorn and Erik de Man

7.1 Introduction 97
7.1.1 The role of culture 98
7.2 Cultural typology of recipient conditions 99
7.2.1 Hofstede's 4D model 99

7.2.2 Potential influence of culture on practices of geospatial data technologies 101
7.2.3 GIS properties 102
7.2.4 Cultural typology of regional groups of countries 103

7.3 Matching recipient culture and GIS properties—an exploration of cultural desirability 103
7.3.1 Culture's impact on the functionality of GIS; social desirability of the technology 104
7.3.2 Desirability of GIS 105

7.4 Feasibility of the introduction and implementation of geospatial data technologies 106

7.5 Conclusion 109

Bibliography 110

Technology framework for GDI

8. The foundation technologies
Wolfgang Kainz

8.1 Introduction 113

8.2 System architecture 113
8.2.1 Stand-alone computers 114
8.2.2 Client–server architecture 114

8.3 Computer networks 115
8.3.1 Transmission media 116
Twisted pair 116
Coaxial cable 116
Fibre optics 116
Wireless transmission 117
8.3.2 Network hardware 118
Local area networks (LANs) 119
Wide area networks (WANs) 119
Internetworks 120
8.3.3 Network software 121
Layer services 121
8.3.4 The OSI reference model 122
Physical layer 122
Data link layer 122
Network layer 123
Transport layer 123
Session layer 124

Presentation layer 124
Application layer 124
8.3.5 The TCP/IP reference model 124
Internet layer 125
Transport layer 126
Application layer 126
8.3.6 Comparison between the OSI and TCP/IP reference models 127
8.3.7 The World Wide Web (WWW) 127
8.4 Database technology 128
8.4.1 Database architectures 129
8.4.2 Data models 130
8.4.3 Distributed databases 131
8.5 Clearinghouses 131
8.6 Conclusion 134
Bibliography 134

9. GDI architectures
Yaser Bishr and Mostafa Radwan

9.1 Introduction 135
9.2 Framework of data sharing 135
9.2.1 Heterogeneity issues 137
9.3 Inter-operability and distributed systems 137
9.3.1 Distributed systems 137
9.3.2 The client–server model 138
9.3.3 Inter-operability defined 139
9.4 Components of GDI architecture 139
9.4.1 Local server 141
9.4.2 Global server 141
9.4.3 Multi-level server 141
9.5 Clearinghouse component 142
9.5.1 Local server 143
9.5.2 The clearinghouse server 144
Graphic user interface (GUI) 145
Metadata 146
Content of metadata 146
9.6 Geospatial Data Service Centre 148
9.7 Conclusion 149
Bibliography 149

10. Conceptual tools for specifying geospatial descriptions

Martien Molenaar

10.1 Introduction 151

10.2 Geographic phenomena and geospatial data modelling 152

10.2.1 Fields, objects, and geometry 152

10.3 Fields and rasters 154

10.3.1 The geometry of rasters 155

10.4 The geometry of geospatial objects 156

10.4.1 Objects represented in rasters 157

10.4.2 The vector structure 158

10.5 Object hierarchies 160

10.5.1 Terrain object classes and generalization hierarchies 161

10.5.2 Aggregation hierarchies 165

10.5.3 Object associations 167

10.6 Object dynamics 167

10.7 Object definitions and context 169

10.8 Conclusion 171

Bibliography 172

Data acquisition and display

11. Spatial referencing

Marco Hofman, Erik de Min, and Ruben Dood

11.1 Introduction 175

11.2 Datums 176

11.2.1 The shape of the Earth 176

11.2.2 Defining a local datum 177

11.2.3 Height references 177

11.2.4 Map projections 178

Introduction 178

Distortion 178

Choosing a projection 180

11.2.5 Other reference systems 181

11.3 The realization of a datum and a summary of geospatial data acquisition methods 181

11.3.1 Realizing the datum 181

11.3.2 Summary of geospatial data acquisition methods 182

11.3.3 Dynamic positioning 182

11.3.4 Towards GPS and a new positioning infrastructure 182

11.3.5 Practical use of GPS 183
11.3.6 Global Positioning System 183
11.3.7 Standard Positioning Service 184
11.3.8 Pseudo-range differential GPS, and the new dynamic spatial referencing infrastructure 184
11.3.9 Carrier wave phase difference 185
11.3.10 Dynamic positioning 185
11.3.11 GLONASS 185
11.3.12 Conclusions on GPS 186

11.4 Combining data sets 186

11.4.1 The problem 186
11.4.2 Description of the cases 187
Case I. Both datums are known, ellipsoids can be transformed into each other 187
Case II. Both datums are known 188
Case III. At least one datum is unknown, and there are corresponding points 189
11.4.3 Choosing a datum transformation 192
11.4.4 Heights 192
Combining or connecting height data 192

11.5 Conclusion 193

Bibliography 194

12. Photogrammetry and remote sensing in support of GDI

Gottfried Konecny

12.1 Introduction 195

12.2 Remote sensing and photogrammetry 195

12.3 Geospatial data infrastructure in relation to photogrammetry and remote sensing 196

12.4 Sensors and platforms 198

12.5 Vector data acquisition by map digitizing and photogrammetry 202

12.6 Raster data acquisition by photogrammetry 206

12.7 Digital elevation models by photogrammetry 207

12.8 Thematic data acquisition by remote sensing 211

12.9 Cadastral photogrammetry 212

12.10 Cost considerations 213

12.11 Conclusion 215

Bibliography 216

13. Access to GDI and the function of visualization tools
Menno-Jan Kraak

13.1 Introduction 217
13.2 Geospatial data before use 219
13.2.1 Maps as indexes 220
13.2.2 Maps for preview 221
13.2.3 Maps to search 222
13.3 Geospatial data during use 223
13.4 Geospatial data after use 224
13.5 Maps at work: online integration 227
13.6 Conclusion 230
Bibliography 230

Human Resources Issues

14. Human resources issues in the emerging GDI environment
David Coleman, Richard Groot, and John McLaughlin

14.1 Introduction 233
14.2 The challenges 234
14.2.1 Economic pressures 234
14.2.2 Technological advances 235
14.2.3 Management and institutional issues 236
14.3 Alternative approaches to meeting the challenges 237
14.3.1 Undergraduate degree courses 237
14.3.2 Trends in post-graduate courses at universities and other degree-granting international institutions 238
14.3.3 College diploma courses 239
14.4 Discussion 240
14.5 Conclusion 242
Bibliography 243

15. Four cases
David Finley, Karen Siderelis, Don Grant, and Arnold Bregt

15.1 Service New Brunswick: the modernization of a land information service 245
David Finley
15.1.1 Background information 245
15.1.2 Developing the information infrastructure and building the database 246

15.1.3 Standards 247
15.1.4 Management structure 248
15.1.5 Financial structure 249
15.1.6 Stakeholder involvement 250
15.1.7 Measures of success 250
15.1.8 Summary 250

15.2 Geospatial data infrastructure in North Carolina 251
Karen Siderelis

15.2.1 Introduction 251
15.2.2 Management structure 252
15.2.3 Stakeholders 253
15.2.4 Purpose of the geospatial data infrastructure 253
15.2.5 Costs and financing 254
15.2.6 Success factors 255

15.3 The Public Sector Mapping Agencies of Australia 255
Don Grant

15.3.1 Meeting a national need 255
15.3.2 Task management 258
15.3.3 Previous mapping procedures 260
15.3.4 Future applications 260
15.3.5 Conclusion 262

15.4 The Dutch clearinghouse for geospatial information: cornerstone of the national geospatial data infrastructure 262
Arnold Bregt

15.4.1 Introduction 262
15.4.2 Pioneer period (1995–1996) 263
15.4.3 Project period (1996–1997) 264
15.4.4 Institutionalizing period (1998–) 265
15.4.5 Conclusion: A personal view 266
Bibliography 267

16. Advancing the GDI concept
John McLaughlin and Richard Groot

16.1 Introduction 269
16.2 Progress to date 269
16.3 Lessons learned 272
16.4 Concluding remarks 273
Bibliography 274

Index 277

List of Contributors

Yaser A Bishr, Dr.
Assistant Professor at the Institute for Geoinformatics, University of Muenster, Germany. Research interests: semantic inter-operability in distributed databases; intelligent agents; mobile geospatial computing; and workflow and performance analyses for the design of component-based architecture.
Institute for Geoinformatics, University of Muenster, Robert-Koch-Str. 26–28, D-48149 Muenster, Germany
Tel: +49 251 83 33724; Fax: 49 251 83 39763; E-mail: bishr@ifgi.uni-muenster.de

Arnold K Bregt, Dr.
Professor of Geo-information Science at the Wageningen University and Research Center in The Netherlands. Following more than 15 years of experience in the field of GIS research and applications, his current areas of interest are spatial data quality, dynamic modelling of land use change, and spatial data infrastructures. From 1996 to 1998 he was one of the project leaders for the development of the national clearing-house geo-information in The Netherlands.
Wageningen-UR, Centre for Geo-information, PO Box 47, 6700 AA Wageningen, The Netherlands
E-mail: arnold.bregt@staff.girs.wag-ur.nl

David J Coleman, Ph.D.
Professor and Chair of the Department of Geodesy and Geomatics Engineering at the University of New Brunswick. Research interests: land information management; geomatics operations management and production workflow; GIS performance in network environments; and technical and policy aspects of spatial data infrastructure implementation.
Department of Geodesy and Geomatics Engineering, University of New Brunswick, PO Box 4400, Fredericton, New Brunswick, E3B 5A8, Canada
Tel: +1 506 453-5194; Fax: +1 506 453-4943; E-mail: dcoleman@unb.ca

Peter L Croswell
Executive Consultant, PlanGraphics, Inc., a GIS consulting firm headquartered in Frankfort, Kentucky, USA. Manages GIS design and development projects for government agencies and utility companies in North America. Emphasizes the appropriate adoption of open system standards in the implementation of spatial information management systems.

PlanGraphics, Inc., 112 East Main Street, Frankfort, KY 40601, USA
Tel: +1 502 223-1501; Fax: +1 502 223-1235; E-mail: pcroswell@plangraphics.com

WH Erik de Man, Dr.
Assistant Professor, Social Sciences Division – International Institute for Aerospace Survey and Earth Sciences (ITC), The Netherlands. Research interests: culture, institutions, and geographic information technology; socio-technical approaches to GIS; geographic information management; and local governance.

ITC, Social Sciences Division, PO Box 6, 7500 AA Enschede, The Netherlands
Tel: +31 53 4874231; Fax: +31 53 4874399; E-mail: deman@itc.nl

Erik J de Min, Dr.
Senior Consultant at the Research and Consultancy Group on Positioning of the Survey Department Rijkswaterstaat. Research interests: quality description of spatial data sets; spatial data analysis; geostatistics; positioning techniques; integration of measurements and modelling; and use of quality information in GIS.

Survey Department of Rijkswaterstaat, Ministry of Transport, Public Works and Water Management, Kanaalweg 4, PO Box 5023, NL – 2600GA, Delft, The Netherlands
Tel: +31 15 2691 467; Fax: +31 15 2135 456; E-mail: e.j.dmin@mdi.rws.minvenw.nl

Ruben GW Dood, Ir.
Consultant at the GIS Consultancy and Research group of the Survey Department of Rijkswaterstaat.
Survey Department of Rijkswaterstaat Ministry of Transport, Public Works and Water Management, Kanaalweg 4, PO Box 5023, NL – 2600GA, Delft, The Netherlands
Tel: +31 15 2691 477; Fax: +31 15 2618 962; E-mail: r.g.w.dood@mdi.rws.minvenw.nl

Mark E Doucette, P. Eng.
President of DataQC, Inc., which provides a wide range of quality control and project management services. DataQC designs and implements quality control procedures for the creation, upgrading, maintenance, and archiving of digital data as well as developing both specifications for data processing and procedures to test for compliance with customer-supplied data specifications.

DataQC, 301 Woodstock Road, Fredericton, New Brunswick, E3B 2H9, Canada
Tel: +1 506 444-8142; Fax: +1 506 444-8125; E-mail: dataqc@nbnet.nb.ca; Web-site: WWW.dataqc.com

David B Finley, M. Eng.
Project Manager – Development, Service New Brunswick. Current interests: web-based GIS and data integration; land information management; and land information infrastructure development.

Service New Brunswick, PO Box 1998, Fredericton, New Brunswick, E3B 5G4, Canada
Tel: +1 506 444-4644; Fax: +1 506 453-3898; E-mail: david.finley@snb.gov.nb.ca

Don Grant, A.M.
Professor and Surveyor-General, New South Wales, Australia. Research interests: land administration; mapping; and project management.

Box 1599, Bathurst NSW 2795, Australia
Tel: +61 63 326-026; Fax: +61 63 322-320; E-mail: grantd@lic.gov.au

Richard Groot, Ir.
Professor of Geoinformatics, Management and Infrastructure, International Institute for Aerospace Survey and Earth Sciences (ITC). Interests: Institutional and organizational aspects of providing routine, purposeful supply of geospatial data; regulatory reform of National Mapping Agencies.

Division of GMI, ITC, PO Box 6, 7500 AA, Enschede, The Netherlands
Tel: +31 53 487 4333; Fax: +31 53 487 4335; E-mail: groot@itc.nl

Marco Hofman, Ir.
Former head of the research and consultancy group on Positioning of the Survey Department Rijkswaterstaat, currently head of the Traffic Safety Division of the Transport Research Centre of Rijkswaterstaat. Research interests: relation of positioning techniques and geodetic computations with GIS; innovation of observation techniques for positioning and deformation.

Transport Research Centre of Rijkswaterstaat, Ministry of Transport, Public Works and Water Management, Boompjes 200, PO Box 1031, 3000BA Rotterdam, The Netherlands
Tel: +31 10 282 5702; Fax: +31 10 282 5646; E-mail: m.hofman@avv.rws.minvenw.nl

Jan Kabel, Mr.
Professor of Information Law. Universities of Amsterdam and Utrecht. Main interest is in information law in the EU with some concentration on geospatial information access and supply.

Institute for Information Law, Rokin 84, 1012 KX, Amsterdam, The Netherlands
Tel: +31 20 525 3452; FAX: +31 20 525 3033; E-mail: kabel@jur.uva.nl; Web: http://www.ivir.nl/medewerkers.kabel.html

Wolfgang Kainz, Ph.D.
Professor and Chair of Geoinformatics, Spatial Information Theory and Applied Computer Science, International Institute for Aerospace Survey and Earth Sciences (ITC), Enschede, The Netherlands. Research interests: theory of GIS; representation of spatial features and relationships; spatial data quality; topology; ordered sets; and fuzzy logic.

International Institute for Aerospace Survey and Earth Sciences (ITC), PO Box 6, 7500 AA Enschede, The Netherlands
Tel: +31 53 4874434, Fax: +31 53 4874335; E-mail: kainz@itc.nl

Gottfried Konecny, Ph.D.
Emeritus Professor, University of Hannover. Current interests: consulting in geoinformatics.

Institute for Photogrammetry and Engineering Surveys, University of Hannover, Nienburger Strasse 1, D 30 167 Hannover, Germany
Tel: +49 511 762-2487; Fax: +49 511 762-2483; E-mail: gko@ipi.uni-hannover.de

Menno-Jan Kraak, Dr.
Professor in Geoinformatics, Cartography and Visualization at ITC. Research interests: web mapping; exploratory cartography; visualization of 3D; and temporal geographic phenomena.

Division of Geoinformatics, Cartography and Visualization, International Institute for Aerospace Survey and Earth Sciences (ITC), PO Box 6, 7500 AA Enschede, The Netherlands
Tel: +31 53 4874463; Fax: +31 53 4874335; E-mail: kraak@itc.nl; Web-site: http://www.itc.nl/~kraak

Lance McKee
Vice President, Corporate Communications and Public Sector Programs, Open GIS Consortium, Inc. Interested in the ways that OGC's enablement of inter-operability will affect the markets for geodata, geoprocessing software, and geoservices, as well as the overall effect on society of increased web access to geospatial information services.

Open GIS Consortium, Inc., 35 Main Street, Wayland, MA 01778, USA
Tel: +1 508 655-5858; Fax: +1 508 655-2237; E-mail: lmckee@opengis.org; Web-site: http://www.opengis.org

John McLaughlin, Ph.D.
Vice-President (Research & International Cooperation). Interests: land administration; land information management; and property studies.

University of New Brunswick, Fredericton, New Brunswick, E3A 5A3, Canada
Tel: +1 506 453-5189; Fax: +1 506 453-3522; E-mail: jdm@unb.ca; Web-site: http://www.unb.ca

Martien Molenaar, Dr.
Professor of Geoinformatics, Spatial Data Acquisition, International Institute for Aerospace Survey and Earth Sciences (ITC). Research interest: spatial object modelling in GIS.

Division of Spatial Data Acquisition, ITC, PO Box 6, 7500 AA, Enschede, The Netherlands
Tel: +31 53 487 4269; Fax: +31 53 487 4335; E-mail: molenaar@itc.nl

Chris M Paresi, M.Sc.
Assistant Professor, Division of Geoinformation Management and Infrastructure at the ITC, The Netherlands. Research interests: land information management; structured information systems development and maintenance; business process re-design; workflow management; and quality management.

Division of Geoinformation Management and Infrastructure, ITC, PO Box 6, 7500 AA Enschede, The Netherlands
Tel: +31 53 4874339; Fax: +31 53 4874335; E-mail: paresi@itc.nl

Mostafa Radwan, Ph.D.
Associate Professor, Division of Geoinformatics Management and Infrastructure, International Institute for Aerospace Survey and Earth Sciences (ITC). Research interests: business process redesign of geoinformatics organisations, workflow management, GDI architecture.

Division of GMI, ITC, PO Box 6, 7500 AA, Enschede, The Netherlands
Tel: +31 53 487 4351; Fax: +31 53 487 4335; E-mail: radwan@itc.nl

David Rhind, Ph.D., D.Sc., FRICS, FRGS
Vice-Chancellor and Principal, City University, London, UK and Chairman of the UK National Geospatial Data Framework. Formerly Director General and Chief Execu-

tive, Ordnance Survey. Research interests: information policy; knowledge creation and transfer; and globalization of universities. He has worked in GIS and computer mapping since 1968.

City University, London, Northampton Square, London ECIV OHB
Tel: +44 20 7477 8002; Fax: +44 20 7477 8596; E-mail: d.rhind@city.ac.uk

Karen Siderelis
Director, Center for Geographic Information and Analysis, Government of North Carolina. Research interests: state-wide GIS development; spatial data infrastructure implementation.

CGIA, 115 Hillsborough Street, Raleigh, NC 27608, USA
Tel: +1 919 733-2090; Fax: +1 919 715-0725; E-mail: karen@cgia.state.nc.us

Willem H van den Toorn, Dr.
Professor of Sustainable Land Resource Planning and Management. Head, Social Science Division – International Institute for Aerospace Survey and Earth Sciences (The Netherlands). Research interests: decision and planning support systems; conflict analysis and management; and culture, institutions, and Information Technology.

ITC—Social Science Division, PO Box 6, 7500 AA Enschede, The Netherlands
Tel: +31 53 4874527; E-mail: toornwh@itc.nl

List of Figures

1.1: Role of a Geospatial Data Service Centre (GDSC) 5
2.1: The tension field in geospatial information policy 15
4.1: Costs and revenues involved in running the Ordnance Survey 47
4.2: Level of funding to national mapping agencies not directly voted by Parliament or Congress 48
5.1: Primary themes of information technology standards 59
5.2: Desktop operating system wars 69
5.3: Geospatial data exchange and inter-operability 71
5.4: Example of metadatabase content 76
8.1: Stand-alone computers without co-operation 114
8.2: Client–server architecture 115
8.3: Coaxial cable 116
8.4: Fibre cable 117
8.5: Network topologies 119
8.6: Wide area network 120
8.7: Layers, protocols, and interfaces 121
8.8: ISO/OSI reference model 123
8.9: ISO/OSI and TCP/IP layers 125
8.10: Database architecture 129
8.11: Models in database design 130
8.12: Top–down distributed database design 132
8.13: Federated database design 133
8.14: Clearinghouse concept 133
9.1: Components of GDI 138
9.2: Distribution possibilities 139
9.3: Inter-operability requires mutual understanding of requests and responses 140
9.4: Example architecture of GDI with a decision-support system at three levels 140
9.5: Conceptual system architecture for the proposed national clearinghouse 142
9.6: Clearinghouse local server 143
9.7: Clearinghouse server 145
9.8: Levels of detail for metadata 147
9.9: Meta-metadata system architecture 148
10.1: Two basic structures for spatial data 153
10.2: Several configurations for linking attribute values to position 154
10.3: The geometric definition of rasters 155

10.4: Three geometric object types 157
10.5: Two geometric structures for terrain objects 158
10.6: Objects represented in a raster 159
10.7: Geometric elements for a vector-structured terrain representation 160
10.8: Several structures for representing a terrain situation in a vector geometry 160
10.9: Several topologic relationships between spatial objects 161
10.10: Tables for two classes of agricultural land use 162
10.11: Diagram representing the relationship between objects, classes, and attributes 163
10.12: Table structure extended for new thematic attributes 164
10.13: Class hierarchy for agricultural objects 164
10.14: The aggregation of agricultural objects 166
10.15: Examples of object associations 168
10.16: Geometric object changes 169
10.17: Changes of the collections of objects 169
11.1: Latitude and longitude 176
11.2: Geoid—ellipsoid—map 178
11.3: Preservation of distance 179
11.4: Conformal projection 179
11.5: Equidistant projection 180
11.6: Equal area projection 181
11.7: Levelling 183
11.8: Fitting new data into old 187
11.9: Lambert and Gauss-Krüger 188
11.10: 2D similarity transformation 190
11.11: Rubber sheeting 191
12.1: Principles of remote sensing 196
12.2: Principles of photogrammetry 197
12.3: Classical and modern geospatial information system 197
12.4: Components of a modern geospatial information system 198
12.5: Aerial survey camera 199
12.6: Electro-mechanical and electro-optical scanners 200
12.7: Resolution and repetitiveness of remote sensing missions 203
12.8: KVR 1000 (DD5) image of Hanover 205
12.9: GIS extracted buildings 209
12.10: Image correlation results for buildings 210
12.11: Automatic image classification 211
12.12: Automatic change detection 213
13.1: The cartographic visualization process 218
13.2: Cartographic visualization strategies, thinking, and communication 219
13.3: The map as an index to other geospatial data 220
13.4: Maps as a preview to the data to be obtained 221
13.5: The map as a location component of a spatial data search engine 223
13.6: Cartographic exploratory tools 225
13.7: Synthesis of the modern mapping process 226
13.8: Relationship between the providers of foundation, framework, and application data, the clearinghouse, and the national atlas 228
14.1: Education and skills framework for individuals working in GDI 240

15.1.1: New Brunswick's electronic *Land Gazette* 248
15.1.2: Types of land information/agency responsibility (after Simpson, 1990) 249
15.2.1: Management structure of North Carolina's GDI 252
15.3.1: The public sector mapping agencies (PSMA) 256
15.3.2: Extracts from PSMA data description tables showing variations in data characteristics for urban roads across all jurisdictions 257
15.3.3: PSMA data categories across Australia 258
15.3.4: PSMA feature list for the census mapping project database 259
15.3.5: Sample of the PSMA data set 261
15.4.1: Project structure in phase 2 264
15.4.2: Search screen of the National Clearinghouse for Geo-Information 265

List of Tables

1.1 Selecting framework data for NGDI 9
3.1 The main legal issues 26
3.2 International legal instruments currently in force 26
4.1 Some possible agendas of the various 'players' and stakeholders in a national GDI 42
5.1 Types of computing standards with selected examples 63
5.2 Layers of the ISO Open Systems Interconnect model 67
7.1 Country cultural typology of recipient conditions 104
7.2 Connotations of cultural indicators *vis-à-vis* functional role 107
7.3 Combined influence of desirability and feasibility on the introduction and implementation of GIS/GDI 108
12.1 High-resolution missions 204
12.2 Status of mapping in the world (1993) 206
12.3 Annual update rate of maps 206
12.4 Summary of costs for technical co-operation projects 214
12.5 Prices of digital map data (Lower Saxony, Germany) 215

List of Abbreviations

ACM	Association for Computing Machinery
AM/FM	Automated Mapping and Facilities Management
ANSI	American National Standards Institute
API	Application Programming Interface
ARPANET	Advanced Research Projects Agency Network (US Ministry of Defense)
ATKIS	Amtliches Topographisch Kartographisches Informationssystem
ATM	Asynchronous Transfer Mode
B-ISDN	Broadband ISDN
CCD	Charged Coupled Device
CDE	Common Desktop Environment
CDPD	Cellular Digital Packet Data
CEC	Commission of the Economic Communities
CEN	Comité Européen de Normalisation (European Committee for Standardization)
CERCO	Comité Européen des Responsables de la Cartographie Officielle
CGI	Common Gateway Interface
CGSB	Canadian General Standards Board
CICG	Canadian Inter-agency Committee on Geomatics
CIG	Canadian Institute of Geomatics
CMIP	Common Management Information Protocol
COM	Common Object Model
CORBA	Common Object Request Broker Architecture
CSDGM	FGDC's Content Standard for Digital Geospatial Metadata
DB	Database
DBS	Data Base Server
DBMS	Database Management System
DCDB	Digital Cadastral Data Base
DCE	Distributed Computing Environment
DCM	Digital Cartographic Model
DCOM	Distributed Common Object Model
DCW	Digital Chart of the World
DDL	Data Definition Language
DEM	Digital Elevation Model
DGIWG	Digital Geographic Information Working Group (sponsored by NATO)
DIGEST	Digital Geographic Exchange Standard
DLM	Digital Landscape Model
DML	Data Manipulation Language

DMR	Department of Main Roads (now known as RTA, Roads and Traffic Authority)
DNS	Domain Name Service
DVD	Digital Versatile Disk
EC	European Community
EGNOS	European Geostationary Navigation Overlay Service
EIA	Electric Industries Association
EPPI	Electronic Procurement Program Interface
ERIM	Environmental Research Institute of Michigan
ERS	European Resource Satellite (also Radarsat, JERS)
ETAC	Electronic Technical Assistance Center
EU	European Union
EUROGI	European Umbrella Organization for Geographic Information
FGDC	Federal Geographic Data Committee, USGS
FTAM	File Transfer Access Management Protocol (ISO 8571–1, 2, 3, 4, 5)
FTP	File Transfer Protocol
GDI	Geospatial Data Infrastructure
GDSC	Geospatial Data Service Centre
GGDI	Global Geospatial Data Infrastructure
GII	Geospatial Information Infrastructure
GIS	Geographic Information Systems
GLONASS	GLObal NAvigation Satellite System
GNSS	Geostationary Navigation Satellite Service
GPS	Global Positioning System
GSDI	Geospatial Data Infrastructure /Global Spatial Data Infrastructure
GUI	Graphical User Interface
HTML	Hypertext Markup Language
HTTP	Hypertext Transfer Protocol
IAAO	International Association of Assessing Officers
IAPRS	International Archives for Photogrammetry and Remote Sensing
IEEE	Institute of Electrical and Electronic Engineers
IGES	Interactive Graphics Exchange Standard
IGN	Institut Géographique National
IMW	International Map of the World at the one to one millionth scale
IP	Internet Protocol
ISDN	Integrated Services Digital Network
ISO	International Organization for Standardization
ITC	International Institute for Aerospace Survey and Earth Science, Enschede, The Netherlands
ITU	International Telecommunications Union
LAN	Local Area Network
LIN	Land Information Network
LIS	Land Information System
MADTRAN	Mapping Datum Transformation Programs (National Imaging and Mapping Agency)
MAN	Metropolitan Area Network
MSAS	Multi-(functional transport) Satellite-based Augmentation Service (Japan)
MSC	Mapping Science Committee, National Research Council, Washington DC

MSL	Mean Sea Level
NAD	North American Datum
NAPA	National Academy of Public Administration
NAS	National Atlas Standard
NATO	North Atlantic Treaty Organization
NAVD	North American Vertical Datum
NFS	Network File System
NGDC	National Geospatial Data Clearinghouse
NGDI	National Geospatial Data Infrastructure
NGII	National Geographic Information Infrastructure
NGS	National Geodetic Survey
NII	National Information Infrastructure
NIMA	National Imagery and Mapping Agency
N-ISDN	Narrowband ISDN
NIST	National Institute of Standards and Technology (USA)
NNTP	Network News Transfer Protocol
NOAA	National Oceanographic and Atmospheric Administration (USA)
NSDI	National Spatial Data Infrastructures
NSFNET	American National Science Foundation Network
ODBC	Open Database Connectivity
OGC	OpenGIS Consortium
OMG	Object Management Group
OSI	Open Systems Interconnection
PID	Property identifier
PPS	Precise Positioning Service (GPS)
RAID	Redundant Array of Independent Disk
RAVI	Dutch Advisory Council on Spatial Information
RDBMS	Relational DataBase Management System
SA	Selective Availability (time and orbit parameter errors in GPS)
SAR	Synthetic Aperture Radar
SCDB	Survey Control Data Base
SDI	Spatial Data Infrastructure
SDTS	Spatial Data Transfer Standard (USA)
SeQueL/SQL	Structured Query Language
SLAR	Sideways-Looking Airborne Radar
SLM	Sustainable Land Management
SMTP	Simple Mail Transport Protocol
SNMP	Simple Network Management Protocol
SPOT	Système Probatoire de l'Observation de la Terre
SPS	Standard Positioning Service (GPS)
TCP	Transmission Control Protocol
TDB	Topographic Data Base
TIN	Triangular Irregular Networks
TSO	The Stationery Office (London, UK)
UDP	User Datagram Protocol
UNCED	United Nations Commission on Environment and Development
URISA	Urban and Regional Information Systems Association
URL	Uniform Resource Locator
USGS	US Geological Survey
UTM	Universal Transverse Mercator map projection

UTP	Unshielded Twisted Pair
WAAS	Wide Area Augmentation Service (USA)
WAN	Wide Area Network
WGS	World Geoditic System
WWW	World Wide Web
XML	Extensible Markup Language

1

Introduction

Richard Groot and John McLaughlin

1.1 The challenge

Over the last three decades, governments and the private sector in the industrial world have invested tens of billions of dollars in the development of geospatial information systems designed largely to serve specific communities (forestry, urban planning, land records management, business geographics, etc.) within a local, national, and even international framework. In the USA, for example, the Mapping Sciences Committee of the National Academy of Sciences reported that annual federal spending on geospatial data activities was in the order US$4.4 billion (Mapping Sciences Committee, 1994).

Now the focus is increasingly shifting to the challenges associated with integrating these systems, building what has come to be called geospatial data infrastructure (GDI). Such infrastructures have been described as information highways, linking environmental, socio-economic and institutional databases ('horizontal highways'), and providing for the flow of information from local to national levels and eventually to the global community ('vertical highways') (Coleman and McLaughlin, 1997).

Proponents of the concept often envision a data infrastructure as being analogous to networks of national highways or vast electric power grids (Branscomb, 1982). The highway metaphor conveys the notion of new or heightened connections to different locations. Further, as with electric power grids, the physical location of the data source is usually less important to the end user than the continuing availability, utility, reliability, and cost of the data itself (Anderson, 1990). However, while these metaphors have contributed to promoting a shared vision of the proposed networking effort, it has been a third metaphor—that of the information market-place (i.e. envisioning all those applications that could be built upon a completed infrastructure)—which has attracted increasing imagination and support. The high expectations in terms of the social and economic benefits to society have been articulated at top political levels, for example, in CEC (1989), CEC (1994), Clinton (1994), and in CEC (1998).

The proposed development of national geospatial data infrastructures (NGDI) has received considerable attention from institutional suppliers, the private sector, and user communities in Canada (McLaughlin, 1991), the United States (Mapping Sciences Committee, 1993; Tosta, 1994), Europe (Brand, 1995; EUROGI *et al.*, 1996), and elsewhere (Onsrud, 1999).

Evolving from earlier data sharing and programme co-ordination efforts, the infrastructure concept has come to encompass the sources, systems, network linkages, standards, and institutional issues involved in delivering spatially related data from many different sources to the widest possible group of potential users at affordable costs.

1.2 The motivation to write this book

In September 1996 an international forum on the 'Emerging Global Geospatial Data Infrastructure' was organized in Bonn under the auspices of the European Umbrella Organization for Geoinformation (EUROGI), the German Umbrella Organization for Geoinformation, the Atlantic Institute, the Open GIS Consortium, the US Federal Geographic Data Committee, and the International Federation of Surveyors. This represented one of the first initiatives to address the challenges and opportunities in building geospatial data infrastructures in a systematic and comprehensive fashion (EUROGI *et al.*, 1996). The forum brought together some very diverse communities: traditional surveying and mapping agencies, telecommunications and networking specialists, information retailers, etc. And this diversity brought with it very different ideas as to what should constitute a global geospatial data infrastructure (GGDI), what the priorities should be in constructing such an infrastructure, who the ultimate customers would be in the resulting information marketplace, as well as to what would be the appropriate roles of the public and private sectors in building and managing the infrastructure. A whole host of institutional and cultural concerns were identified, but, without a common framework or language, it was virtually impossible to articulate an agenda to address them in any meaningful fashion.

The conference has followed through with its agenda in subsequent meetings in 1997 in Chapel Hill, North Carolina (USA) and 1998 in Canberra (Australia). See *http://www.eurogi.org/gsdi*.

However, immediately following the first conference in 1996, the authors concluded that there was a real and immediate need for a publication which could contribute to the process of developing a shared understanding of the nature of, and need for, geospatial data infrastructures. The goal would be: (1) to lay out the key elements of the infrastructures, (2) elaborate on designing, building and sustaining such infrastructures, and (3) identify some of the best practices associated with GDI implementation. While clearly overly ambitious, we hope that this book will go some way towards serving this need, and that it will be of value both to practitioners and students.

1.3 What is geospatial data?

Geographic information, geographic data, spatially related data, spatial information, and geospatial data, are all terms used more or less interchangeably. For the purpose of this book we have decided to use the term 'geospatial data', meaning data defined spatially (in location) by four dimensions (geometry and time) related to the Earth. Geospatial data is spatially referenced in some consistent manner, for example by means of latitude and longitude, a national co-ordinate grid, postal codes, electoral or administrative areas. It often has a temporal dimension as well when it signifies how some feature has changed over time. Examples of geospatial data collected by government organizations (including national mapping agencies, census and national statistical agencies, etc.) include topographic, hydrographic, Earth and atmospheric science data, soil and forest survey inventories, property and boundary records, and a large variety of social and economic data such as population characteristics, labour force surveys, etc. Governments use this data for their own purposes in legislative and policy development for the allocation and management of natural resources, for defence and public safety purposes, in support of a variety of regulatory activities, and in promoting a better understanding of the physical, economic and human geography of the nation. A vast array of private agencies also collect geospatial data for a wide variety of commercial, social and environmental applications.

A key characteristic of geospatial data is its potential for multiple applications. Especially since the advent of Geographic Information Systems (GIS) technology, it has become much easier to correlate and integrate different data sets to provide new and useful insights into the interaction of many geographic phenomena. For example, recent advances in understanding the need for, and approaches to, sustainable development (addressing concurrently the economic and environmental dimensions of resource allocation and management) have benefited significantly from the applications of GIS technology. A necessary but not sufficient condition for this integration is that the data must be collected and spatially referenced in a consistent manner. See for example Smith and Rhind (1999).

1.4 Towards the Geospatial Data Infrastructure concept

Beginning in the late 1970s many national surveying and mapping agencies recognized the need to create strategies and processes for standardizing the access to, and applications of, geospatial data. Initially, the requirements for standards were seen in narrow technical terms. Over time, however, the scope of the resulting inquiries was expanded to incorporate a host of institutional and organization issues. Examples included the Major Surveys Review in the

Federal Government of Canada (Canadian Government, 1986), the Report on the Handling of Geographic Information of the Government of Great Britain (Department of the Environment, 1987), the report of the National Research Council of the USA (Mapping Science Committee, 1990), the Netherlands Council for Geographic Information (RAVI, 1995), and similar reports from Australia and New Zealand (ANZLIC, 1996).

The notion of a data infrastructure as a mechanism for providing more effective access to geospatial data first emerged in the early 1980s in Canada. The federal and provincial surveying and mapping organizations were under intense pressure at that time to justify the huge investments that had been made in constructing geospatial databases and to develop new business cases for sustaining their activities (with reducing duplication a priority). The infrastructure concept had immediate intuitive appeal, appearing to provide a useful way of moving beyond the focus on computer and communication technology per se to providing a framework to address more comprehensively the wide array of technical, legal, financial, organizational, and other issues at stake in providing effective data access. But as invariably seems to be the case, the infrastructure label brought with it its own history and complexity.

The term infrastructure has a long history, dating back to at least the mid-nineteenth century. The French railways, for example, have used the term infrastructure for more than a century in reference to fixed installations such as the permanent way of track and bridges. It came into widespread use in English in the early 1950s when it was applied within the North Atlantic Treaty Organization (NATO) to distinguish shared common infrastructure, e.g., fixed installations such as airfields, telecommunications, pipelines, and ports, which might be used by the forces of any ally and were therefore financed by a central NATO fund. Since then the term has acquired a still more general meaning to envelop the basic capital investment of a country or enterprise, and it now includes factories, roads, telecommunications, schools, etc.

These descriptions have emphasized common hardware facilities to be shared by participants in some kind of common endeavour. Over time, however, it has been difficult to separate the facilities and hardware from the institutional, regulatory, and financing elements required for the effective design, creation, maintenance, and actual use of such facilities. Others have expanded the notion of infrastructures even further to incorporate processes that promote or facilitate broad social participation.

1.5 Terminology

For the purpose of this book we will use the term Geospatial Data Infrastructure (GDI), together with the related concept of a Geospatial Data Service Centre (GDSC), for enterprise-wide or domain-oriented activities. National Geospatial Data Infrastructure, Regional Geospatial Data Infrastructure, and Global Geospatial Data Infrastructure are special cases and will be defined in terms of what gives the GDI its regional, national, or global character.

Geospatial Data Infrastructure encompasses the networked geospatial databases and data handling facilities, the complex of institutional, organizational, technological, human, and economic resources which interact with one another and underpin the design, implementation, and maintenance of mechanisms facilitating the sharing, access to, and responsible use of geospatial data at an affordable cost *for a specific application domain or enterprise.* The application domain or enterprise suggest a large degree of common semantics in the definition of geospatial entities which facilitates optimal data sharing.

A Geospatial Data Service Centre (GDSC) is a facility or organization which is the intermediary ('broker') between the data users and the suppliers for the applications in the enterprise or domain (see Figure 1.1). It facilitates the integrity of access to the required data by ensuring system technical services as well as the administrative, data security, and financial services necessary to broker between data suppliers and data users within the information policies governing the GDI. It will facilitate and oversee data standardization activities within an application domain or enterprise and hopefully ensure that optimal data sharing is being pursued economically and with integrity. In several jurisdictions GDSCs are evolving from clearinghouse activities.

The National Geospatial Data Infrastructure seeks to support the sharing of data in the national context by means of a set of standards, such as: *national spatial reference systems, a national topographic template, a national elevation model, any other standardized spatial data set of national scope such as geographical names, administrative boundaries, certain thematic data sets (soils, hydrology, vegetation population, etc.), and meta data standards to describe in a consistent manner each of the GDI holdings.*

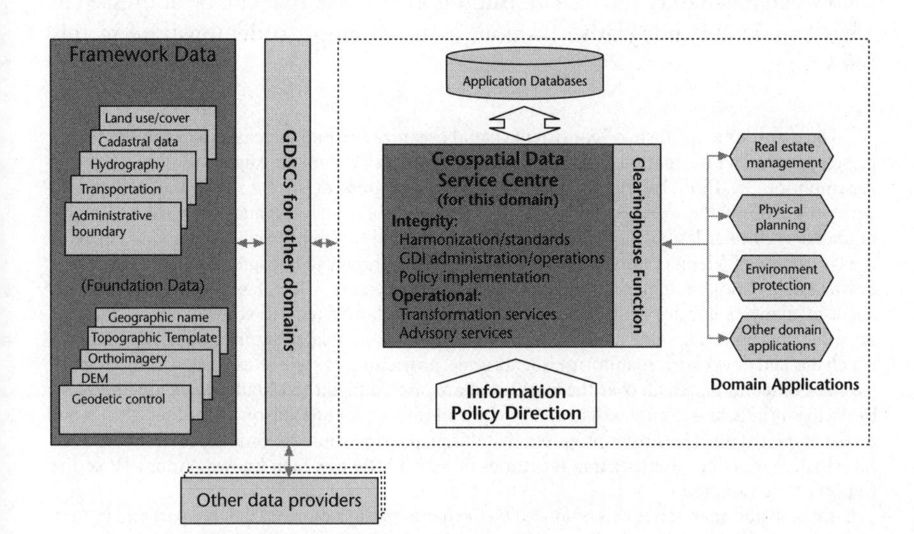

FIGURE 1.1 Role of a Geospatial Data Service Centre (GDSC).

Many enterprise- and domain-oriented GDIs will have used foundation data which adhere to these standards. Linking them into an NGDI is then not necessarily onerous as long as standardized data descriptions are being applied. However, this requires the application of national meta standards—in most cases a time-consuming retrofitting exercise.

Regional GDI initiatives refer to areas crossing administrative (including international) boundaries associated with a particular application domain, for example, the environmental assessment of the catchment region of the River Rhine. The associated regional GDSC will have to standardize the data from relevant NGDIs to be meaningful for the regional infrastructure and application, and operate within an information policy which must be in harmony with the national ones (see, for example, Mounsey, 1991).

Global GDI concepts relate to the infrastructure necessary to support global-domain or enterprise-wide applications. The foundation will be global referencing systems, including topographic templates at various levels of detail, elevation models, and global thematic data sets. As in the case of RGDI it will have to adhere to the information policies of the data suppliers which may not be easy.[1]

1.6 Direction and contents of the book

The challenge of designing, building, operating, and maintaining a geospatial data infrastructure lies in the purposeful orchestration of a substantial number of different disciplines and the examination of a huge array of issues. It is a complex, messy task fraught with difficulties in sustaining a shared language, a shared sense of purpose, and reliable financing (as the chapters which follow will make only too clear!). But it is also a task that can be addressed in a systematic and integrative fashion, as we attempt to demonstrate in this book.

[1] The quest for some form of consistent spatial organization and increased accessibility to dispersed geospatial information is not new. In Bern, Switzerland, on 10 August 1891, the General Assembly of the Fifth International Geographical Congress decided to take the initiative to develop a map of the world at the scale of 1 : 1,000,000 of which the maps would be regularly bounded by lines of latitude and longitude. In his proposal to this project, Professor A. Penck of the University of Vienna refers to the vast volume of geographical/cartographic information laid down in many books and maps of different scales and projections which has become available as a result of many exploratory expeditions. 'Various circumstances tend to considerably reduce the scientific and practical value of this enormous mass of accumulated information. The maps on which it is laid down are not uniform either in scale, projection, or style of execution; they are published at different places all over the world, and are often difficult to obtain. Some are only published in journals, and cannot be had separately, while others, for one reason or another, are not put on the market at all.' He further proposed that this map series would become the depository for all future surveys of the countries and territories of which little was then known (Annex IV of the Report of the Congress).

Ultimately this map series was produced by the participating countries in large part and in 1948 it also became the basis for the International Civil Aviation Organization's World Aeronautical Chart at the 1 : 1,000,000 scale, which in digital form became the Digital Chart of the World.

The book begins with a discussion by Lance McKee, of the Open GIS Consortium, on the GDI transition from a 'technology-push' to a 'market-pull' orientation and identifies a series of specific near-term applications of GDI. This is followed by a section on A Policy Framework for GDI comprising five chapters which examine some of the major issues affecting design, implementation, and sustained operation. This invariably is a complicated subject with often conflicting demands on the organizations involved in the acquisition, development, custody, and dissemination of geospatial data. There is the issue of political accountability and operational responsibility for a GDI initiative. In addition, there are significant issues related to public access, copyright and other intellectual property rights, data protection and security, liability, and privacy. Fitting a domain-oriented or an enterprise-wide GDI initiative into an existing legal and regulatory environment can be an especially demanding challenge, with significant consequences for sustained GDI financing.

Jan Kabel, a professor of law at the Universities of Amsterdam and Utrecht, addresses these legal issues mostly from a European context in Chapter 3. Citations are also provided on the Anglo-American perspective.

The economics of GDI (including financing, pricing strategies, etc.) is a large and complex subject, which only a few organizations have yet had to address. David Rhind, immediate past Chief Executive Officer of the Ordnance Survey of Great Britain (an agency which has examined the subject in depth), provides a comprehensive review of the subject in Chapter 4, and draws some interesting conclusions with respect to the viability of financing NGDI implementation in countries with a liberal free market political philosophy.

There are a plethora of standards relating to the operational governance of a GDI. These include the fairly stable and broadly applied standards relating to the core computer and communication requirements within the domain of organizations such as the International Telecommunications Organization. However, there are also more specific data model standards, exchange standards, standards dealing with inter-operability, and quality management standards which need to be examined in some detail. Peter Croswell provides a general examination of the subject in Chapter 5; Mark Doucette and Chris Paresi examine it from a quality management perspective in Chapter 6.

There are also strong cultural dimensions to any GDI initiative (a topic of special importance given the American dominance of the Internet to date). Willem van den Toorn and Erik de Man discuss this cultural dimension in Chapter 7, with emphasis on a method to help anticipate cultural factors affecting GDI development and implementation.

The next section, A Technology Framework for GDI, examines several of the key technologies required to construct and effectively use the infrastructure. For our purposes we have distinguished between foundation technologies, i.e. those required to build and link the relevant databases (and which are increasingly associated with web-based paradigms), and the more specifically geospatial-related technologies. The latter include those tools and processes required for the positioning, acquisition, modelling, and visualization of geospatial data.

Wolfgang Kainz presents an overview of the foundation technology in Chapter 8, giving special emphasis to the Internet. He also introduces the clearinghouse concept, as developed by the US Federal Geographic Data Committee. This is followed by a discussion of potential GDI architectures by Yaser Bishr and Mostafa Radwan in Chapter 9.

The need to facilitate and promote the sharing of data once collected is at the heart of the GSDI concept, and raises the question as to what extent data collected for one application can be effectively used in another. Some data are clearly more sharable than others, having been collected with a broad audience in mind from the outset. Examples could include the framework data contained in topographic, cadastral, hydrological, hydrographic, soil, vegetation, and regional geological maps and databases (see for example Table 1.1).

But other data sets are application specific, and raise all sorts of issues when horizontal or vertical integration is attempted. Assume, for example, an application in agricultural extension related to selected improvements of farming practices. We might be asked to determine whether the investments in extension, such as enlarging the size of the farming fields or the farm units themselves, are making a difference. Answering this question may require organizing a wide variety of data collected at various operational levels: individual farm fields, the farm as a whole, or groups of farms operating as an integrated system. Beyond this, several farm systems may together take on special administrative significance for the management of the agricultural economy. At the farm level, the data definitions need to be designed in such a manner that the aggregated data necessary for the applications at the higher levels can be computed from the data surveyed or collected at the field level without having to collect separate data at the higher levels. This is a matter of economics and of internal consistency.

The point is that for a particular application within an application domain the information requirement at all levels must be defined as much as is practically feasible before data collection takes place. Hence data classes, abstraction levels and requirements of accuracy and resolution need to be known in advance. Thus decisions having a significant economic dimension, about which data to draw from existing sources and which to survey and develop, from the ground up so to speak, can be made. Martien Molenaar reviews these aspects of the GDI agenda in Chapter 10 within the context of semantic modelling to facilitate optimal data collection and 'sharability'.

The next three chapters deal with data acquisition and display. In Chapter 11 Marco Hofman, Erik de Min, and Ruben Dood review key concepts and terminology associated with the geometric component of geospatial positioning/geographical referencing. In particular, they discuss the problems of matching data referenced by traditional technologies and specific reference systems with new data collected using global positioning system (GPS) technology.

TABLE 1.1 Selecting framework data for NGDI

Country	*1*	*2*	*3*	*4*	*5*	*6*	*7*	8	9	10	11	*12*	13	14	15	16	17	18	19	20	21
Australia	***V***	***V***	***V***	***V***	***V***	***V***	***V***		v		v	***V***									
Canada	***V***	***V***	***V***	***V***	***V***	***V***	***V***	v		v	v	***V***				v					
Colombia	***V***	***V***		***V***	***V***	***V***	***V***					***V***	v								
Cyprus	***V***	***V***		***V***	***V***	***V***	***V***			v											
Finland	***V***	***V***			***V***									v	v						
France		***V***		***V***	***V***											v					
Germany	***V***	***V***	***V***	***V***	***V***																
Greece	***V***	***V***	***V***	***V***	***V***			v			v	***V***					v				
Hungary	***V***	***V***	***V***	***V***	***V***							***V***	v		v			v			
India	***V***																			v	
Indonesia	***V***	***V***		***V***	***V***				v			***V***									
Japan	***V***	***V***	***V***	***V***	***V***	***V***	***V***		v												
Kiribati		***V***			***V***																
Malaysia	***V***	***V***			***V***																
Mexico	***V***	***V***	***V***	***V***		***V***	***V***		v			***V***	v					v			
The Netherlands		***V***			***V***		***V***	v		v		***V***	v						v		
New Zealand	***V***	***V***			***V***		***V***				v			v							
Northern Ireland	***V***	***V***		***V***	***V***	***V***	***V***			v		***V***						v			
Russian Federation	***V***	***V***	***V***	***V***	***V***	***V***	***V***		v												
South Africa		***V***	***V***	***V***	***V***	***V***	***V***			v											
Sweden	***V***	***V***	***V***		***V***	***V***				v		***V***		v		v				v	v
United Kingdom	***V***	***V***	***V***	***V***	***V***	***V***	***V***	v		v	v	***V***				v					
United States	***V***	***V***	***V***	***V***	***V***	***V***	***V***		v												
Total (23)	***19***	***22***	***12***	***16***	***21***	***12***	***13***	4	6	7	5	***11***	4	3	2	4	1	3	1	2	1

V	selected for framework data of NGDI
v	not selected

Data types

1. ***geodetic***
2. ***land surface elevation/topographic***
3. ***digital imagery***
4. ***government boundaries/administrative boundaries***
5. ***cadastral/land ownership***
6. ***transportation/roads***
7. ***hydrography/rivers and lakes planimetric***
8. ocean coastlines
9. bathymetry
10. physical features/buildings
11. place names
12. ***land use/land cover/vegetation***
13. geology
14. real estate price register/land valuation
15. land title register
16. postal address
17. wetlands
18. soils
19. register of private companies
20. gravity network
21. zoning and restrictions

(*Source*: Onsrud [1999] URL *http://www.spatial.maine.edu/harlan/gsdi/GSDI.html*)

Gottfried Konecny (Chapter 12) assesses recent advances in acquiring geospatial data using both spaceborne and airborne platforms, emphasizing the relative operational performance of each. The continuing importance of ground-based surveys is also acknowledged in this chapter with a short review of recent advances in GPS technology. Visualization, which is evolving quickly at the intersection of cartography, multi-media and animation technology, is playing a growing role in facilitating access to and use of data in a GDI. Menno Jan Kraak looks at the opportunities and state of the art of visualization in the Internet environment in Chapter 13.

As the GDI concept evolves, new functions at the managerial and operational levels are emerging, to which the traditional survey engineering and geography schools have been slow to respond. Chapter 14 provides a brief review of these human resources requirements.

Chapter 15 presents four brief case studies, which can contribute to building a best practice literature in the field. While every GDI initiative will address jurisdiction-specific needs and priorities, there is much of a technical, administrative, and even institutional nature which can be usefully shared with others.

Remarking on the expectations at a recent Global Data Infrastructure conference in Australia, Michael Brand (then president of EUROGI) noted that when his paper had been proposed 'it was hoped that it would lead logically to an optimum solution which would appear almost magically as systematic consideration was given to all the issues as they were identified. In practice, as consideration was given to the issues, delving deeper into the detail the number of questions multiplied, most of which can only be answered by debate, and the consensus that we hope will ensue.' See *http://www.eurogi.org/gsdi/canberra/theme html* page 8. In the light of the original objectives set for this book, we can identify with this observation. What can be said in practical terms about GDI development and operational maintenance after completing this book is certainly not as detailed and definitive as we had hoped, simply because such conclusions cannot be substantiated yet by a mature case literature and research.

Tentatively, the quantification of economic benefits appears to become the decisive factor in the sustained financing of GDI. The purposeful development of enterprise or application domain-oriented GDI has a clarity of focus, difficult to replicate in the development of NGDI. In the former, the number of players is limited and usually under one command. Under those circumstances the economic motivation can usually be well quantified and funding responsibilities assigned. To create similar success conditions for an NGDI is much more complex. Its development involves many more diverse stakeholders while the economic benefits to individual departments and society as a whole are very difficult to quantify, as David Rhind has concluded in Chapter 4. The sustained financing and acceptance of NGDI and, for example, the required development and maintenance of core data therefore becomes a non-trivial challenge.

If the definition of NGDI cannot be kept elementary, implementation may require costly and probably unacceptable retrofitting of data models, reference systems, access, use, and pricing policies, as well as quality management processes in domain and enterprise GDIs that are already in place. In the 1980s and early 1990s the complexity of most NGDI initiatives in the industrial world assumed that central governments had a much greater power to impose standards for example, than was realistic in light of the multitude of dispersed initiatives either in development or already functioning at the domain or enterprise level. These are tied up in the inertia of institutional history and mandates of government departments, and the imposition of such additional costs may be resisted, unless the benefits are very explicit. On the other hand the political vision of the societal benefits of the information society recognizes the need for a high degree of efficiency and effectiveness in national, regional, and global geospatial data infrastructure.

Hence, we must conclude that the top–down approach from the political level must be accompanied with a programme of making stakeholders sensitive to the longer term benefits of NGDI and involving them in the development of standards. Especially if the benefits can be quantified, the necessary readiness may be achieved to participate and make a commitment to the initial investments of implementing various aspects of standardization in the geospatial data supply environment.

This is one of several related conclusions and practical guidelines for advancing GDI which have been elaborated in Chapter 16 by the editors.

Bibliography

ANDERSON, N. M. (1990). 'ICOIN infrastructure', *Proceedings of a forum on the Inland Waters, Coastal and Ocean Information Network (ICOIN)*, June 1990, The Champlain Institute, Fredericton, New Brunswick, Canada, pp. 4–17.

Australia New Zealand Land Information Council (ANZLIC) (1996). Spatial Data Infrastructure for Australia and New Zealand, a discussion paper.

BRAND, M. J. D. (1995). *The European Geographic Information Infrastructure (EGII).* Paper presented at the ESIG Conference, 27–29 September 1995, Lisbon, Portugal.

BRANSCOMB, A. (1982). 'Beyond Deregulation: Designing the Information Infrastructure', *The Information Society Journal*, 1/3: pp. 167–90.

Canadian Government (1986). *Management of Government: Major Surveys.* A Study Team Report to the Task Force on Program Review, 31 July 1985. Canadian Government Publishing Centre, Ottawa, Canada.

CEC: European Commission (1989). *Guidelines for Improving the Synergy Between the Public and Private Sectors in the Information Market*, Office for official publications of the European Communities.

CEC: European Commission (1994). *Europe and the Global Information Society. Recommendations to the European Council*, Office for official publications of the European Union.

CEC: European Commission (1998). *Public sector information: A key resource for Europe.* Green Paper on public sector information in the information society. Office for official publications of the European Communities. *http://www.echo.lu/legal/en/access.html*

CLINTON, W. (1994). *Coordinating geographic data acquisition and access to the National Spatial Data Infrastructure.* Executive Order 12906, Federal Register 59, 17671–4. Washington, DC, 2 pp.

COLEMAN, D. J. and MCLAUGHLIN, J. D. (1997). 'Information Access and Network Usage in the Emerging Spatial Information Marketplace', *Journal of Urban and Regional Information Systems Association*, 9/1: pp. 8–19.

EUROGI, DDGI, AI, ILI/ILA, OGC, FGDC and FIG (1996). *Proceedings of the Conference on Emerging Global Spatial Data Infrastructure.* Dr Martin Bangemann, Patron. Konigswinter, Germany. Distributed through the European Umbrella Organization for Geographic Information. Mr Michael Brand, President, September.

MCLAUGHLIN, J. D. (1991). 'Towards National Spatial Data Infrastructure', *Proceedings of the 1991 Canadian Conference on GIS, Ottawa, Canada,* Canadian Institute of Geomatics, Ottawa, Canada, March, pp. 1–5.

Mapping Sciences Committee (1990). *Spatial Data Needs: The Future of the National Mapping Programme.* Washington, DC: National Academic Press.

Mapping Sciences Committee (1993). *Towards a Coordinated Spatial Data Infrastructure for the Nation*, Washington, DC: National Academy Press.

Mapping Sciences Committee (1994). *Promoting the National Spatial Data Infrastructure through Partnerships*. Washington, DC: National Academic Press.

MOUNSEY, M. (1991). 'Multisource, multinational environmental GIS: lessons learnt from CORINE', in MAGUIRE, D., GOODCHILD, M., and RHIND, D. (eds.), *Geographical Information Systems: Principles and Practices*. London: Longman, pp. 185–200.

ONSRUD, H. (1999). URL http://www.spatial.edu/harlan/gsdi/GSDI.html

RAVI (1995). *The National Geographic Information Infrastructure (NGII)*. Amersfoort, The Netherlands: RAVI Netherlands' Council for Geographic Information.

SMITH, N. S. and RHIND, D. W. (1999). 'Characteristics and sources of framework data', in LONGLEY, P., MAGUIRE, D., GOODCHILD, M. and RHIND, D. W. (eds.), *Geographical Information Systems: Principles, Techniques, Management, and Applications*. New York: John Wiley, p. 655.

TOSTA, N. (1994). 'Continuing evolution of the national spatial data infrastructure', *Proceedings of the 1994 GIS/LIS Conference*, pp. 769–77.

2

Who wants a GDI?

Lance McKee

2.1 Introduction

The foreword to this book refers to the expected evolution of the information infrastructure in general, its social and cultural impact, and the growing significance of the geospatial data infrastructure component of this information infrastructure. This chapter explores the technology push and market pull which are at the heart of this evolution, and discusses the possible effects on society. Predicting what humans will do in the future has never been an exact science, but some observations are inescapable, and some predictions about the growth of the GDI are, in my opinion, hard to deny. Technology push and market pull are driving each of the following developments.

2.1.1 Synthesis of geospatial information with other information

It is highly unlikely that current information technologies will fall into oblivion; they will probably continue to combine and recombine as they evolve individually, and as new technologies appear. What is in analogue form can be represented digitally and is thus capable of extraordinary manipulation by software and miscible through software interfaces that map one form into another. So digital maps, digital verbal directions, and other kinds of geographic information will merge into the collection of information devices, sources, and services comprising information environments in the twenty-first century.

2.1.2 Mobile, wireless, and location-aware devices

Based on the technical feasibility and profit potential of: (1) the microminiaturization of circuits, and (2) the expansion of wireless communication networks, it is safe to predict that we will see smaller, cheaper, more numerous, more mobile devices on wireless computer networks. These will often be embedded in and performing several functions in information appliances such as televisions, telephones, watches, vehicle instrument panels, radios, and even cameras, binoculars, and glasses. Mobile information appliances will establish their geographic positions using the global positioning system (GPS) or via an index into a geographic network of wireless communication cells, and these same appliances will query Internet-resident databases that hold geospatial data relating to their positions.

2.1.3 Commerce in geospatial information products and services

The information technology infrastructure being built today will encourage the information services sectors to grow and become elaborately structured. This commerce will certainly spawn a family of geospatial information products and services which will be a part of the general information infrastructure. How significant will they be? If we admit that geodata has so far been isolated from mainstream information systems because of its complexities, and the limitations of the technology to manipulate this data, and if we consider that 'everything happens somewhere, and everything and everyone is somewhere', it is hard to imagine that geographic information will not become a very significant part of information infrastructure.

2.1.4 Geospatial becomes mundane

In conventional Geographic Information Systems (GIS) and remote sensing applications such as environmental monitoring, agriculture and forestry, transport planning, facilities management, defence, urban planning, and civil engineering, we will see progress towards automated processes, specializations, and richer markets for specialized geo-information and geo-processing. But the benefits of GDI will rapidly extend to other possibly as yet unexpected domains in society, and special training will not be required for most of the GDI beneficiaries.

Perhaps police officers too will be able to see at a glance where their fellow officers are, and where their help is needed. Perhaps the location transmissions will be routinely encrypted. Similarly, geospatial tools will become part of many human activities in which GIS and related technologies were unknown or only experimental a few years ago. It is remarkable at first, but most people will not think about it much once such systems are in place.

2.1.5 The progress will be chaotic

The roll-out of geospatial technologies will be driven by a wide variety of partially overlapping, or at times barely overlapping, agendas. Proponents of a global GDI aim at 'substantial development in both rich and poor countries'. This will be partly achieved by concerned officials investing public funds in data co-ordination and other essential components of GDI, although these may not be immediately or even necessarily profitable. Development will also occur as both large and small companies seize on business opportunities that appear in the rapidly evolving geospatial information and geo-processing technology environment.

Pro-GDI forces in the government and private sectors will support each other, and also frustrate each other, in various ways. At the same time, anti-GDI forces will appear as people become concerned about the shifting individual and institutional roles, and as individuals grapple with the technology's implications for privacy and personal freedom. Figure 2.1 illustrates the tension field concerning the acceptance of policy with respect to GDI development.

When professionals who have worked extensively with GIS think of local, national or global GDI, they are usually thinking in terms of the world they know. They envisage the era-specific combination of early and limited market, early and limited pre-World Wide Web technical capabilities, an important government role, and significant user expertise requirements that defined the old GIS industry paradigm. What GDI will look like by 2010 will certainly be different from what we see today, and it will probably be both better and worse than we now hope.

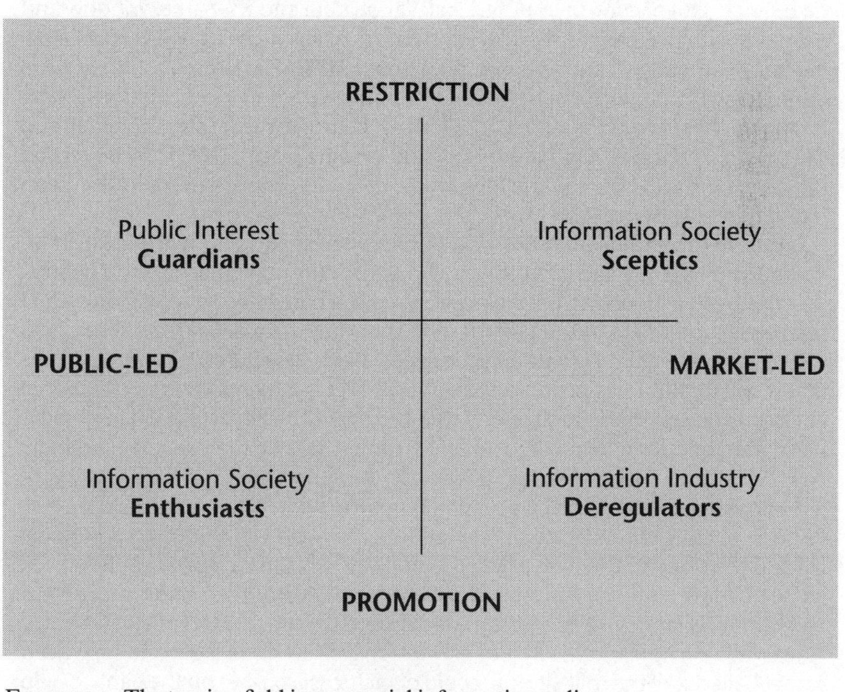

FIGURE 2.1 The tension field in geospatial information policy.
Source: Adapted from fig. 5.1 in Dutton and Vedel (1992, p. 72).

2.2 Two examples

'Why GDI?' is easily and eagerly answered by anyone who has worked in traditional GIS, remote sensing, or automated mapping and facilities management (AM/FM) application areas. Such a practitioner might dream of a scenario like this:

I am a landscape architect in the City Planning Office, working with a team to develop a small city park on what is currently an abandoned machine tool factory site. It will be located next to the mill pond and spillway that powered the factory's early mill 130 years ago. I can sit at my desktop computer, call up the City Assessor's property boundary map on my web browser, and rubber band a rectangle around the neighbourhood that includes the future park site. I select a menu option on the browser that takes the Earth co-ordinates of the rectangle, goes to the server that serves the website of the City Engineer's office, and retrieves the street engineering drawings for that area. (The meter on the City Engineer's data server does not charge me because I work for the city council.) This drawing segment neatly and automatically overlays itself on the street map, showing me where the curbs and pavements lie relative to the property boundary. I repeat the procedure to obtain digital elevation data from the surveying contractor's topographical study for this site, which is available on the surveyor's data server. Out of curiosity, I put that elevation data into a separate window and overlay it with the inexpensive elevation data available from a commercial Earth-imaging company. That costs me only about US$0.75. Hmm . . . , they have done this with images and automated orthophotogrammetric techniques, and it is roughly the same as the surveyor's work. Of course, it misses the stone wall and the trench, and the elevations are 50 cm off in some places. (No data conversion has been necessary, though the four source data sets were created and stored on four different software systems.)

I retrieve the engineering drawing of the redesigned spillway from the Engineers' Corps website and it automatically overlays the other map layers. The spillway has been widened to relieve pressure on the dam after heavy rainfall. Can I still put a soccer field in the park despite the widened spillway? A full size field would be too large so I create a junior soccer field, slide it into position, and see that it will fit, but I will need to calculate fill. This is a menu option on a special landscape design tools menu on my web browser. (The fill calculation function is performed on a remote application server somewhere on the network. I willingly pay US$5.27 for my time on that server, which is automatically added to my project expense account.) The fill situation is not too bad. If I lower the field surface 0.25 m we can grade the site to create the soccer field without using additional fill, and the slope to the path along the spillway will also be less steep. But it will not be so low as to create a blind spot that will worry the security-conscious neighbouring landowners. Good. I will propose the junior soccer field. Maybe if I lower it just a little more, I'll have enough soil left to make a bank to help keep soccer balls out of the spillway. We could plant fuchsia bushes on the bank to help retain the balls.

In fact, the landscape architect's computing environment might not be so dependent on the World Wide Web. The city might have a more tightly controlled geospatial data warehouse on which most of this data is kept, virtually on one server (though in reality it is on multiple distributed servers), with different departments responsible for different map layers. Or another Intranet approach might be used. But the essence is that the data will reside on diverse systems on the network. Network-resident data and processing services will be accessible as if on one's local system, because the software, on both the client side and server side, will have interfaces that conform to a globally used standard specification (the OpenGIS specification). There will be no need for data conversion except for certain archival purposes.

Geo-processing inter-operability software interfaces are necessary for this scenario, but not sufficient. Such interfaces can enable communication between different geospatial databases that have different spatial reference systems, different internal data formats, different units of measurement, and different geometric representation methods, and they can provide access to diverse geo-processing services, but they cannot solve the problems of different naming schemes and metadata schemas (see Chapters 5 and 10). For example, if those who collect data register a 'dam' only where they see a concrete spillway, and if I then search the database for all the dams in the city, I might miss the small earthen dam with a spillway made with a metal culvert pipe. Comparisons can be 'apples to apples' and not 'apples to oranges', only if the data have been prepared according to data content standards established through data co-ordination efforts. Metadata schemas, too, need to be co-ordinated. Geospatial catalogues similar to web text search engines such as Yahoo and AltaVista will require indexed geospatial information to be described by metadata which is consistent across databases.

Technology has a role to play in data co-ordination: this could be supported by use of special software that constrains data collection to established data content and metadata standards. And 'metadata translators' could be configured to make the best possible use of legacy data collected before the acceptance of metadata standards, or data collected by disciplines that have partially overlapping data content and metadata requirements. Such software could also report on quality, accuracy, likely areas of confusion, etc. But such tools only make sense in the context of an inter-professional, inter-agency, intergovernmental, public/private data co-ordination 'culture' that recognizes the importance of a coherent data framework.

Most current GIS practitioners should be able to relate to the above landscape design example. It is a desktop application, the data sources are known and understood by the user, and the user is an application expert. Most proponents of GDI have this professional mindset. They argue simply that many professionals' work will be much more efficient when they are working with a coherent GDI that exhibits consistency within and across local regions, and with software whose inter-operability interfaces eliminate much of the need to convert and manipulate whole data sets. This efficiency will help governments deliver services more effectively to their citizens, and many businesses will also benefit from these efficiencies.

But there is more to the picture we can paint of the future GDI. The new geospatial capabilities will become part of everyone's daily life. Current geospatial applications are relatively purpose-specific and affect only a few peoples' lives immediately. You could compare them to the stock ticker machines that preceded the telephone. The stock ticker was familiar to only a few people, but the telephone became part of everyday life for the average citizen. So, too, will geospatial technology evolve: rubbish collections, family life, boating and hiking, car travel, and neighbourhood life will be affected, along with other human activities which are not predictable at our current 'point of beginning'.

The next example highlights the mundane quality of the consumer side of GDI, and it sheds further light on how the information economy aspects of a GDI might develop. Dr Clifford Kottman, Vice President and Chief Scientist for the Open GIS Consortium, Inc., emailed this example to someone who works for a national mapping agency.

> Imagine a future where you can say to your personal appliance (maybe on your wrist) 'I'm hungry, show me the nearest pizzeria.' Imagine that the appliance performs speech recognition (or wirelessly connects to a service that does). By wireless networks, a trader in restaurant information is found and queried, and through it a database of pizzerias is accessed. The personal appliance reports your GPS position to the database, and the nearest three pizzerias are located. These locations, and your location (and perhaps other information, such as the map projection and scale, and a symbol library you understand) are handed to a map generation service (somewhere on the web), which creates a digital map showing three routes to three restaurants. This map is sent to your appliance where it is displayed for you. Perhaps the display also gives information on today's specials, and the approximate waiting time at each pizzeria.

Technologically, this example is not far-fetched. By the time this book is published, readers will probably have seen magazine articles about new cell phones with graphical displays, cell phones with voice recognition, cell phones that strap on to your wrist, and cell phones that know and can transmit their location. By then the public will have read that telecommunications companies have begun to put powerful servers on line to provide their mobile customers with many kinds of information. The market may take a few years to bring all the pieces together, but it will probably offer these capabilities faster than it brought us fax machines and answering machines. Certainly these technologies will reach the general public and achieve major economic significance faster than electricity bridged the gap from Edison's and Tesla's laboratories to the wall socket. Dr Kottman's message continued:

> Geospatial data in the emerging business market is different from map sheets, one-degree cells, and today's commerce in geodata. Today we are in what we can call a pre-wholesale era. Large blocks of data are for sale, perhaps from organizations like a national mapping agency, or from commercial data providers.

They are not really intended for sale to the general public, because the public has no tools to open and browse such data. The public wants information and knowledge (that is, answers to specific questions) not data. The general public does not want to browse geodata any more than it wants to read telephone books. In many countries, if you want a phone number, you can dial a directory number, give the name of the party you want to call, and be given the number. In the US this costs US$0.40 or less per service on your phone bill. Access to geodata is going along the same route, except the access will be more automated.

The future market for geospatial data and information will be similar to the telephone number directory service. It will have both wholesalers, e.g., France's IGN or the US Geological Survey's National Mapping Division selling information to a company that provides the map generation service imagined above, and retail outlets. An example of a retail outlet is contained in the relationship between the map generation service imagined above and you. How much would you pay to have a map or directions to the nearest pizzeria? US$0.05? US$0.50? Let me ask a related question: how much do you pay now for a year of Internet access? Internet allows access to information at the retail level. Most of it is free now, but the installation of information metering 'Internet cash registers' has begun. Or, maybe the pizzerias will absorb the cost into their advertising budget (like your automobile association does with the hotels it lists). Another analogy shows it is one thing to have a dictionary, and another to sell the whole 'book' on CD-ROM. It is yet another option to offer a dictionary service, and to sell access to it on a word-by-word basis. I think both visions will apply to geospatial information (and for geospatial data services, too) in the not-too-distant future.

Dr Kottman's example provides a good view of technology push (telecommunications infrastructure supporting communication between computers that can share geodata) and market pull (people who want to know the answer to the question 'Where?').

This discussion of mundane consumer use of digital geospatial information is not meant to suggest that the traditional GIS applications will disappear from view. On the contrary, practitioners in environmental monitoring, agricultural analysis and planning, municipal planning, public safety, disaster management, defence, and AM/FM, etc., will all experience significant improvements in ease of accessing, acquiring, packaging, managing, updating, and distributing data, and in the general usefulness of geographic information. Such applications will become easier to use because expertise in data conversion will not be a requirement for most users, so more people will have access to these tools. Many specialist businesses will develop to support these types of work. But at the same time, mundane applications will work their way into everyday life, and the mundane and the professional will overlap and mix in unpredictable ways.

2.3 Shaping the inevitable commercial development of the GDI

National mapping agency policy makers, packagers and end users of geospatial information, GIS software executives, state and municipal geographic information co-ordinators, private and government data producers, and others interested in geography are having lively discussions about the relative roles of government and private sectors in providing geospatial information to 'the market'. In each country, the governments, institutions, customs, laws, and legacy information systems are different, but every country faces commercial/technological challenges to the status quo. The new technologies, delivered by commercial interests and enabling radically new methods of commerce in geospatial information, threaten existing arrangements (which may be good or bad) and offer the possibility of new arrangements (which may turn out to be good or bad). Now is certainly a good time for policy debates, because policy can shape the outcome, even though the market will also shape the outcome. (For discussions on policy framework for GDI, see Chapters 3–7.)

Here are some policy considerations:

- Policy makers should seek to understand the evolving technical issues and carefully consider the public interest. This is difficult. It is easier to simply wish that all these wonderful technological improvements might work their way into our lives in the best way possible. Indeed, some people believe that markets alone lead to the best of all possible worlds. I would suggest that geospatial information is similar to both the physical infrastructure and the activities that such infrastructure supports. Some of these systems need to be designed and constructed with the public interest in mind, to promote safe and orderly daily life, and to promote wholesome commerce. However, some of these systems belong to the commercial sector, and should be left largely to entrepreneurs, their financiers, and their customers.
- Policy makers and geodata producers and users should converge as soon as possible on a 'baseline' set of standard data layers. Some data layers, such as cadastral and wetlands in the US at least, seem to cause too much controversy to be established soon. But by establishing those data layers that are not particularly controversial, we would immediately have a basis for much useful data sharing and data commerce, and we would also establish a precedent for how to agree on data content standards and metadata standards. We could thus get started on the great geospatial data infrastructure enterprise. The remaining foundation data layers would be standardized through continuing negotiation in formal data co-ordination proceedings. Hopefully, such proceedings would respect the agreements already achieved by the International Standards Organization (ISO).

Ways need to be devised to fund coherent geospatial data infrastructures. Bruce Cahan, President of Urban Logic Inc., a New York non-profit organization, suggested to me in a personal communication that significant local public sector funds are spent on building physical infrastructure such as roads, subways and waste treatment plants, and on planning and administering social programs. These public funds come from a combination of: (1) local taxes, (2) federal grants, (3) the proceeds of municipal bonds sold to mutual funds, and (4) royalties and returns from public–private partnerships such as airport facilities, cable television franchises, and other infrastructure. Cahan also suggested that cities should try to tap the same four sources that fund physical infrastructure for building 'community data warehouses' or other sharable information infrastructure, such as distributed information systems that receive, hold, and give access to the cities' essential geospatial data layers. Once cities and their public and private sources of funding see how much time and money cities (and state and federal governments) can save with data content standards and inter-operable software, the money will flow. Everyone—citizens, city agencies, federal agencies, contractors, integrators, software vendors, value-added data suppliers, and bankers—will benefit. Cahan thinks the key to financing is to show that these emerging standards will help underwrite appropriate investments in information infrastructure as capital assets for the communities that the information describes. See also Didier (1990).

Kottman's email message went on to predict an interesting market phenomenon:

> The Open GIS Specification will enable inter-operability between different speciality products. This means the same map service used in the pizza example could be used for navigation, for emergency response, for 'call before you dig' applications, and so on. These additional applications would require additional speciality databases (like the restaurant database), but that is exactly the point: all these databases will inter-operate. How do you predict the relationship between geospatial commerce and e-commerce and the related aspects of the larger market place: the wireless services, the database services, the trader services, the security services, the fee collection services, the personal appliance services, etc.?
>
> Imagine some of the hard questions facing a national mapping agency: What is the role of the agency with respect to the scenarios described above? Does the agency want to be in the retail business? Does the agency want to sell their data in the wholesale market, advertising 'Up to date! Accurate! Authoritative! Inexpensive!'? What partnerships should the agency form now to hold that market niche open?

There must be a demonstrated market for the information services to justify the infrastructure investments needed to create the open access environment, which is a condition for the exploitation of government and private geodata. Hopefully, people creating geo-processing standards and geodata content and metadata standards—whether consensus standards or de facto standards—will understand that they are creating standards for broad, general, multi-stakeholder, multi-purpose use. Many of the markets that will use this infrastructure cannot be imagined now, and in fact may never materialize if the standards infrastructure is too specific or confining.

People who seek to eliminate barriers to GDI ought to be encouraging or participating in the development of the Open GIS Specification, because market development is greatly enhanced by the availability of commercial, off-the-shelf software that inter-operates easily with other software over networks. But GDI proponents also ought to be attempting to implement institutional solutions by participating in the following activities:

- Learn from each other and seek to avoid duplication of effort. Education is perhaps the most critical element in the early phase of GDI development.
- Acquire knowledge, find models to apply, and find experts to consult.
- Establish inter-agency and inter-government co-operation. Such efforts help to instil better communication and co-operation, even as they put technology in place to support better communication and co-operation. One key goal of such co-operation is the development of an essential set of databases 'the Framework' (Federal Geographic Data Committee, 1995) with standardized feature names and metadata formats.

Data co-ordination efforts are a strategic investment, not an investment with a return that can be measured in a calendar year.

2.4 Effects on society

Rather than offering a list of speculations here, I think it is more useful to explore ways of thinking about the effects of this technology. Wegener and Masser prepared a document titled 'Brave New GIS Worlds' for the US National Research Council Commission on Geosciences, Environment, and Resources Mapping Science Committee, which appeared as chapter 2 in Wegener and Masser (1996) (see *http://www2.nas.edu/besr/223a.html*). In that chapter they comment on the pro-technology bias of 'technology diffusion' and explore the possible effects of the rapidly advancing GDI. They generate four scenarios, each of them a projection of one possible evolution of the uses of GIS and their impact on society.

These scenarios are too long to repeat here but they are worth reading. They suggest several ways of thinking about the technology's possible benefits and dangers. New technologies enable and induce so many new activities that society strains its previously adequate institutions, such as for protecting privacy, or for generation and distribution of culture, or for labour relations. Designing new institutional frameworks requires looking ahead to envision the new conditions, the new pitfalls and opportunities. No other kind of design struggles with such amorphous, dynamic, and slippery design elements.

Most people would agree that there is a strong need for policy, but there are more businesses than policy makers. Businesses have ways and means to influence policy makers, and, in this decade of the market, the market dance of the businesses and their new technologies is almost too fast to follow. This dynamic commercial and technological activity and the policy-making that frenetically tries to keep pace are shaping the technology environment of the individual.

Of course in this respect the enthusiasts tend to be somewhat euphoric about the potential for the good of mankind of these new developments. Lummaux (1997) highlighted a caveat of some potential dangers. 'A global spatial data infrastructure accessible by means of electronic networks, which allows citizens to evaluate the choices made on their behalf by those who govern and thus exercise in a well informed manner their democratic right of control, would indeed be a formidable instrument of democracy.' But he then continues: 'if it merely serves to emphasize the chasm between those who have and those who don't, if it would just be another model for the former to profit at the expense of the latter, and if it would just aggravate the technological and economic dependencies which already exist, this infrastructure would indeed become a terrible instrument of oppression and retreat.' Those are valid concerns and they emphasize the need for the international community to keep the subject of the possible adverse effects on humankind on the agenda of its relevant institutions.

If we step back from the exciting activity in the political and economic arenas, there are equally interesting and, perhaps, more fundamental ways in which technology, in this case geospatial data processing technology, affects individuals, society, and the Earth. It may happen that personal and public policy will be based on consideration of these effects and how they impinge on, or help us toward 'the way things are supposed to be'. Conversely, it may happen that most of these effects, though real and operative, are too subtle or indirect to matter to most people.

McLuhan (1964) wrote about media (technologies) as 'extensions of man'. He suggested that the technology itself, not only its information content, shape our perceptions and behaviours. As you read this sentence, you are silently decoding linear arrangements of about a hundred visually perceived symbols (letters, rather like a programming interface). They reference the English subset of sound symbols we can make with our voices (rather like another level of programming interface) These, in combinations, relate to a set of compound auditory symbols (words) that reference a roughly common experience of physical reality and culture. This linear visual process and simultaneous decoding process produce a completely different mode of thinking than live storytelling or watching TV. Over time, this subtle and unrelenting influence of the medium can be more important than most of the soon forgotten messages it carries.

Every tool extends our physical or mental abilities to meet an existing need more efficiently but later and more importantly uncovers new, as yet unperceived needs. GIS extends our mental ability to put together an understanding of our world so that we can shape not only our world but also our behaviour, because ultimately our survival depends on our directed individual and collective behaviour.

Maps are older than alphabets, so it is reasonable to hypothesize that GIS will play a big role in our future use of digital information systems. Through GIS, everything looks like overlay thematic maps that can be combined to provide new thematic maps. This technology instils a mental habit of analysing spatially, temporally, visually, above the fray, non-linearly, aware of boundaries but unconstrained by them, and aware of the overlaid human and natural elements of the planet. It is a kind of analysis that synthesizes and integrates rather than divides, and thus may influence our values in a direction which is more in line with the needs for a sustainable life on Earth.

Bibliography

Didier, M. (1990). *Utilité et valeur de l'information géographique*. Paris: Economica.

Dutton, W. H. (ed.) (1996). *Information and Communication Technologies: Visions and Realities*. Oxford: Oxford University Press.

—— and Vedel, T. (1992). 'The dynamics of cable television in the United States, Britain and France', in Blumler, J. G., *Television and the Public Interest: Vulnerable Values in West European Broadcasting*. London: Sage, pp. 70–93.

FGDC (Federal Geographic Data Committee) (1995). *Development of a National Digital Geospatial Data Framework*. USGS: Washington (*http://www.fgdc.gov/framework.html*).

Lummaux, J. C. (1997). *GDI, la meilleure ou la pire des choses.* (*http://www.eurogi.org/GDI/GDI3.html*. French version, for English summary see: *http://www.eurogi.org/GDI/jclpaper.html*).

McLuhan, M. (1964). *Understanding Media, The Extensions of Man*. New York: McGraw Hill.

Wegener, M. and Masser, I. (eds.) (1996). *GIS Diffusion: The Adoption and Use of Geographical Information Systems in Local Governments in Europe*. London: Taylor and Francis, chapter 2.

already available and do not contain obligations to process the requested information. Lastly, some jurisdictions make it possible to refuse public access to information if it is to be used for commercial purposes.

Thus, if a public body considers using public information commercially, the first question to be answered must always be whether this information is subject to freedom of information regulations. To answer this we must look at the object (is the information to be considered covered by the scope of the law in question?), at the way in which the information must be made available (if it is not already available in electronic form, we could consider commercialization in that form, giving access to the raw materials at the same time), and finally at the purposes for which access is requested. National legislation may forbid access to public sector information if it is to be used for commercial purposes.

3.2.4 Democratic access rights used for commercial purposes: different views

In the Framework of the American Paperwork Reduction Act of 1980 (Title 44 United States Code, Section 3501), as amended in 1996, there are guidelines on the commercialization of Federal Government information, that is 'information created, collected, processed, disseminated, or disposed of by or for the Federal Government'. The regulations on this point are clear and are an absolute ban on commercialization by the Federal Government. The authorities may collect, produce and disseminate information, but only in so far as these activities are necessary for a just execution of their public task (Section 8a). It is not permitted to license information on an exclusive basis if the availability of that information to other persons is thereby restricted. Charges for the re-use of information are also forbidden. The Federal Government should not be in a position to claim copyright on government information. Charges may only amount to the costs of distribution and must not cover the processing of the information, since the latter costs have already been incurred in the public task.

However, there are exceptions which seem to cloud the clarity of the system. Some public agencies, such as the Bureau of the Census, at the request of commercial companies, may produce additional information if its production could not be financed from federal resources. In these cases the companies must pay the total costs, but the information thus supplied must at the same time be kept available for the general public. However, this kind of commercialization may only be practised if it is consistent with Federal Government tasks. It should not merely be a commercial activity. Agencies should not expand public sources to fill needs that have already been met by others in the public or private sectors.

The French Circular of 14 February 1994 (Maisl, 1994) also expressly forbids the commercial use of government information but, in contrast to the US Federal Government approach, this ban on commercialization is based on the principle that private bodies should not profit from government resources since these have been financed by public means. This principle has been heavily criticized by the European Commission and will probably be abolished on a European level in the near future. The documents available on this issue (Stockholm Conference, 1996) represent the Commission's view that access to public sector information could be a key to economic growth. Free access to public sector information is necessary to be able to compete with the Americans. In particular, the information market makes its profits through the production and use of content. This content is mainly provided by the government.

This vision assumes, however, a clear delineation of the public task. In general, it will require the government to be modest, the solution being for everyone to stay within their own territory: the government shall do nothing more with the information than is needed to perform its assigned public task, and it is up to the private sector to process the information further. It is exactly this requirement of modest government functioning in the processing of information which nowadays calls for serious attention. Public bodies seem to be infected by the desire to behave in a client-orientated and cost-efficient manner; in short, to become more commercial.

3.2.5 A level playing field: unfair competition and commercialization of public information

Developments in the fields of liberalization of public services and privatization of public agencies have given rise to the question of whether public bodies should be subject to the laws of competition or of unfair competition. The general opinion is that there should be a level playing field in the market for both public and private competitors, but views diverge as to how this should be executed. One end of the spectrum advocates a complete economic and legal unbundling of public and private activities; at the other end, the existing laws of competition and of unfair competition are considered sufficient to suppress possible abuse.

Abuses may occur when public bodies are able to use the benefits that are inherent to their public status to compete with private competitors. The competitive advantage may be due to secure public funding, tax exemptions, legal positions (such as data ownership), profiting from the public image, equipment or personnel normally used for the execution of the public task. The use of all these facilities, separately or combined, which are not available to the private sector, could result in distortions of competition to the detriment of private competitors. Nevertheless, in most countries there are no specific regulations prohibiting commercial activities by public bodies, even if these activities transgress their public task.

With respect to access to public geospatial data by private entities and similar bodies, some national decisions have already been taken: note, for example, the French Meteo case (information about weather conditions) or the Dutch Vierhand case (information about companies). The decisions in these cases have proved to be more or less unsatisfactory and have not led to clear guidelines. On the European level, however, there are clear guidelines on at least one major point: the transparency of accounting and cross-subsidization.

The EC Commissions Transparency Directive (1980) obliges state-owned companies to use a transparent accounting system so that it is possible to control whether these companies comply with the regulations on state aid (Section 92 of the EC Treaty). These regulations have been elaborated on in some recent cases. In its decision of 27 February 1997, the EC Court of First Instance set out detailed regulations for the cross-subsidization of public bodies in relation to commercial activities. If a public body receives less from the state for the accomplishment of its public tasks than the costs related to the fulfilling of this task, then the subsidization by that public body of commercial activities is not to be considered as a cross-subsidization which could result in distortions of competition. If, on the other hand, there is a credit balance which is used to undertake commercial activities, the use of these resources could be challenged under Sections 92–94 of the Treaty. Note that the concept of state aid is not restricted to subsidies but also includes other interventions having the same effect of mitigating the charges normally included in the budget of an undertaking.

3.3 Protection of investments in geospatial databases

3.3.1 Methods of protection

Geospatial information has an economic value. The question therefore arises of whether this kind of information can be protected against retrieval and secondary use by others. Without such protection, the building of geospatial databases is a risky business, since the necessary investments can easily turn into losses when competitors are free to use the data. Four different legal methods of protection are considered here: copyright law, unfair competition law, the new right of extraction provided for by the EC Directive 96/9/EC on the Protection of Databases, and contractual law. The first three methods do not require prior activities: an appeal on copyright law, the extraction right, or unfair competition law do not depend on formalities. These methods of protection have the advantage that they can be upheld against anyone, not only against contractual parties.

Before the introduction of the extraction right, the main problem connected with the protection of geospatial information was the character of the copyright protection and the unfair competition law. Copyright does not protect economic investments as such. Copyright protects intellectual achievements which show a certain originality and that condition, legally speaking, is not applicable to facts. Geospatial data are mostly related to facts. The condition of originality could be fulfilled if the way in which these facts are organized in a database or on a map show a certain intellectual activity that could be characterized as different to the standard ways of organizing such material.

But that still leaves the 'facts' free to everyone. Free riders could only be stopped if the way in which they collect these facts constitutes an act of unfair competition, for example, if a competitor bribes an employee of the database owner to obtain the data. Copying a whole database without the organizing features is considered fair in many countries. The freedom to use other's achievements is restricted by statute (competition) law or by the unfairness of the methods used in competition, but not by making use of competitors' achievements as such.

3.3.2 The extraction right

From the legal situation described here, it is clear that database producers required a right that protected the facts in the database and the investments connected with the collection of these facts. This right was introduced by the European Commission by means of a *sui generis* right:

> a right for the maker of a database which shows that there has been qualitatively and/or quantitatively a substantial interest in either the obtaining, verification or presentation of the contents to prevent extraction and/or re-utilization of the whole or of a substantial part, evaluated qualitatively and/or quantitatively, of the contents of the database.

A database, according to the Directive on the protection of databases, is 'a collection of independent works, data or other materials arranged in a systematic or methodical way and individually accessible by electronic or other means'. This definition is applicable to geospatial databases. The maker of such a database is accorded the extraction right for a period of fifteen years following the completion of the database. (Databases completed after 1 January 1983 may profit from this protection until 1 January 2003.) It is interesting that a factual or contractual protection which exceeds the limits of the protection accorded by the Directive is not valid. The maker of a database made available to the public may not prevent a lawful user of the database from extracting and/or re-utilizing insubstantial parts of its contents for any purpose whatsoever.

3.3.3 The other side: freedom of information versus protection of investments

The dark side of this new extraction right relates to the legal position of the customers or the users of databases in the geospatial sector. The free flow of information will be substantially hampered, since the greatest producers of geospatial databases have been public bodies. There is no distinction in the Directive between databases held by public bodies or private bodies, so there is nothing to prevent the maker of a public database from claiming an extraction right. Situations may vary depending on national law. Copyright law in some countries provides for free reproduction and publishing of works published by a government agency, provided the copyright is not expressly reserved. Database protection by way of extraction rights, however, is not a copyright protection, so exceptions in one field cannot simply be considered as exceptions in another field. As we have seen, freedom of information acts may prevent claims on exclusive rights. The question remains whether access to a substantial part of a public database could be formulated in terms of freedom of access to public information. If the database concerned is not to be considered as part of public information governed by freedom of information acts, specific regulations on specific public databases, such as the land register, could also prevent claims on exclusive rights. Nevertheless, investments by a public body in derivative information products, which leave the public functioning of the original database intact, could be protected by database rights.

Access problems cannot be solved by compulsory licensing since the Directive provides no regulations on this kind of compulsory access. Compulsory licences compel the database maker to provide the data at reasonable prices. A compulsory licensing system was provided for in one of the drafts of the Directive but was dropped at a later stage. To reach a result which is comparable to a compulsory licensing system, one has to appeal to the regulations on economic competition, as provided for in Sections 85–86 of the EC Treaty or in national regulations on economic competition. In its decision on the Magill case, the EC Court of Justice has made it clear that owners of one-source databases could be compelled to license their data to third parties at reasonable prices if refusing a licence impedes the introduction of a new information product which would fulfil consumers' needs.

More fundamentally, the marketing activities of public bodies could be attacked under the regulations of unfair competition for public bodies, a problem already discussed. It must be noted that the synergy guidelines of the European Commission (1989) expressly prescribe that contracts or other arrangements of public sector bodies with private sector database providers or host services should not grant exclusive rights if they lead to a distortion of competition.

3.4 Privacy and protection of personal data

3.4.1 Ratio and scope of protection

Geospatial data may be of a personal nature, i.e. information related to identified or identifiable individuals. In the 1970s, the German Supreme Constitutional Court in its fundamental Census decision upheld that if society is organized in such a way that a citizen is not aware of who know facts, personal data or details of his or her personality, this situation has to be qualified as a contradiction to the right of self-determination and to a legal structure which wants to establish democracy. Anyone who cannot be sure that a deviating attitude will not be registered for a certain period, and could then be used or transferred in a non-transparent way, will try not to reveal such attitudes.

Of course, this consideration is particularly relevant to databases held by public authorities, since these authorities are able to determine people's social behaviour in a far more stringent way than private bodies. Nevertheless, the German Court's consideration also applies to private and semi-private bodies which, in fact, have powers to determine social attitudes in much the same way as the government. Insurance companies, financial institutions, or employers are a few examples. These agents clearly give rise to the greatest threat to the protection of privacy. And the existing regulations on the protection of personal data make no difference: everyone who processes personal data is subject to these regulations. They are therefore relevant to every geospatial data collection which could be defined as a collection of personal data and which, in some way, are the object of processing.

The European definition of personal data as stated in Directive 95/46 is characterized by its broad scope. Personal data are 'any information related to an identified or identifiable natural person'; an identifiable person is one who can be identified directly or indirectly, in particular by reference to an identification number or to one or more factors specific to his physical, psychological, mental, economic, cultural, or social identity. The Directive is applicable if these data are the subject of processing, which means any operation which is performed upon personal data, whether or not by automatic means, such as collection, recording, organization, storage, adaptation or alteration, retrieval, consultation, use, disclosure by transmission, dissemination or otherwise making available, alignment or combination, blocking, erasure, or destruction.

This wide range of activities subject to the regulations of data protection necessitates a clear definition of personal data. The criterion that personal data also consist of data on persons who can be identified indirectly leads to the conclusion that anonymous data, under certain circumstances, could be considered as personal data. This conclusion implies the data protection regulations apply to many geospatial databases which, at first view, seem to fall outside the scope of these regulations. This conclusion may be mitigated by statement (26) of the Directive, which states that to determine whether a person is identifiable, account should be taken of all the means, *likely or reasonably*, which could be used in identifying that person. Nevertheless, when one thinks of the technical possibilities which are available today, the condition of reasonability should be easily fulfilled. For example, this brings geo-segmentation systems which are based on postal codes and not on personal addresses into the area of data protection. In short, the fact that the data seem to be anonymous does not always make a database immune to these regulations.

It must be noted that most national laws contain specific regulations on the protection of personal data for specific databases in the public sector. These regulations have to provide at least the level of protection of the Directive where EC members are concerned.

3.4.2 Principles of protection

The Directive's main principles for the protection of personal data are already adhered to in many countries. They refer to the transparency of data processing, to limitations on the collection, use and disclosure of personal data, to security of data processing, to the rights of the data subject (such as individual access), of being informed about the processing of one's personal data, and the right of correction. A new right is that of the data subject to object to automated individual decisions based on profiles of data subjects.

3.4.3 Specific problems

Specific problems with respect to the geospatial sector mainly concern the secondary use of personal data, the collection of data without the knowledge of the subject, and the use of statistical data which could refer to subjects. If intermediaries are involved in the processing of personal data, their responsibility for the data being accurate and complete has to be assessed. The latter question will be dealt with later, together with other problems concerning the liability of intermediaries.

The secondary use of personal data which are processed for certain primary purposes will, in principle, collide with the finality principle. This principle restricts the use and disclosure of personal data. According to the Directive, personal data must be collected for specified, explicit and legitimate purposes, and may not be further processed in a way that is incompatible with those purposes. This means that the purposes for which data are collected must be defined and announced in advance of the data collection and processing. Generally speaking, personal data collected for one specific purpose, say the registration of land ownership, cannot be used for other purposes, e.g. direct marketing, unless specific conditions are met, such as the consent of the data subject or the specific conditions of the Directive concerning the use of personal data for direct marketing purposes.

If personal data have been collected without the knowledge of the subject, the subject must at least be informed of the identity of the data controller, together with the intended purposes of the data processing. This obligation is, of course, the basis for the execution of the rights, assigned to the subject, such as the right of access to his/her data, the right to correct data, and the right to object to decisions based on fully automated evaluations of a person's characteristics. The right to correct data is of specific importance in cases where statistical data are used which are statistically correct but not necessarily correct when a specific person is concerned.

3.4.4 Free flow of personal data

The importance of the EC Directive is to be found in its free-flow principle. Once personal data has been processed according to the provisions of the Directive, Member States shall neither restrict nor prohibit the free flow of personal data between Member States for reasons connected to the protection of personal data. The international geospatial information sector therefore has an interest in abiding by the law. National laws may be stricter than the Directive, but these restrictions can only apply to nationals and not to the transfer of personal data from other Member States. If, however, personal data are transferred to the EC territory from a country outside the European Union, the principle of the free flow of personal data no longer applies and the same is true for the transfer of data to third countries. The latter case is expressly provided for in the Directive and requires an assessment of the level of protection afforded by a third country. If the country in question does not ensure an adequate level of protection, the transfer of personal data undergoing processing or intended for processing after transfer, is prohibited. Any country that trades personal data with the UK, France, Germany, or any of the other EU States will be required to embrace European standards for the protection of personal data. See also Recommendation No R (91).

3.5 Liability of intermediaries

Geospatial information is not only processed by the initial producer, but also by intermediaries between producers and the end-users. As these intermediaries generally do not help compile the information, they are not in a position to check if it is protected by specific intellectual property rights for instance, or if it infringes specific personal data protection rights. Nor can intermediaries know if the information is incorrect or inaccurate, which could give rise to compensation for damages if decisions are taken based on that information. In most countries, intermediaries are only liable on the basis of a *with-fault liability*, i.e. the intermediary must have actual knowledge of the infringement or of the incorrect content of the information, and must have the possibility of blocking its delivery. Under European product liability law, the intermediary will not be held liable for damages if it can identify the actual producer of the information product. This principle is also found in US liability law and bears a certain logic: it would be senseless to hold an intermediary liable where another party is clearly more directly liable.

Intermediaries have, on the basis of the with-fault liability, a duty of care to check that the information they collect and distribute is correct and cannot be challenged on the basis of intellectual property or personal data protection rights. It is still sensible to provide for detailed contractual clauses with the suppliers of information.

3.6 Conclusion

The legal and regulatory environment of GDI is in a stage of dynamic development. The issue of the right to access government-owned data and information is regulated in most cases at the national level but, at least in Europe, also increasingly by directives at the supra-national level. In addition, the respective roles of governments and the private sector in the commercialization of public-owned data is far from settled. Again, national governments may have views on this, but supra-national policy may direct a different course.

The characteristics of information and communication technology mean we should expect increasing attention to be paid to the international environment with respect to access and use, data protection, and protection of the individual. Similarly, more attention will be paid to the position of intermediaries between data owners and users with respect to their liability for the quality and completeness of the information supplied. Although the current situation is fluid, there are enough legal positions in existence that can be applied. In any case, this fluidity should not be a barrier to municipalities or other organizations starting their local GDI in support of particular missions. In the next few years, case law and jurisprudence will contribute to the developing regulatory environment of GDI, and while the development of the 'geospatial information economy' may not proceed at the rapid rate expected by today's politicians, it will at least proceed at a rate in which adequate consideration can be given to maintaining a number of important institutions, including the role of governments, the protection of the individual, and the role of the private sector.

Acknowledgement

The author would like to thank Mrs Mireille van Eechoud of the Institute for Information Law, University of Amsterdam, for her valuable observations.

Bibliography

DE TERWANGE, C., BURKERT, H., and POULLET, Y. (eds.) (1995). *Towards a Legal Framework for a Diffusion Policy for Data held by the Public Sector*. Kluwer Law and Taxation Publishers, Computer Law Series.

FEENSTRA, J. (1998). 'Should Bodies Governed by Public Law be Subject Without Any Exemption to the General Regulations of the Law of Competition or the Law of Unfair Competition? To What Extent do the Regulations of (Unfair) Competition Operate to Suppress Possible Abuse', *International Review of Competition Law*, no. 2, 26–36.

MAISL, H. (1994). 'La diffusion des données publiques. Ou la service public face au marche de l'information', *L'actualité juridique—Droit administratif*, 20 May 1994.

VAN ECHOUD, M. M. M. and HUGENHOLTZ, P. B. (1997). 'Legal protection of geographical information: copyright and related rights. Bottlenecks and recommendations'. Study commissioned by RAVI (Netherlands Council for Geographic Information) on behalf of EUROGI. Conducted by the Institute for Information Law, University of Amsterdam.

EUROPEAN UNION

CEC: European Commission (1989). *Guidelines for improving the synergy between the public and private sectors in the information market.* Commission of the European Communities. Directorate-General for Telecommunications, Information Industries and Innovation, Luxembourg. Office for Official Publications of the European Communities.

Commission Directive 80/723/EEC of 25 June 1980 on the transparency of financial relations between Member States and public undertakings, *Official Journal L 195*, 29 July 1980: 35–7.

Council Directive 90/313/EEC of 7 June 1990 on the freedom of access to information on the environment. *Official Journal L 158*, 23 June 1990: 56–8.

Directive 95/46/EC of the European Parliament and of the Council of 24 October 1995 on the protection of individuals with regard to the processing of personal data and on the free movement of such data, *Official Journal L 281*, 23 November 1995, 31–50.

Directive 96/9/EC of the European Parliament and of the Council of 11 March 1996 on the legal protection of databases, *Official Journal L 077*, 27 March 1996, 20–8.

EC (1998). Public Sector Information: a key resource for Europe. Green Paper on Public Sector Information in the Information Society. COM (1998) 585.

Recommendation No R(91) on the Communication to Third Parties of Personal Data held by Public Bodies. Adopted by the Committee of Ministers of the Council of Europe. 9 September 1991.

Stockholm Conference on Access to Public Information, 27–28 June 1996, Documentation available at: *http://www2.echo.lu/legal/stockholm/welcome.html*

CASES

Census Decision, BundesVerfassungsGerichtsHof, 15 December 1983, 65, 1.

Guerra and others v. Italy European Court of Human Rights, 19 February 1998 (116/1996/735/932).

Halstead v. United States (1982), 535 F. Supp. 2782 (D.C. Conn. 1982).

RTE & ITP v. European Commission, European Court of Justice, 6 April 1995, joint cases C-241/91 and C-242/91 (Magill).

4

Funding an NGDI

David Rhind

4.1 Introduction

This chapter outlines the sources and balances of resources currently used to create and maintain NGDI—in so far as these are known. It partitions the costs into those associated with collecting NGDI information (the largest of the identified costs), the human resources costs, those involved in enhancing software, and those incurred in providing the physical infrastructure. For illustration purposes, relevant data from the USA and the UK will be used. We examine the different models for funding NGDIs, with particular reference to 'pay as you go'/'user pays' models operating in both the private and public sectors, and to the 'taxpayer pays' model. The advantages and disadvantages of each model are outlined.

A discussion of the need to measure the benefits of creating and maintaining an NGDI follows. This is necessarily brief because very little real work—as opposed to assertion—appears to have been carried out in this arena. Finally, we outline the additional complexities of a global GDI (GGDI).

Throughout the chapter the underlying assumption is that benefits have to be tangible and valued by society if the distribution of public resources is to be modified and the private sector is to be involved in a significant way—both of which are essential to the substantial enhancement of NGDIs.

4.2 The definition of NGDI

The definition of what is included in a National Geospatial Data Infrastructure determines how much it costs and which parties are likely to be involved. Moreover, the origins, form and relative magnitudes of these financial contributions have differed greatly over time. And the situation will inevitably differ considerably from one country to another. It follows that this chapter can only be illustrative and, in the examples given, describe nothing more than a 'snapshot' at a particular moment in time.

It is assumed here that an NGDI (see also Clinton (1994) for an early enunciation of the scope of the US National GDI) irreducibly consists of:

- a topographic template, acting as a geographical framework (see *http://fgdc.er.usgs.gov/framework/frameworkintroguide/*; Smith and Rhind, 1999) which extends over the whole of the nation state;[1]
- other spatial information which is nationally complete and is fitted to the 'template'. In combination, these other sets of information may well be nationally specific;
- a cohort of suitably trained people who ensure the key information is collected or collated, validated, made available, publicized, and used, plus who can offer advice to others when required;
- training and education facilities to enhance and develop the skills base as needs change;
- sets of laws, protocols and standards to ensure that the infrastructure operates safely, effectively, and efficiently;
- the necessary hardware and software in terms of computers, communications networks, and geographical information system (GIS), and other software as appropriate; and
- financial resources to fund the totality of the exercise (though these are very unlikely to originate from one source).

It is important to point out, as Tosta (1997) and others have done, that we already have geospatial data infrastructures since most or all of the above exist, even if they are sub-optimal in many respects in some countries. The 'real world problem' is to ensure that the NGDI components can meet rapidly changing needs, can support greater reliability, can become more readily accessible (typically in digital form) and are maintained up to date.

We already have such NGDIs without having had some previous, umbrella and/or coherent planning to establish them. Thus a coherent vision of an NGDI cannot have been essential hitherto. It can reasonably be argued, however, that the extending nature of NGDI, the shortcomings cited above, and some market failures ensure that at least a national vision is now a necessary condition.

4.3 The NGDI players

According to the numerous papers on this subject (e.g. Grelot, 1997), there are many groups of individuals or organizations which believe they can (or do)

[1] This template includes the national geodetic reference system.

make a material contribution to the national equivalents of a GDI. These include:

- the public sector as data providers, as consumers of information and advice, and as service providers (N.B. their current role varies greatly between different countries);
- not-for-profit organizations;
- science and social science research organizations;
- international aid/development organizations;
- educational organizations;
- private sector information/content providers;
- private sector service providers;
- private sector software vendors;
- private sector hardware vendors; and
- individual citizens.

In some countries, notably The Netherlands (Masser, 1998: see also *http://www.ngdf.org.uk/whitepapers/mass7.98.htm*), there is strong central co-ordination of NGDI-related matters with the Secretary of State for Housing, Spatial Planning, and the Environment named as the individual having a cross-government role. This role is, however, essentially one of persuading, rather than instructing, colleagues to behave in a co-ordinated fashion. It is self-evident that not all of the players on a wider stage will have the same agendas. Table 4.1 suggests some likely motivations and agendas of a selection of these players. This inescapable disparity of views needs to be factored into any proposals for corporate enhancement and financing of an NGDI.

4.4 The distribution of expenditures in an NGDI

How much does it cost to run an NGDI? Indeed, can such estimates ever be made meaningfully? It seems astonishing that such questions have not apparently been asked before. For instance, the National Academy of Public Administration, a body affiliated to Congress, carried out a major study on the desirable changes to the GI 'industry', concentrating substantially on the US NGDI. They published their findings in a voluminous report entitled 'Geographic Information for the 21st Century: Building a Strategy for the Nation' (NAPA, 1998). Yet nowhere in their ten-page Executive Summary—setting out the background to the study and all their key findings—did they mention funding, other than (rarely) in recommendations such as 'develop co-ordinated goals, strategies, performance measures and budgets for federal agency GI programs and activities'.

This situation is absurd. At present, we do not know with any great precision what is included within an NGDI. And we have no real idea how much related activities currently cost, how much more funding might be needed, or what improvements are desirable—at least in any form of business case. This chapter attempts to initiate the process of creating such business cases. Inevitably,

TABLE 4.1 Some possible agendas of the various 'players' and stakeholders in a national GDI

'Player'	Alternative short-term agendas
Central Government organizations	• avoid expense—'hide data'? or • maximize use inside and outside department or • maximize revenues and minimize costs subject to equality of treatment and fairness rules
Local Government organizations	• avoid expense—'hide data'? or • maximize use inside and outside department or • maximize revenues and minimize costs • *plus* conform with national requirement for administrative/ statistical data
Commercial sector—information trader/publisher	• trade profitably and have positive cash flow • minimize costs of getting data from elsewhere • minimize risks by pre-publication agreements to purchases • disaggregate markets and appropriate as much as possible of customer value, prevent arbitrage • make project of all activities to measure costs/benefits/ write-offs
Commercial sector—hardware/ software vendor	• trade profitably and have positive cash flow • obtain data required by customers at minimum cost and bind in to system to provide solutions, preferably on exclusive basis
Commercial sector conglomerate	If focused as one business, 'gestalt' aspects of business (e.g. data helps sell equipment), dictate decision-making and agenda. More normally, individual enterprises within the conglomerate are judged first on their own 'bottom lines'
Non-governmental/ Not-for-Profit organizations (e.g. charities)	• obtain data, software, and hardware at minimum cost so available funds can be devoted to organizational objectives • disseminate information widely to help meet objectives
Academic sector	• produce published papers on the basis of research or observation • challenge the 'taken for granted' views of others • teach knowledge, use of tools, and foster understanding of GIS/GI amongst students
All individuals	• altruism • obtain career, finance, or status benefits e.g. through networking • enhance personal skills, competence, and knowledge development

we have to start with some previously quoted estimates and assume for first approximation purposes that the minimal cost of running the existing NGDI is as described below. (All figures cited are in US dollars, converted where necessary on the basis of the exchange rates pertaining in September 1998.)

4.4.1 The cost of raw data capture or maintenance

It has been estimated by the US Office of Management and Budget that the US federal government spends around US$4 billion per annum collecting and managing spatial information across its agencies (FGDC, 1994, 2). This sum is believed to include staff and all other direct costs and local overheads (it is possible that these proxies for NGDI-related expenditure have been exaggerated through inclusion of some element related to the 'outside US' expenditures by the military mapping organization). There also appears to be a consensus that state and local government in the US may spend as much as US$6 billion annually. As in the federal activities, an unknown but substantial proportion of this is probably devoted to 'usage' or routine operations exploiting the NGDI. Though these figures seem to include heroic estimates, we have no better ones available. In addition to these sums spent by different levels of government, US utility companies spend significant amounts on basic mapping and data conversion—let us (probably reasonably) assume these to be of the order of US$2 billion per annum. The expenditures of other private companies in this arena are difficult to estimate but, in aggregate, cannot be less than US$500 million. Further—in the absence of any real evidence—let us assume that about one-third to one-half of this total of US$12.5 billion—say US$5 billion—is spent on information that is properly required to make the US NGDI function, as opposed to operational use of it. Note that there is no mention here of data maintenance—it is assumed that these significant costs are also included in these 'ball park' estimates.

4.4.2 The cost of the physical infrastructure

Let us also assume there are 100,000 individuals active in GIS in the USA who routinely and effectively use it full time, or to whom it is an essential part of their business. Given the number of GIS licences and other evidence, this seems a reasonable guess. The costs of purchase and maintenance of PC equipment, relatively simple software, and use of standard telecoms, suitably weighted to cope with organizational overheads and the cost of money, can reasonably be assumed to be of the order of US$5,000 per person per annum, i.e. about US$500 million per annum.

4.4.3 The cost of people

Suppose further that the same number of people spend 50 per cent of their time on GDI-related activities, that their average total annual salary is some US$30,000 and the organizational overheads raise this by about 50 per cent.

Thus the total salary bill would be of the order of US$225 million. In this calculation, no account of the cost or value is put on the time of casual users, such as citizens who access information via a terminal on a very occasional basis.

There are, however, inescapable and quantifiable costs of training new staff and continuous professional development for existing staff members. Including attendance at courses and conferences, these cannot be less than a further US$25 million per annum in the USA alone (and could well be much higher). For instance, approximately 8,000 people attend the annual ESRI User Conference in San Diego alone. Software development also goes on continuously. Summing up all of this type of development in the USA, we can only guess that it costs no less than US$250 million per annum.

4.4.4 Other costs

The costs of other factors, such as regulation, development of standards and publications, are probably modest in comparison with the above. We will simply assume them to be of the order of 5 per cent of the total, i.e. up to another US$700 million. Included in this figure is the publicly identified 'one off' co-ordination funds for NGDI in the USA, including the annual sums or so granted by the federal government to the Federal Geographic Data Committee (FGDC; see below).

4.4.5 So how much does *NGDI cost currently?*

The first order approximations cited above amount to some US$5 billion to US$6 billion per annum for a relatively narrow description of the specific cost of creating and maintaining the NGDI, including building tools and creating essential skills. Taking the wider definition, which includes these sums but also expenditures of some of the businesses which depend upon existence of the NGDI, the total expenditure seems likely to be no less than US$15 billion. The narrow definition amounts to about US$20 for each US citizen or a little over US$1,200 per square mile of the US land area. The wider definition is about three times this expenditure.

If we assume that these figures are of the correct order of magnitude—a brave statement in the absence of detailed studies—then some interesting conclusions follow. It would seem that these average costs are actually rather moderate—equivalent to less than a packet of cigarettes every week for every adult US male. It is also clear that, on this analysis, the cost of collecting information is by far the largest contribution to the whole enterprise. Finally, the bulk of the NGDI expenditures come from long-established sources: the US$7.2 million annual funds specifically established by the US federal government to facilitate enhancement of the NGDI through the United States Geological Survey (USGS) are extremely modest in relation to the total expenditures involved, though they may have important leverage or seed-corn effects. Of this amount, some US$1.6 million goes into the USGS's National Mapping Division's operational programmes focused on framework data collection, standards activities, etc. The remainder is devoted to the FGDC, where US$2.2 million is used for staff work and some clearinghouse development and maintenance; and US$3.4 million is spent on grants and other external programmes.

It seems wise to cross-check these figures by reference to a different country, especially one where the government structures and processes are rather different. In the UK, total expenditure on GDI-related matters has not been investigated in any serious way. Nevertheless, the expenditures of certain central government bodies—the two Ordnance Surveys (OS), the Military Survey, the three cadastral organizations, the Meteorological Office, and the British Geological Survey, government executive agencies such as the Environment Agency, and various research bodies—total more than £2 billion (about US$3.3 billion) annually. It is unclear what proportion of this can be properly attributed to the creation and maintenance of the existing UK NGDI, and what parts relate to those operations based on the NGDI and other operations which do not use it. Here, based on a 'best guess' and for consistency with the US example, it is assumed that about 30 per cent is attributable to the creation element. Less money is probably spent by local government in Britain than in the USA since some functions (such as the cadastre) are functions of the central state. Nevertheless, a first guess would be that something of the order of £1 billion is spent by local government on collecting information; again, perhaps 30 per cent of this can be construed as underpinning or creating the NGDI. The utility companies have also incurred major expenditures in the last decade in building detailed spatial databases.

Some minor differences emerge between the US- and UK-based analyses. These include a higher unit cost of hardware and software and a smaller than pro-rata number of individuals involved of around 10,000 people in the UK. Overall, however, the 'ball park' narrow expenditure figure on GDI-related activities in the UK seems likely to be in excess of £1 billion for information collection and people involved in its collection and maintenance. In addition, other expenditures occur of about £50 million on infrastructure as defined earlier, some £150 million on human costs of users, about £5 million on training, and perhaps £30 million on software development. Adding 5 per cent to cover other, unidentified costs, the grand total expenditure on a narrow definition amounts to some £1.3 billion (US$2 billion). With a wider definition, the GDI costs approximate to at least 50 per cent, and perhaps 100 per cent more. The narrow definition equates approximately to US$35 per citizen or about US$20,000 per square mile of the UK territory per year.

In no sense can these figures for the USA and the UK be compared in any detail. The uncertainties, not least what proportion can be safely attributed to creation and support of the existing NGDI, are such that these can be nothing better than educated guesses. Nevertheless, they are very similar per head, giving some reassurance of correctness or, at worst, of consistent errors. The hugely greater expenditure per unit area in Britain largely reflects the much greater population density in that country (approximately eight times that of the USA). Other comparisons support these 'ball park' figures. Work by CERCO (the association of national mapping agencies in Europe) and by its equivalent for cadastres shows that the combined expenditures of their members within the fifteen countries of the European Union probably total in excess of US$3 billion. In all known cases, the domination of NGDI-related costs by those within the information collection agencies, the greatest proportion of which are public sector, is noteworthy.

So far as direct expenditure on the NGDI is concerned in the UK, the absolute and relative amounts spent on the National Geospatial Data Framework (Nanson and Rhind, 1998) have been rather smaller than those spent on the NGDI in the USA. After initial efforts based on voluntary contributions by data providers, the Ordnance Survey announced in September 1998 that some US$370,000 would be made available annually from the National Interest in Mapping Service Agreement funds for NGDF work. Even this will only approximate to about 0.2 per cent of the 'narrow definition' annual cost of creating and maintaining the NGDI. The equivalent figure for the USGS-mediated federal NGDI funds in the USA is of the order of 0.15 per cent. It follows that making any significant changes or improvements to the NGDI in both the USA and the UK (and probably elsewhere) is largely about influencing the distribution of 'old money', or about generating completely new revenues.

4.5 Who pays and how is the money raised?

In the USA, the most easily identified funding comes from the federal government. This clarity reflects the line budget approach operated in that government. Nevertheless, anecdotes and some evidence suggest that other parties make enormously important contributions. Identifying them can be difficult because some are indirect: for example, some are provided in kind through the time of staff already on pay rolls which are not properly attributed, or in returns to government through the tax system.

The identifiable contribution to the NGDI from the private sector is modest. The sources of evidence include statements by leaders of industry—always to be treated with caution as potential propaganda. More reliable are verifiable commitments such as contributions to organizations such as the OpenGIS Consortium (total in the order of US$1–2 million per annum at present) and investments and taxes on profits reported in annual reports (in the case of public companies).

Increasingly, however, a contribution is being made in both the UK and USA by end-customers for information and for services which result, at least in part, from the existence of the NGDI. In the USA, this is mostly mediated through the private sector since federal government is not, in general, allowed to charge more than the cost of copying for the information. There are some curious anomalies, however: for instance, the charges by the USGS for satellite imagery are about an order of magnitude greater than those made by the same organization for digital map data of equal volume and complexity. In the UK (and in various other countries; see Rhind, 1997*a*), revenues from end-users are

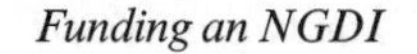

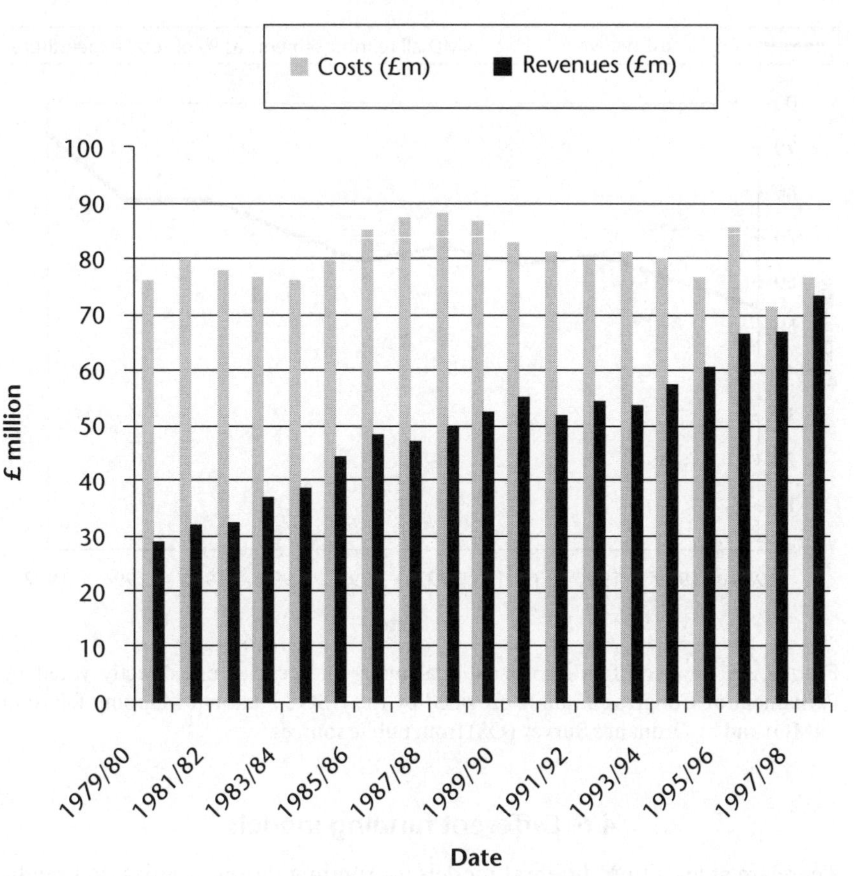

FIGURE 4.1 Costs and revenues involved in running the Ordnance Survey; expressed in constant 1996 £ (US$1.65 = £1.00). *Source*: Ordnance Survey Annual Reports modified with the Retail Price Index to remove inflation effects.

collected through both the private and public sectors. Thus a survey published by the UK central government showed that it benefited annually by no less than US$330 million annually from sales of information and services based upon its work (TSO, 1998); this figure would be significantly larger if revenues from services such as land registration transactions were included.

One of the most spectacular examples of this relates to the Ordnance Survey, the national mapping agency for Great Britain (i.e. England, Scotland, and Wales). Figure 4.1 shows how information and services revenues for this have grown since 1979 whilst Figure 4.2 shows a comparison of the levels of cost recovery achieved by the OS with those from the equivalent US federal body, the USGS National Mapping Division. In essence, these show the level of moneys obtained and used by these organizations which were not voted directly by Parliament or Congress as appropriate.

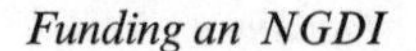

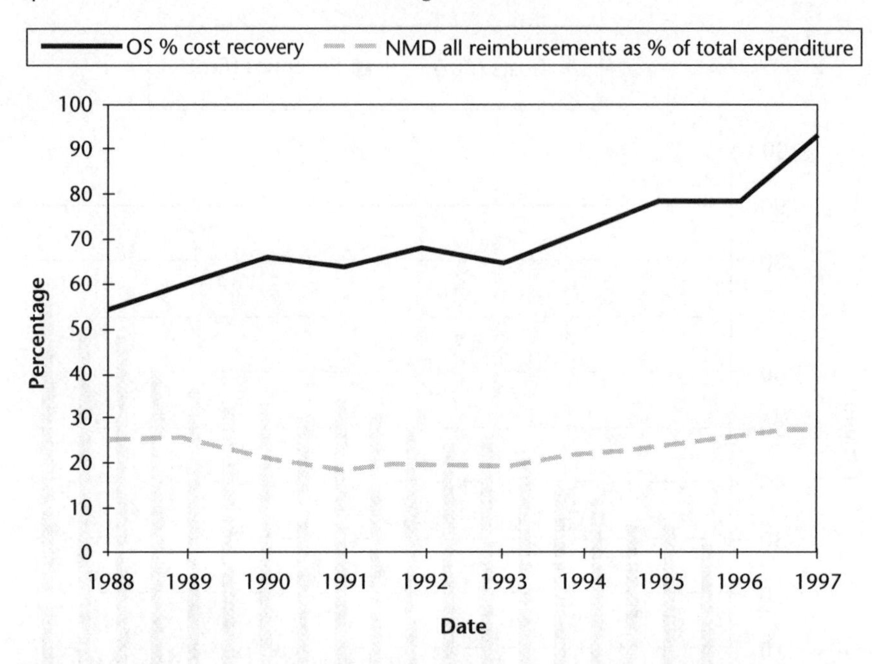

FIGURE 4.2 Level of funding to national mapping agencies not directly voted by Parliament or Congress. Figures provided by the USGS National Mapping Division (NMO) and by Ordnance Survey (OS) from public sources.

4.6 Different funding models

There are at least four different models for funding those activities that might be ascribed as creating or maintaining an NGDI. These are:

- funding by central or local government using appropriated funds derived from taxation and paid out for defined services or activities;
- funding through payments made by customers, collected by the private sector;
- funding through payments made by customers, collected by the public sector; and
- funding on the basis of sponsorship, advertising or other indirect methods. This could be taken to include near-market applied research and development. The creation and part of the operations of Microsoft's TerraServer (see *http://terraserver.microsoft.com*) for instance, seems to have been funded on a basis of 'proof of concept', brand sustenance, and demonstration of competitive advantage.

As we have seen earlier, the great bulk of NGDI expenditure at present seems to derive from the first of these. However, as Grelot (1997), Rhind (1997*b*), Sandgren (1997), and others have shown, the growth of income-generating 'geospatial information' organizations within the public sector has been a feature of the last five years. These also make some significant contribution to the NGDI. Moreover, there are a number of private sector bodies trading in the area of geospatial information which are linked synergistically (or even parasitically) to government bodies, such as those exploiting US Bureau of Census TIGER files. The web of dependencies and beneficiaries is therefore becoming ever more complex.

Perhaps, in its simplest form, the arguments about funding models for an NGDI can be reduced to three questions:

1. Does everyone in society benefit greatly from collection of a certain set of information and its widespread and unrestricted provision at low cost, together with the provision of skills to utilize the information?
2. Would this happen if left solely to the private sector?
3. Are the benefits so gained more important than those from competing calls on the public purse (assuming finite resources)?

The answers to these questions are not self-evident. It is well known for instance that vested interest groups of experts are prone to spend other people's money on projects they see as vital to the (untested) 'national benefit'. Equally, it is strange that the purchase of telecoms, software and (increasingly) of the university-level education components of an NGDI are taken as being acceptable, whilst certain types of geospatial information are held to be a public good and one which governments have a duty to provide. Finally, it is clear that factors other than effectiveness and efficiency sometimes operate when public sector decision-making is concerned, such as political influence peddling.

On the other hand, there are considerable geographical and theme-specific restrictions to what the private sector would provide unaided. The small number of customers for rural data and their often low purchasing power ensure that some parts of the template or framework element of the NGDI—and hence most other information—would be collected rarely and at low levels of detail if the funding was left solely to normal commercial considerations. In part, this has political and philosophical implications: is there a societal right to have equity of access to geospatial information? This may appear to be an absurd suggestion, yet ignoring it could lead to breaches of accepted social justice. For instance, non-availability of up-to-date geospatial information may prevent provision by the emergency services of an agreed service to the seriously ill and those in other acute danger.

Aside from the social justice and market failure aspects, the practical issue of whether the nation state can ever afford to fund the totality of the creation and maintenance of an NGDI is a vexed one. It is noteworthy that, where mixed private and public sector information gathering and exploitation take place, the total sums spent are larger than where all this is simply a public sector matter. It is also evident that, where revenues are returned directly to the nation state or its geographical subsidiaries, these bodies are also much more supportive of continued and increasing investment. A prime example of this is also shown by the Ordnance Survey: after five years of increasing revenues from customers (including other government bodies) and increasing expenditures in parallel, the UK government agreed to boost its annual contribution to the Ordnance Survey in mid-1998 by about 500 per cent! The reasons for this included a recognition of the geographical cross-subsidy cited above which was being funded by customers, the demonstration that users valued the core information greatly, plus an understanding that a coherent, up-to-date national 'template' or framework coverage was essential. As always, success also depended on good presentations by the Ordnance Survey staff to Her Majesty's Treasury and through strong public support for the change.

Despite this success, there are many other factors to be taken into account in assessing whether the end user rather than the taxpayer should fund the creation of those parts of an NGDI which are presently funded by the latter. Special pleading and 'hype' normally complicate the situation: historical analogies such as the economic benefits generated by the creation of interstate highways are spurious, in part at least. A further set of arguments centres on whether the public sector should charge market rates for the products and services it creates and what its relationship should be to the private sector. In reality, there is no one logical answer for policies on matters as diverse as competition, state security, budget balancing and procurement all interact. The advent of new technologies clearly also has the potential to change the advantages and disadvantages of different courses of action. Micro-financial transactions facilitated by Internet- (or Intranet-)based services may well revolutionize the current situation, or they may lead to communities of the information-poor, mimicking current purchasing power and population densities as commercial firms tailor services solely to demand. Yet these changes will not necessarily occur in ways predicted by technologists (for instance, brand images and confidence in information quality, integrity, and institutional longevity seem to have a significant effect in the geospatial information supply 'business').

Some of the advantages and disadvantages of the different courses of action in terms of raising funds for an NGDI or its components are set out below (taken from Rhind, 1999). The italicized text offers a response to each contention. It should be stressed that virtually all state bodies generate some direct revenue returns (see, for example, Figure 4.2): the big question is, what proportion of the total activities of the institutions involved should that part become?

4.6.1 Arguments for cost recovery by public sector bodies

- Charging which reflects the cost of collecting, checking, and packaging data actually measures 'real need' and forces organizations to establish their real priorities. (*But equal charges to all do not necessarily force setting of priorities since not all organizations have equivalent purchasing power, e.g. utility companies can pay more than charities.*)

- Users exert more pressure when they are paying for data and, as a consequence, data quality is usually higher and the products are more 'fit for purpose'.
- It is equitable since the number of data users is presently small compared to the number of taxpayers. Hence, cost recovery minimizes the problem of subsidy of some individuals at the expense of the populace as a whole. (*Some users are acting on behalf of the populace, such as local governments.*)
- Empirical evidence shows that governments are more prepared to part-fund data collection where users are prepared to contribute significant parts of the cost. Hence full data coverage and up-dating is achieved more rapidly where cost recovery at a reasonable level can be achieved. (*But once the principle is conceded, government usually seeks to raise the proportion recovered inexorably.*)
- It minimizes frivolous or trivial requests that may well require much resource and detract from the specified functional objectives of the government body concerned. (*What is the real role of government?*);
- It enables government to reduce taxes in comparison to the level they might otherwise have reached.

4.6.2 Arguments for dissemination of data at zero or copying cost by public bodies

- The data are already paid for, hence any new charge is a second charge on the taxpayer for the same goods. (*This is unlikely to be true if the data are constantly being up-dated; moreover, there are generally many more taxpayers than users.*)
- The cost of collecting revenue may be large in relation to the total gains. (*Since the latter, including benefits of prioritization, are unmeasurable, this cannot be proved even though it may be true. It does not seem to be true in the private sector.*)
- Maximum value to the citizens comes from widespread use of the data through intangible benefits or through taxes paid by private sector added-value organizations. (*The first of these is unmeasurable and the second may fall into the same category, though possibly capable of being modelled; as a result, the contention cannot be proved. In any case, this point is irrelevant where costs and benefits are reckoned by government on an organization-by-organization basis rather than as integrated national accounts.*)
- The citizen should have unfettered access to any information held by his/her government. (*This is a matter of political philosophy rather than economics so is not considered further here.*)

4.7 Measuring the benefits of an NGDI

If few studies have been carried out on the costs of NGDI components, even fewer meaningful attempts have been made to assess the benefits of having an

NGDI. A rare exception is Price Waterhouse (1995), who claimed a substantial tangible cost-benefit from investing in NGDI-like components (see also Smith and Thomas, 1996). Coopers and Lybrand (1996) laid out the basis on which such assessments could be carried out and made some observations on what seemed likely to be the case for public sector investment.

4.8 The global GDI

If defining and then arguing the case for an NGDI is non-trivial, the equivalent actions for a global one are somewhat more complex. This has not, however, prevented much discussion (see Baranowski, 1996, and *http://www.gov.state.nc.us/gGDI97*). The need for some degree of international harmony in the existence and quality of geospatial information is evident (see, for example, Brandenberger and Ghosh, 1990; Collins and Rhind, 1997; Estes, Lawless, and Mooneyhan, 1995; Estes, Kline, and Labonne, 1997; Htun, 1997).

But if there is a plurality of players, all with vested interests, at the national scale, the situation is multiplied greatly at the international level. In education, for instance, there is now a large and growing international market with many universities seeking to increase their number of students from overseas countries. In so doing, they are in active competition with other similar bodies and, in some respects, with private sector software providers and information brokers.

It is also clear that national pre-occupations ensure that the NGDIs differ in some significant ways. The French approach and concerns (Grelot, 1998) are clearly very different to those in the USA (see *http://fgdc.er.usgs.gov*), the UK (Nanson and Rhind, 1998: see also *http://www.ngdf.org.uk*), and in Asia and the Pacific (see Majid, 1997 and *http://www.permcom.apgis.gov.au/index.html*). Given this, no simple melding of the national schemes will obviate the identified problems. This is even more the case when the needs of the different players are taken into consideration. For example, the business and strategic plans of the US military National Imagery and Mapping Agency place great emphasis on meeting the needs of their primary customers (the 'warfighters' of the USA and allied forces) to the exclusion of meeting the needs of others (NIMA, 1997*a*, 1997*b*). Indeed, there is vital competitive advantage in ensuring that key information does not leak out to other parties.

It was pointed out in an earlier section that the great bulk of NGDI funding is provided by a number of public sector organizations and changes can only be made by influencing their dispositions of expenditures. Clearly this is even more difficult in the international arena. Bodies like the United Nations have very limited discretionary funds and their primary ability to influence is through obtaining agreement on particular policies, which then may not be universally implemented (consider, for instance, the agreement on data sharing embedded in Agenda 21: see UNCED, 1992).

In practice, the situation is even more complex for private sector players who are set to achieve a greater uniformity of influence at the global than at the individual national levels. The prime manifestation of this is the likely advent of high resolution satellite imagery for almost anywhere in the world on a speedily available basis. Such a development has profound ramifications for traditional GDIs and for the 'information poor'—their lot may become even worse. Such ramifications would be even greater if some of the multi-national players were to take over a key supplier of GIS software, education and training facilities, becoming in effect a quasi-monopoly, as a vertically integrated supplier of a global GDI.

4.9 Conclusion

From our fairly broad brush estimation of the cost of maintaining the NGDI we cautiously conclude that the identified costs per person are probably of the same order of magnitude in the USA and UK. Funds specifically provided by governments to foster NGDI seem to be tiny fractions of the total existing expenditure in the geospatial information infrastructure arena: different portions of these expenditures have been justified on quite different grounds.

However, the assumptions we had to make in this analysis have also revealed that we know very little about how much money and other resources are actually being spent on maintaining the existing national GDIs, let alone on creation of enhanced versions of them, or who is providing these resources. In broad terms, we do not know whether these resources are being applied wisely. It would seem helpful, therefore, to carry out some sound accounting of this expenditure: arguments for adding to it or for using it more effectively or efficiently are unconvincing if we do not know the current practice. This will require, of course, a widely agreed and used definition of what is subsumed by the term 'National Geospatial Data Infrastructure'.

The initial evidence and calculations also suggest that funding has highly pluralistic sources and is therefore difficult to track in totality. But it seems virtually certain that the bulk of expenditures are currently provided by governments using moneys generated through taxation. Considerable variations in the pattern exist and the trend is towards more of a 'user pays' philosophy. That said, this situation is largely true in the areas of information collection and provision. It is also largely true in the arena of laws, but private sector-driven initiatives are increasingly influencing the standards element of NGDI. In the field of education and training, governments have some influence but diminishing control as market-led decisions by individuals become more common. In the case of the software and physical infrastructure components, the private sector is the key provider, at least in the Western nations, and the public sector influence is indirect through provision of incentives, support of national or regional industries, etc.

In terms of achieving some form of global GDI, additional complexities mostly of a non-technical but rather cultural character need to be considered. The absence of accepted institutions which could authoritatively deal with the associated issues is just one of the barriers to this. For institutional, financial, political, historical, and other reasons, this will be more difficult than the enhancement of existing national GDIs. It is possible, however, that commercial developments might become the driver for a much more consistent and widely available global GDI.

It is often argued that no one can own the national GDI (see, for example, Tosta, 1997). That is true in one important sense but the key players—and, in a practical sense, the owners—are those who invest the bulk of resources in creating and maintaining that entity. Some of these players can be influenced by government edict, some require financial incentives. At the global level, however, the decision-making forums are many and various: there is over-lapping responsibility for GGDIs, and few of the organizations involved have disposable resources akin to the global private sector players which, driven entirely by commercial agendas, are appearing on the scene.

Simplistic conclusions may be drawn from observation of the 'hype' underlying national GDIs to date and the complexities of the global scene. Many busy individuals will conclude that committing organizational and personal resources to such embryonic 'things'—which operate in complex environments, where traditional management is impossible and where rewards seem at best to be measured in approving words by politicians—are difficult to justify. Yet that would probably be a mistake: there are substantial benefits that could arise from having enhanced NGDIs, both nationally and at a global level. It is necessary to demonstrate the likely consequences of the required investments and to convince the investors that suitable returns can be generated in whatever currencies they value. Despite persuading key individuals like the Secretary of the US Department for the Interior, or the UK government's minister in charge of the Ordnance Survey, of the importance of NGDIs, we are nowhere near the necessary rigorous assessment of the business merits, with 'business' viewed in a wide sense, of GDI-related developments. There is no well-articulated and intellectually and financially convincing case for them—and hence they are not yet sustainable in the longer term. Once enthusiastic individuals move on, we are likely to find others with different enthusiasms. Thus, until such time as better justifications are produced, and widely and convincingly adopted, the resources available for building NGDIs will remain to be drawn serendipitously from a multiplicity of sources. As a result, any consistency and coherence of a national or global GDI plan will be an accidental result. And accountability for the GDI will thus be impossible, with obvious consequences. It is time to create formal business plans for (N)GDIs.

Acknowledgements

Thanks are due to Steve Guptill, Bryan Nanson, and Karen Siderelis for their comments and suggestions on a draft of this chapter; however, responsibility for any errors rests with the author.

Bibliography

BARANOWSKI, K. (1996). *Proceedings of the Emerging Global Spatial Data Infrastructure Conference*, Bonn, Germany 4–6 September (see also *http://www.eurogi.org/gGDI/index.html#documents*).

BRANDENBERGER, A. J. and GHOSH, S. K. (1990). 'Status of world topographic and cadastral mapping', *World Cartography*. New York: United Nations, vol. xx, pp. 1–102.

CLINTON, W. J. (1994). Executive Order 12906. *Co-ordinating Geographic Data Acquisition and Access: The National Spatial Data Infrastructure*. Washington, DC: 2 pp.

COLLINS, M. and RHIND, J. (1997). 'Developing global environmental databases: lessons learned about framework information', in RHIND, D. (ed.), *Framework for the World*. Cambridge: GeoInformation International, pp. 120–9.

Coopers and Lybrand (1996). *Economic Aspects of the Collection, Dissemination and Integration of Government's Geospatial Information*. Southampton, UK: Ordnance Survey.

ESTES, J., LAWLESS, J., and MOONEYHAN, D. W. (1995). (eds.), *Report on the International Symposium on Core Data Needs for Environmental Assessment and Sustainable Development Strategies*, United Nations Development Programme/United Nations Environment Programme, New York.

——KLINE, K., and LABONNE, B. (1997). *Proceedings of the International Seminar of Global Mapping for the Implementation of Multinational Environmental Agreements*, Department of Geography, University of California, Santa Barbara.

FGDC (1994). *The 1994 Plan for the National Spatial Data Infrastructure: Building the Foundations of an Information Based Society*. Reston, VA: Federal Geographic Data Committee, USGS.

GRELOT, J.-P. (1997). 'The French Approach', in RHIND, D. (ed.), *Framework for the World*. Cambridge: GeoInformation International, pp. 226–34.

——(1998). 'Infrastructures de données spatiales: situation en France'. *Proceedings of Canadian Geospatial Data Infrastructure Conference*, Ottawa, July.

HTUN, N. (1997). 'The need for basic map information in support of environmental assessment and sustainable development strategies', in RHIND, D. (ed.), *Framework for the World*. Cambridge: GeoInformation International, pp. 111–19.

MAJID, D. A. (1997). 'Geographical data infrastructure in Asia and the Pacific', in RHIND, D. (ed.), *Framework for the World*. Cambridge: GeoInformation International, pp. 206–10.

MASSER, I. (1998). *Governments and Geographic Information*. London: Taylor and Francis.

NANSON, B. and RHIND, D. (1998). 'Establishing the UK National Geospatial Data Framework', *Proceedings of the Canadian Geospatial Data Infrastructure Conference*, Ottawa, July (see also *http://www.ngdf.org.uk/whitepapers/*).

NAPA (1998). *Geographic Information for the 21st Century: Building a Strategy for the Nation*. Washington, DC: National Academy of Public Administration.

NIMA (1997*a*). *Business Plan 1997*. Washington, DC: National Imagery and Mapping Agency.

NIMA (1997*b*). *Strategic Plan 1997*. Washington, DC: National Imagery and Mapping Agency.

Price Waterhouse (1995). *Australian Land and Geographic Infrastructure Benefits Study*. Canberra: Australian Government Publishing Service.

RHIND, D. (1997*a*). (ed.), *Framework for the World.* Cambridge: GeoInformation International.

——(1997*b*). 'Facing the challenges: redesigning and rebuilding Ordnance Survey', in *Framework for the World.* Cambridge: GeoInformation International, pp. 235–46.

——(1999) 'National and international geospatial data policies', in LONGLEY, P., GOODCHILD, M., MAGUIRE, D. and RHIND, D. (eds.), *Geographical Information Systems: Principles, Techniques, Management and Applications.* New York: Wiley.

SANDGREN, U. (1997). 'Merger, government firms and privatization? The new Swedish approach', in RHIND, D. (ed.), *Framework for the World.* Cambridge: GeoInformation International, pp. 235–46.

SMITH, M. and THOMAS, E. (1996). 'National spatial data infrastructure: an Australian viewpoint', *Proceedings of the Emerging Global Spatial Data Infrastructure Conference*, Paper 7, Bonn, Germany, 4–6 September, 16 pp.

SMITH, N. and RHIND, D. (1999). 'Characteristics and sources of framework data', in LONGLEY, P., GOODCHILD, M., MAGUIRE, D., and RHIND, D. (eds.), *Geographical Information Systems: Principles, Techniques, Management and Applications.* Chichester, UK: Wiley.

TSO (1998). *Crown Copyright in the Information Age.* A consultation document on access to public sector information. Command 3819. London: The Stationery Office.

TOSTA, N. (1997). 'National Spatial Data Infrastructures and the roles of National Mapping Organizations', in RHIND, D. (ed.), *Framework for the World.* Cambridge: GeoInformation International, pp. 173–86.

UNCED (1992). *Agenda 21.* New York: United Nations Commission on Environment and Development.

5

The role of standards in support of GDI

Peter L. Croswell

5.1 Introduction

In the 1990s the popularity surrounding the topic of standards in geospatial information systems has contributed to a greater awareness of their purpose, but it has also resulted in considerable confusion about standards implementation (see Croswell and Ahner, 1990; Moyer and Niemann, 1993; and Wellar, 1972). The wisdom in the industry today says that a user organization must adopt standards to deploy and support a successful geospatial data infrastructure, but many questions remain: What types of standards are appropriate? Which specific standards are current? How can they best be implemented to support and encourage GDI development and use rather than inhibit it?

The geospatial information user community is rife with examples of problems and inefficiencies resulting from a lack of standards or ineffective use of standards. Consider these examples:

- A growing local government establishes no clear standards on local or wide area network infrastructure or protocols. The demand for GIS and other information systems increases, and multiple incompatible networks spring up making interoperability and connectivity very difficult. Satisfying user needs for computing capability and providing access to different computer platforms require a complex array of communication hardware and software to link systems together, and an increasing support staff to keep the network operating.
- A utility company has no clear standards on operating systems and computer hardware to guide procurements. Individual divisions in the organization purchase systems to meet specific needs without a full overview of long-term information sharing requirements. The growing patchwork of

PCs, servers, and mid-range systems are expensive to maintain. The situation inhibits effective integration and presents obstacles to a systematic approach to computer upgrades.

- A regional agency with growth planning responsibilities is gathering land use information from multiple agencies (e.g. local government organizations) within the region to support short-term and long-term planning decisions. The regional agency finds that a wealth of land use information is collected and updated by multiple organizations, but it is captured according to multiple, incompatible classification schemes. A great deal of time is needed to resolve these differences and provide a basis for comparison and data integration.
- A large local government spends big sums of money on a GIS database development project to serve multiple departments. Lack of standards and procedures to document the characteristics, quality, and availability of this database presents obstacles to its effective access and use, and to its ongoing maintenance.
- A national government has multiple agencies that gather and collect geospatial information for their programmes. Mapping activities supporting environmental protection, transportation, natural resources, social services, demographic statistical collection, and defence result in overlapping activities and redundant data compilation, thereby wasting large sums of money. A lack of standards in database design and poor co-ordination limit sharing or re-use of the data.
- A multi-national military peacekeeping body fails to adopt a unified convention for map compilation and symbology, frustrating operations among units from different countries.

These examples illustrate cases where effectively applied standards would ease real problems in the efficient use and maintenance of geospatial information infrastructures. These problems range from computer system issues to the design of GIS databases and the products generated from them. Standards can facilitate the sharing of information and computer resources *within* an organization, and *between* organizations. As underscored in Tom (1988), URISA (1998), Ventura (1993*a*, 1993*b*), and Wellar (1973), standards are not an end in themselves but the foundation to help make information systems and databases easier to use and maintain.

The value of wisely chosen standards for geospatial information users is reflected in three primary themes illustrated in Figure 5.1:

1. *Portability*, with the concept of 'interchangeable parts' implies an ability to use and move data, software, and custom applications among multiple computers and operating system environments without re-tooling or reformatting.

2. *Inter-operability and information access* impact computers and networks, and the users' ability to connect and retrieve information from multiple systems.

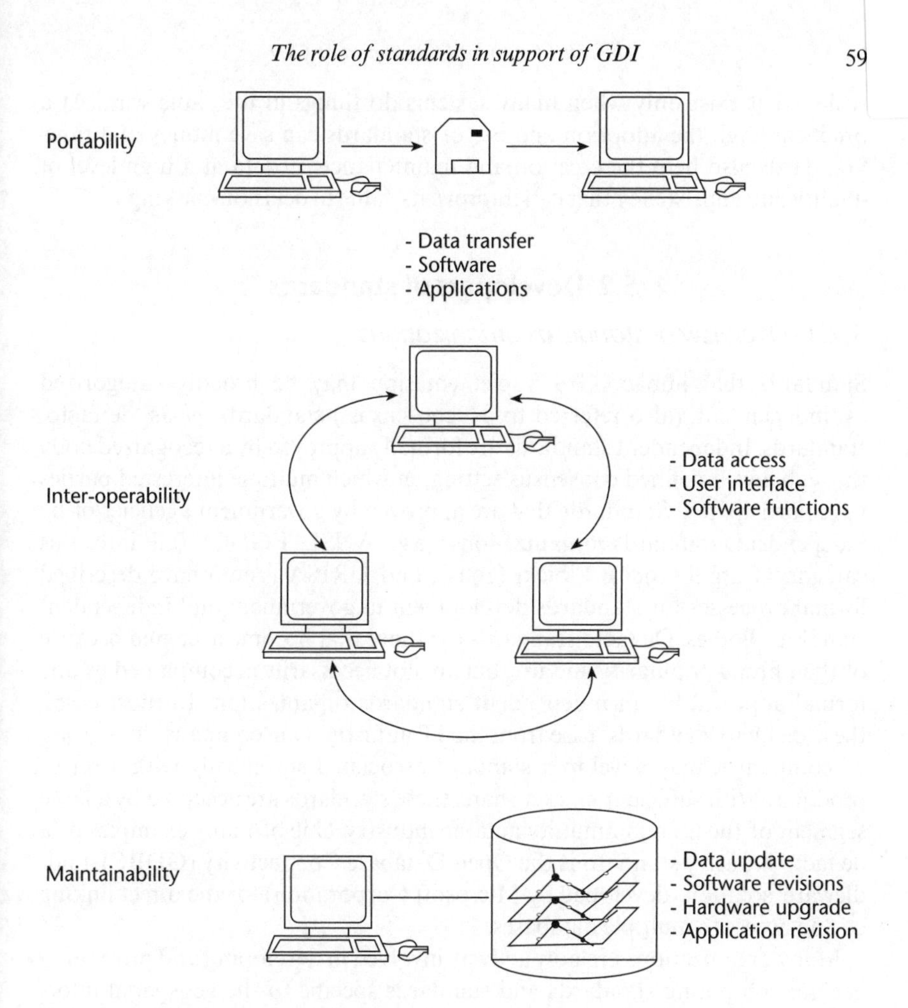

FIGURE 5.1 Primary themes of information technology standards.

3. *Maintainability* addresses the use of standards to promote long-term and efficient updating, upgrading, and the effective use of computer systems and databases.

Real examples of the effective use of standards to facilitate information sharing have been described in Dueker and Vrana (1995), Bamberger (1995), and Levinsohn (1997). For designers and developers of geospatial information systems, the question is not whether standards should be adopted, the challenge is to choose suitable standards and a sensible approach for their implementation, to facilitate sharing of information and to make systems easier to support and maintain.

Libicki (1995: 7) responds to the question, 'What are standards good for?' in this way: 'Standards solve particular problems, such as how to represent data efficiently or manage a communications system, and they create benefits—interoperability, portability, ease of use, expanded choice, and economies of scale—that exist only when many systems do things in the same way.' At a practical level, the adoption and use of standards can save money and time. Standards also help the creation and maintenance of data at a high level of quality and consistency that can improve its value in decision-making.

5.2 Developers of standards

5.2.1 Overview of standards organizations

Standards that impact GIS implementation may be broadly categorized as 'independent' (also referred to as 'consensus') standards, *or* as 'de facto' standards. Independent standards are formally approved by a recognized body through a well-defined consensus setting, in which multiple interested parties have participated. Standards that are approved by government agencies or by independent standards organizations (e.g. ANSI, FGDC) fall into this category. Cargill (1989), Libicki (1995), and URISA (1998) have described formal processes for standards development in government and independent standards bodies. De facto standards are those that become accepted because of their broad popularity and use, but are not necessarily accompanied by any formal approval by an independent standards organization. In most cases, these de facto standards arise from the IT industry. One company, or a group of companies, may develop a standard associated specifically with a set of products. With sufficient market share, these standards are accepted by a large segment of the user community and the industry. One of many examples of a de facto industry standard is the Open Database Connectivity (ODBC) 'middleware' solution (developed by Microsoft Corporation) for the direct linking of and access to multiple databases.

Many organizations are now actively involved in developing and promoting general computing standards and standards specific to the geospatial information technology industry. Government organizations establish and enforce standards for more consistent communication among agencies and international bodies, and with multi-national representation, seek to encourage communication standards globally. Several independent professional associations and industry consortia are also heavily involved in the standards movement. In many cases, organizations in the categories described below work together on the development and approval of formal standards.

National government organizations

These are national government agencies in all industrialized nations with specific responsibility for approval of information system standards (and other types of standards) for use by their constituencies. Examples include the National Institute of Standards and Technology (NIST) in the USA, and the Canadian General Standards Board (CGSB). These groups and similar organizations elsewhere address all types of standards, including information system standards in general and specific standards impacting geospatial data and software. Some national governments have created bodies that focus solely on geospatial information standards with a goal towards developing national GDI. The Federal Geographic Data Committee (FGDC) in the USA and the Canadian Inter-Agency Committee on Geomatics are two examples.

Independent standards bodies

These formal standards bodies work in a consensus building process to adopt and promote formal standards. They include representation from government agencies, professional organizations, and private companies. Independent standards bodies have open policies for membership (with rules for participation) and formal committee structures and procedures for standards development, review, and approval. Examples include the American National Standards Institute (ANSI), the Institute of Electrical and Electronic Engineers (IEEE), the International Telecommunications Union (ITU), the European Committee for Standardization (CEN) and the International Organization for Standardization (ISO).

Industry consortia and trade associations

These are formal or informal associations, predominantly made up of information system product or service companies with missions for joint definition, development, and promotion of standard-based products for their customer base. Many of these organizations, some with limited life spans, are created to address specific market niches and demands. Examples include the Open Group (formerly X/Open), the Object Management Group (OMG), and the OpenGIS™ Consortium (OGC).

Professional organizations

There is a range of professional organizations with missions involving education, interaction between members, and a review of proposed standards. Examples include the Urban and Regional Information Systems Association (URISA), the International Association of Assessing Officers (IAAO), and the Association for Computing Machinery (ACM). In some cases, professional associations formally participate in government and independent standards bodies.

Standards groups in the USA, Canada, Europe, and the Far East have become very active in the past five years in developing standards for geospatial processing and management. At the international level, the International Organization for Standardization (ISO) is the focus for standards development. In 1993, a GIS Committee was formed within the ISO to address a wide range of standards that impact GDI. This Committee, now called the Geographic Information/Geomatics Committee (Technical Committee 211), includes participants from most of the developed and developing nations. The Committee has established Working Groups to address standards that cover such topics as data transfer, metadata, data classification, data quality, and other related topics (see URISA, 1998; and Ostensen, 1996).

5.3 Types of standards impacting GDI

Standards of importance to geospatial information users range from the details of computer hardware and networks to the design of databases and map products. Standards may be categorized into low-level and high-level categories (URISA, 1998).

The low-level standards cover detailed technical concerns in the following categories:

- hardware and physical connection standards
- network communication and management standards
- operating system software standards.

Low-level standards are vitally important for the interoperability of computer systems and provide the basic computing and communication infrastructure for all system integration and information sharing. Development of hardware and software products that comply with these low-level standards is the domain of the computer industry.

High-level standards deal primarily with the following database design, data exchange, and presentation topics:

- user interface standards
- data format/data exchange standards
- programming and application development standards
- user design standards.

Developers of geospatial information systems generally have a great deal of influence over the shape of high-level standards and how they are applied within their organizations. The process of database design, establishing procedures for data exchange, and the development of custom applications requires wise decisions about standards to encourage consistency and information standards. Table 5.1 presents a list of standards categories with a description and examples.

5.3.1 Hardware and physical connection standards

This class of standards addresses important concerns relating to the basic architecture, physical connection, and cabling of hardware devices. Many accepted standards for cabling types, electrical interfaces, and cable connectors are now widely adhered to in the computer industry, although physical standards continue to emerge as communication technology advances into higher speed communications using fibre optic and wireless communication technology (see Slone, 1998). The computer hardware industry, with support from independent standards bodies such as the IEEE and the Electronics Industry Association (EIA), has been prompt in establishing physical connection standards. Also included in this category are physical format standards for

TABLE 5.1 Types of computing standards with selected examples

Standards categories	Explanation and examples
Hardware and physical connection standards	
Cabling and couplers	Cabling types (e.g. twisted wire, coaxial, fibre optic), physical couplers, and connectors. Standards approved by the Electric Industries Association (EIA), the Institute of Electrical and Electronic Engineers (IEEE), and the International Organization for Standardization (ISO).
Electrical interfaces	Voltage and frequency standards for data communication lines. Standards approved by the Electric Industries Association (EIA), the Institute of Electrical and Electronic Engineers (IEEE), and the International Organization for Standardization (ISO).
Computer hardware design	Standards governing computer architecture (processor chips, memory, internal bus, monitors). De facto standards developed by leading computer companies (Intel, IBM, others).
Storage media format	Physical format for data storage on tapes, magnetic disks, and optical storage media. Examples include such industry standards (from individual companies or industry consortia) as DAT and QIC for tape, CD-ROM and Digital Versatile Disk (DVD) for removable optical disks, and redundant array of independent disk (RAID) for magnetic storage.
Network communication and management standards	
Local network protocols	Protocols supporting communication on network connecting devices in close proximity over direct cabling schemes to support speeds of 10 megabits per second to 1 gigabit/second (Ethernet, Fast Ethernet (100BaseT), Asynchronous Transfer Mode, Gigabit Ethernet). Some approved by the International Organization for Standardization (ISO) and the American National Standards Institute (ANSI).
Wide area network protocols	Protocols supporting communication among widely spaced devices using remote communication media (e.g. Frame Relay Asynchronous Transfer Mode (ATM)).

TABLE 5.1 (*cont.*)

Standards categories	Explanation and examples
Wireless data communications	Industry standard protocols to support digital communication over the airwaves (radio frequencies, microwave). An example is the Cellular Digital Packet Data (CDPD) standard developed by an industry consortium of data communication vendors.
High-level protocols	Communication protocols to support high-level services such as file transfer, e-mail, and file encryption. Examples are simple mail transport protocol (SMTP), FTAM, and FTP file transfer protocols.
Communication protocol suites and models	ISO Open Systems Interconnect Model, TCP/IP protocols and services.
Operating system software standards	
Operating systems	Portable operating systems allowing more flexible networking and applications which run without modification on multiple platforms (UNIX standardization through the Open Group; Windows 95 and Windows NT as industry standard).
Distributed network management	Capabilities and software products to provide effective monitoring and administration of networks. Products based on such standards as Desktop Management Interface Profile, X.500 directory services standards, the Common Management Information Protocol (CMIP), and the Simple Network Management Protocol (SNMP).
Object management architectures	Standards and compliant products for object management and communication. Examples include the Common Object Request Broker Architecture (CORBA), the distributed Common Object Model (DCOM), and interoperability standards being developed by the OpenGIS™ Consortium.
User interface	
Industry standard graphic user interfaces	Standard GUI packaged with the operating system or off-the-shelf application. MOTIF (UNIX) and Microsoft Windows as basic standards.

TABLE 5.1 (*cont.*)

Standards categories	Explanation and examples
Custom application GUI design	Need for consistency within organizations for custom interface look-and-feel.
Data format standards	
Geospatial data formats and exchange (vector)	De facto industry and government graphic data standards (SIF, DXF, DLG, etc.) and formal intermediate exchange formats for GIS (SDTS, DIGEST).
Geospatial data formats and exchange (raster)	Standard file formats for raster image and raster document data, including TIFF, CCITT, JPEG, GIF. SDTS Raster Profile for raster data exchange.
Attribute data exchange and access	Exchange of attribute data among disparate systems and interactive access supported by standards as ODBC and SQL.
Inter-operability and transparent geospatial data access	OpenGIS™ Specification for interoperability. Growth of RDBMS-based GIS software products encourages open access.
Programming and application development standards	
Open application development tools	Industry standard tools with access to GIS functions at executable level (Visual Basic, Delphi).
SQL spatial extensions	Work in ANSI and ISO towards approval of SQL/MM spatial standard.
Internet and World Wide Web	Increasing use of Internet for spatial data queries, analysis, and data distribution with popularity of Java language.
User design standards	
Database schemas	Attribute data file parameters such as field length, format, and other characteristics of data elements. One example is standards for formatting address information (e.g. US Census Bureau).
Geospatial data coding and classification	Classification and coding schemes for data elements providing consistent reference by multiple users (e.g. standard land use or zoning codes).
Geospatial metadata	Databases storing information about geospatial data sets. Example: The Federal Geographic Data Committee (FGDC) *Content Standard for GeoSpatial Metadata*

TABLE 5.1 (*cont.*)

Standards categories	Explanation and examples
Map design	Consistent selections of feature content by layer; placement of annotation; and use of symbols, line types, shading, and colour.
Map accuracy	Standards governing the accuracy of survey control points and the horizontal and vertical positional accuracy of maps. Example: Subcommittees of the Federal Geographic Data Committee (FGDC) are developing a standard called the 'Geopositional Accuracy Standards'.
Map presentation	Standards and mapping for an organization or discipline governing scale and sheet format, symbology, colours, annotation style, and placement for specific types of maps. Standards exist for such map types as USGS Topographic Maps (US Geological Survey), water utility maps (American Waterworks Association), parcel maps (state agencies and the International Association of Assessing Officers), and engineering maps and drawings (state Transportation Departments).

mass storage of data on tape and disk. GDI users will not normally be directly involved with the details of industry standards in this hardware/physical connection category. Those involved, however, in system design and support should be aware of the status of these industry standards and the ways in which computer hardware manufacturers are using them in their products.

5.3.2 Communication and network management standards

A data communication network provides the physical infrastructure for sharing information in a GDI. Communication of data simply involves the transfer of binary digits (bits). For devices and software to make sense of these bits, however, they must be 'packaged' in a specific format. This is the role of a particular communication 'protocol'—to describe how the bits are arranged so that they can convey all the information necessary for effective communication. Protocols describe the format of transmission to accomplish the following tasks:

- transfer data from one device or node to one or more destination nodes on a network;
- control routing of messages between multiple networks;

- provide information for error checking and connection;
- convey data to support network monitoring and management.

Incompatibilities in protocols among multiple computer systems will frustrate interoperability and system integration. National and international standards organizations have made significant progress in drafting standard protocols that have been adopted by many computer hardware and software vendors to promote network integration (see Annitto and Patterson, 1995; Microsoft Press, 1998; and Slone, 1998).

The most ambitious attempt at comprehensive standardization in communication protocols is represented by the Open Systems Interconnect (OSI) model defined by the International Organization for Standardization (ISO). It is structured as a series of 'layers' for communication between devices (see Table 5.2). Lower levels of the OSI model address very basic needs for connectivity between devices on networks. Higher levels respond to increasingly sophisticated requirements for network communication. The OSI Model does

TABLE 5.2 Layers of the ISO Open Systems Interconnect model

OSI layer	Explanation
7. Application layer	Organizes information to support specific applications such as e-mail, file transfer.
6. Presentation layer	Provides applications with the means to interpret information exchanged on a network and provides specific data formatting rules. Character coding (e.g. ASCII) and data compression standards are examples.
5. Session layer	Provides services for high-level monitoring, synchronization, and error control in support of message transport.
4. Transport layer	Establishes routing paths for message and for transparent end-to-end flow of data from origin to destination.
3. Network layer	Provides services for creating packets for transmission through a network reserving space for address and error control information handled by Layer 4. Standards guiding high-speed communications services such as Frame Relay and Asynchronous Transfer Mode (ATM) apply here.
2. Data link layer	Establishes protocols for proper transmission of messages between two network nodes and detects errors that occur at Layer 1. The ISO has approved Ethernet and Token Ring standards for this layer.
1. Physical layer	Defines electrical interfaces, cabling, and physical connection standards. Many IEEE, ITU, and EIA standards apply at this level.

not define specific protocols, but protocols addressing the requirements of this series of layers are evaluated and accepted by the ISO if they comply with the tenets of the model.

The computer industry has been slow to develop products that fully comply with all levels of the OSI model, but it has provided a workable model to promote interoperability on data communication networks and the development of products with at least partial OSI compliance. For example, the Ethernet standard, accepted by ANSI and the ISO, is a local area network standard that addresses Layer 2 of the OSI model.

5.3.3 Operating system software standards

Core operating system issues

Traditionally, the core software or 'operating system' that directs all fundamental operations of a computer system was unique to particular makes and models of processing units (e.g. IBM MVS, Digital Equipment Corporation VMS, etc.). Figure 5.2 shows the current industry competition for operating system control that makes standards decisions difficult for users.

Over the past 15 years, the UNIX operating system has been the primary focus of GIS vendors and users alike. This was in part due to the promise of its 'portability', or its ability to run on different computer models from different vendors. UNIX portability, however, has not been fully realized because of the proliferation of many UNIX varieties developed by computer manufacturers for their own particular models. UNIX's status as the choice of GIS vendors and users has been challenged by the great popularity of Microsoft Windows™—particularly Windows NT™ and Windows 95™. Windows NT entered the market by offering a less-expensive option for use on desktop personal computers. NT has by no means replaced UNIX, rather it has offered GIS users an alternative. At the time of writing, UNIX systems still hold the edge over NT in performance and system administration tools, but GIS software built to run in an NT and Windows 95/Windows 98 environment is increasing in popularity, with NT becoming a de facto standard operating system for GIS. The growth of the Internet as an environment for geospatial information exchange may threaten the growing market share of the Windows environment by providing non-proprietary JAVA-based applications.

One of the most critical low-level standards decisions that an organization must make is the choice of operating systems. This should be viewed as a long-term decision that will guide procurements and system development. Effective GDI will require access to information on many different systems. It is therefore important that the operating system standards take into account both overall information system standards in the organization and trends in the computer industry. The latter drive market share and the availability of software products compatible with the operating system. Wise decisions about operating system deployment will promote interoperability and portability (Croswell, 1993).

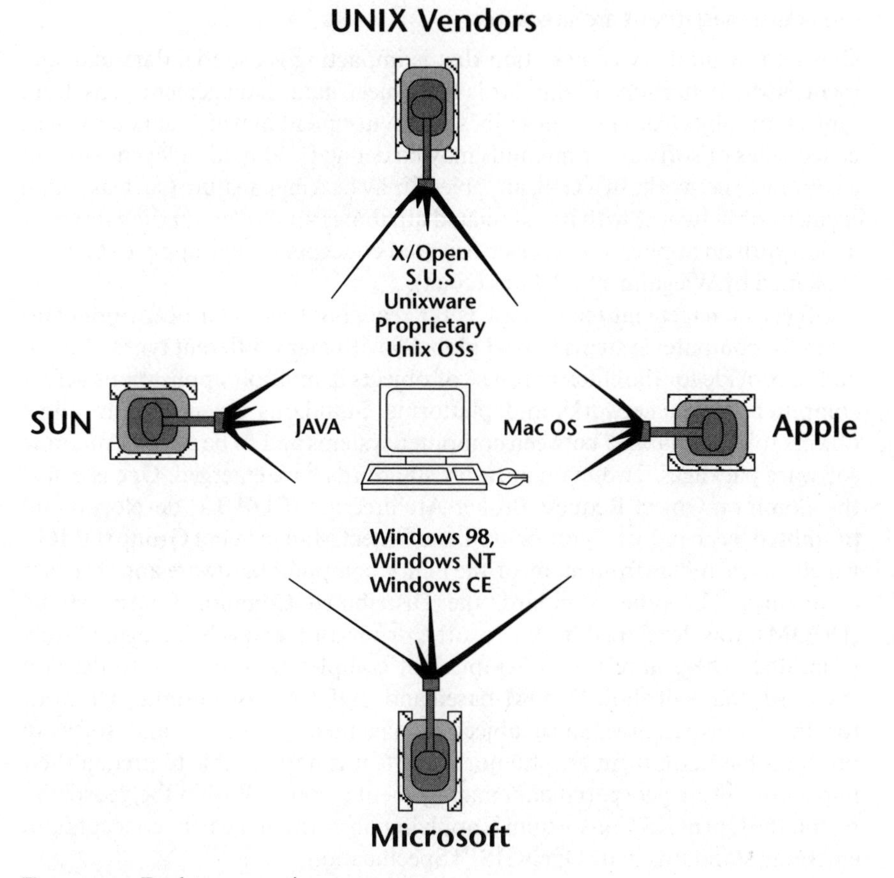

FIGURE 5.2 Desktop operating system wars.

System and network management

In complex computer networks supporting many servers and workstations, there is an important need to manage the network. This includes such functions as: (1) monitoring and reporting on system use, (2) directing and monitoring network traffic, (3) problem diagnosis and correction, and (4) establishing access security and such high-level management tasks as automatic network mapping, hardware/software inventory, data back-up, software distribution, and software licence management.

Most computer vendors provide capabilities addressing many of these needs as a standard part of their operating system, and many add-on software packages support these important system administration functions. The computer industry has responded to this need with a wide variety of software and hardware products, which are, in part, based on independently approved and industry-accepted standards. *It is vitally important that GDI managers adopt sound procedures and standards for system and network administration.*

Object management architectures

One area of industry competition that is impacting geospatial data management is the definition of standards for object data management. The term 'object' or 'object-oriented' describes an environment in which data and associated rules or software commands may be defined and used independently in a computer network. In a GIS, an 'object' may be a map feature (such as a pipe segment or network) with its associated attributes and 'rules' guiding its interaction with an application. Object-oriented concepts as they apply to GIS are explained by Wiegand and Adams (1994).

Object management technology is far reaching because it opens opportunities for computer systems to deal globally with many different types of data, and to provide for the efficient re-use of objects in multiple applications across computer system networks and platforms. Standards are needed to allow objects to be exchanged between computer systems and to be used by multiple software packages. Two main industry standards have emerged. One is called the Common Object Request Broker Architecture (CORBA) developed and promoted by an industry consortium, the Object Management Group (OMG), which has members from many of the major computer hardware and software companies. The other standard, the Distributed Common Object Model (DCOM), was developed by Microsoft Corporation and is being aggressively promoted. There have been attempts, not completely successful, to develop standards that will allow DCOM-based and CORBA-based products to work together transparently. Since object management standards and software products based on them are still not mature, it is not possible to predict their impact on future geospatial data management systems. Within the geospatial realm, the OpenGIS Consortium is applying object management concepts and emerging standards in its OpenGIS™ Specification.

5.3.4 User interface

A popular topic in data processing today is the graphical user interface (GUI), which is the most common vehicle for user interaction with the computer systems (usually through a mouse). In addition to providing a flexible development environment for programmers, GUIs provide an efficient way for users to invoke commands and access portions of the database (as opposed to entering commands from a keyboard). A standard GUI presents a common 'look-and-feel' for users and programmers, and supports the concept of portability if GUI standards are used with different software packages.

Recently, a number of industry-standard graphic user interfaces have been launched; each has gained considerable market share but none, to date, has become predominant. These include MOTIF and the Common Desktop Environment (CDE) developed by the OSF consortium (recently merged with X/Open to form the Open Group), and Open Look developed by the UNIX Systems Laboratory (this organization no longer exists). Industry forces, driven by the large number of users of Microsoft's Windows operating systems, have made the Windows look-and-feel a de facto standard in the GIS industry.

Many GIS applications specific to a particular organization require customized user interfaces designed and developed by users. These interfaces apply GUI widgets (drop down menus, pick lists, buttons, slide bars, etc.) that have become standard in the industry. Along with the need for industry-wide standardization of GUIs, it is important that standards be set within organizations to ensure consistency in the look-and-feel for all applications and users in the organization.

5.3.5 *Data format, exchange, and access*

This class of standards covers the logical and physical structure of geospatial data and capabilities for access and exchange of geospatial data among multiple computer platforms and applications. Daniel (1993) and Kottman (1992) explain that specific formats for storing vector and raster data are almost as numerous as vendors offering GIS or automated mapping software. While there has been movement in the industry towards more open geospatial data formats, proprietary data structures predominate, often necessitating batch file translation and loading for moving data from one system to another.

Geospatial data exchange

GIS developers and users have focused on the need for the exchange of data (and the challenges to implement it) for many years (see Friesen and Sondheim, 1994; Jiwani, 1988; Kindrachuk, 1992; Lee and Coleman, 1990; Peled and Adler, 1993; and Wellar, 1973). This exchange has been addressed through two different strategies, as shown in Figure 5.3.

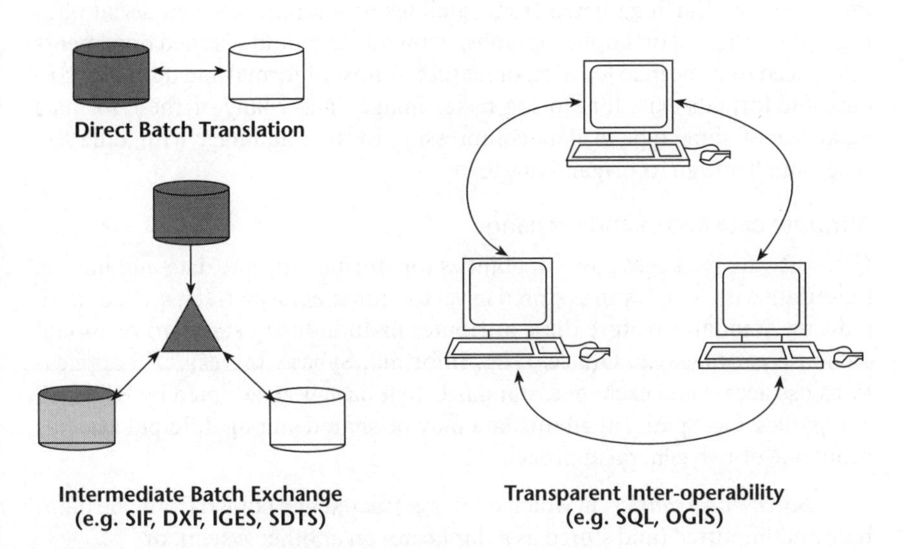

FIGURE 5.3 Geospatial data exchange and inter-operability.

1. Development of specific programs to translate formats *directly* from one proprietary structure to another.

2. Translation of a specific vendor format to an accepted *intermediate* exchange format followed by translation to another vendor's structure.

The second approach is most appropriate when data must be exchanged between several different software formats. There are several industry standard exchange formats, including:

- Interactive Graphics Exchange Standard (IGES)
- ISIF (developed by Intergraph Corporation for the exchange of vector graphic data)
- DXF (developed by Auto Desk, Inc., for the exchange of vector graphic data).

Several government agencies and international bodies have developed neutral exchange formats encompassing not only the exchange of vector graphics but more comprehensive GIS database content and format information. These may include attributes associated with map graphics, complex map feature structures, map symbology, and metadata (see Coleman and McLaughlin, 1992; Fegeas *et al.*, 1992; Moellering and Hogan, 1997; Kottman, 1992; and Lazar, 1996). Several examples of comprehensive GIS data exchange standards that have been formally accepted by standards organizations include: the Spatial Data Transfer Standard (SDTS) developed by the US federal government and the Digital Geographic Exchange Standard (DIGEST) developed by a multi-national Digital Geographic Information Working Group (DGIWG), sponsored by NATO.

Most GIS software packages have capabilities for storing and accessing documents and images in raster form. These may include remote-sensing images of the Earth gathered from satellites or aircraft, scanned aerial photographs or digital orthophotographs, scanned maps, and scanned documents referenced to some map location or feature. A host of formal and de facto standard file formats exist for storing raster image data. Many of these formats make use of some type of data compression to store an image with a file size much smaller than its original raw form.

Attribute data access and exchange

GIS software packages have capabilities for storing attribute data and linking these data with map features stored in vector or raster format. It has become an industry standard to store these attributes in an industry standard relational database package (e.g. Oracle, DB2, Informix, Sybase, Ingres). GIS applications use access and exchange, standards that have been adopted by RDBMS companies. Geospatial attribute data may be shared among different systems using one of two general approaches:

1. *Batch File Transfer*, in which data is extracted and copied from one database and imported (and stored as a duplicate) on another system, or

2. *Interactive Access*, whereby an interactive session is established between two or more databases (perhaps on separate computer platforms) to allow a GIS application or user query to automatically and transparently access the external data. This latter approach does not involve any copying or batch transfer of data.

Two standards facilitate interactive access:

1. SeQueL (SQL)—This standard, sometimes referred to as 'structured query language', is accepted by the American National Standards Institute (ANSI) and has been adopted by many software and database vendors. It provides a standard dialogue to access data, perform queries, and select records from a relational or table-based database.
2. Open Database Connectivity (ODBC)—This is a new industry standard, developed by Microsoft Corporation, which establishes an application programming interface (API) for transparent interactive access among disparate databases which have ODBC compliance. Because of its popularity, most major database software vendors are providing full compliance with ODBC.

Interest has been considerable in developing 'open' GIS environments to give users more flexible access to software and data environments. Work carried out by the OpenGIS Consortium (OGC) and by several individual GIS software companies has helped encourage interoperability and portability. The OpenGIS™ Specification, the umbrella standard being developed by the OGC, provides a basis for the connection and interoperability among distributed computing platforms.

5.3.6 Programming and application development standards

Most GIS software packages provide macro languages of varying levels of functionality, and some provide more powerful 'fourth generation' languages for the customization of applications. These proprietary languages work well with the particular software package but offer minimal consistency or standardization regarding functionality, commands, syntax, and user interfaces. Currently, the high-level application-programming environment for GIS is still dominated by proprietary development languages, but this is changing rapidly. Several industry standard development languages such as Visual Basic, Visual C++, and Delphi have been adopted by users and GIS software vendors as 'open' development environments. Companies providing serious support for these industry standard programming languages allow their GIS functions to be accessed directly as program code objects in a standard program. In this way, a Visual Basic program, for instance, could directly access GIS functions and data. Critical issues and developments in software and data interoperability are discussed in Zajac (1994), Doyle (1995), Glover (1995), and Trudeau (1996).

Interest in standard languages for user commands and database queries has been evident since at least 1993 in the GIS community. The SeQueL language

(SQL) has become a de facto standard by many GIS and database management software developers. Since 1993, there has been active work towards developing standardized 'spatial extensions to SQL'—an extended set of SQL commands that handle geospatial query and analysis operations on a consistently defined set of geospatial data types. A working group within the ANSI Database Languages Committee (X3H2), in co-ordination with an ISO joint technical committee, is in the final stages of approving such a comprehensive SQL set called 'SQL/MM Spatial'. This could establish a stronger basis for a common geospatial data manipulation language and portability of GIS software and applications (National Institute of Standards and Technology, 1993).

A greater degree of 'openness' in the application-programming environment will provide several benefits to developers and users of GDI. First, it will create an opportunity to apply the same 'open' languages for geospatial applications that are being used for information systems development in general. This will help maximize the efficiency of staff resources in application development. Secondly, the applications developed with these languages will be inherently more portable. Finally, common development languages will increase opportunities to integrate data from multiple systems and databases.

5.3.7 User design standards

There are many standards issues that should influence the design and implementation of GDI. In any automation project, the creation of a database should be preceded by a design phase that describes the content and format of the database and the products to be generated from the system. These decisions impact such design factors as: (1) attribute data schemas and coding rules, (2) map compilation and map accuracy standards, (3) quality control procedures, and (4) map design criteria, all of which are largely independent of a particular software or hardware environment.

Database schemas

Schemas (physical file format design) for storing geospatial attribute data should adhere to existing standards within an organization as a whole and, if applicable, to a broader community of users. The schema specifies field lengths, data element formats (e.g. integer, character, etc.), and other characteristics of the database. Use of a consistent schema, whether approved through an independent standards body or via an agency-wide policy, can greatly reduce the problems encountered in exchanging attribute data between systems. Users should take sufficient time to review existing schemas and comply with an existing government agency, or a standards body, if such standards exist. When no formal database schema standards are available, database administrators should evaluate existing informal standards or conventions inside the organization to avoid internal inconsistencies that could limit integration of databases and their maintainability.

Geospatial data coding and classification

Alpha or numeric coding and classification schemes for data are used for many types of data stored on computer systems. Classifying data into categories and coding these categories appropriately make it easy to retrieve and analyse data at different levels of detail. For example, land use or land cover classification schemes normally use some hierarchical classes, each having alpha or numeric codes assigned (e.g. code of '100' for urban with a subclass code of '110' for residential).

A large amount of geospatial data is stored in coded form, and it is easy to see why consistent use of coding and classification standards can facilitate exchange of data among systems. Unfortunately, all too often database design is done in a vacuum, and standard coding schemes are not followed. A full discussion of specific data coding and classification standards and initiatives now underway is beyond the scope of this chapter, but many standards (not all universally accepted or applied) exist for specific disciplines and topic areas. Designers and users of geospatial information systems should adhere to accepted coding and classification standards within their own organization, as well as standards established by government and professional groups, at national and international levels.

Geospatial metadata

With the diverse sources from which geospatial databases are built, it is extremely important to maintain information about the content, quality, source, and lineage (history of use and changes) of the data. There has been considerable activity in the development of metadatabases—tabular databases. These hold information about geospatial databases to support: (1) queries about the availability of data, (2) finding and accessing data, (3) determination of the suitability of the data for particular applications, and (4) programs for data maintenance (see Newman *et al.*, 1991; and Onsrud and Rushton, 1995). A comprehensive metadatabase will contain the types of information depicted in Figure 5.4.

A number of standards organizations have developed, or are in the process of developing, standards for storing and maintaining metadata. The most mature of these is the US Federal Geographic Data Committee's *Content Standard for GeoSpatial Metadata* to address the requirements for geospatial 'data sets'. A data set may be a full GIS database, a digital map database, or a tabular database of geographically related information. The FGDC Metadata Content Standard defines metadata elements and their logical format and domains (see Federal Geographic Data Committee, 1995).

Map compilation and map accuracy standards

Along with concerns for standards about the way data is formatted and coded, some standards issues impact the compilation of geospatial data. These

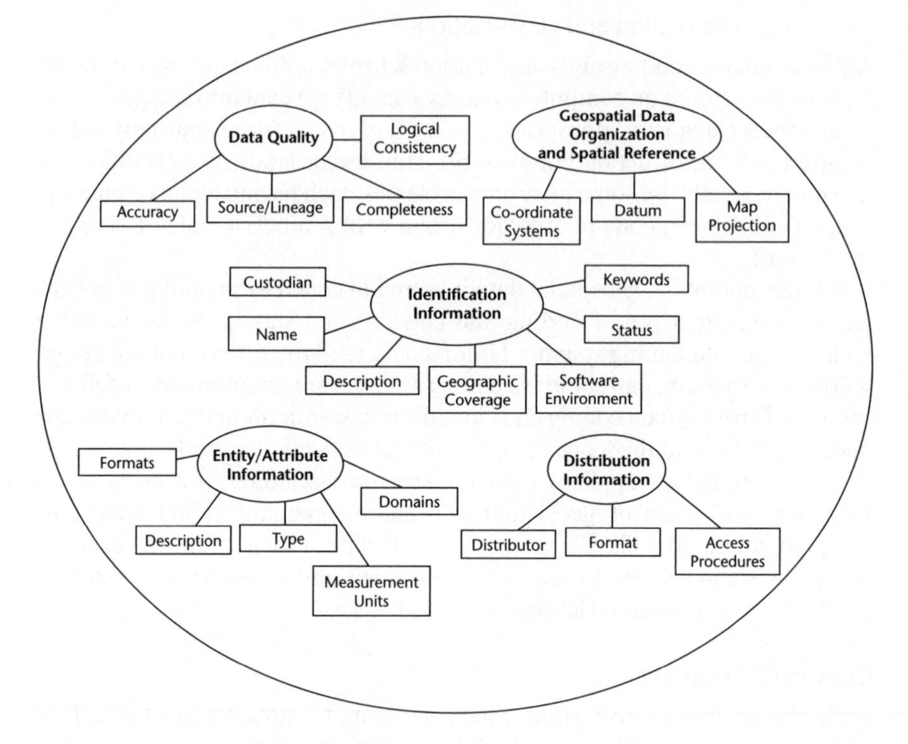

FIGURE 5.4 Example of metadatabase content.

standards address such areas as: (1) survey control and geodetic datum, and (2) map positional accuracy.

Survey control standards establish specifications for the accuracy of different classes of survey control points and documentation of monumented control. The Geodetic Sub-committee of the Federal Geographic Data Committee (formerly the Federal Geodetic Control Committee) in the USA and Geomatics Canada are the primary government agencies that set and oversee these standards. Also included in this category of standards are the geodetic datums on which surveying and mapping operations are based. The North American Datum (NAD) 1983 and the North American Vertical Datum (NAVD) have been adopted for mapping in North America. Similar activities in Europe are dealt with through CERCO, the European committee of heads of national mapping agencies, for example, through the Euref project.

Selecting suitable accuracy levels for the compilation and automation of map layers in GIS is a perplexing issue for most users. The objectives of map accuracy standards are to explicitly define and measure levels of accuracy, not to dictate what accuracy is appropriate for particular users. Accuracy standards address both the horizontal and vertical placement of map features and describe maximum errors of displacement relative to their actual position.

Map presentation standards

The graphic presentation of maps for soft copy display or hard copy generation includes such factors as scale, map sheet, map margin design and content, map legends, map feature symbology and colours, annotation styles and placement, and graphic design considerations. While general cartographic design criteria provide a general basis for sound map presentation design, few universal standards govern the design of specific types of maps. This is not surprising when we consider the number and types of maps and uses to which they are put. Professional associations and government organizations have, however, developed presentation standards for certain types of map which are in frequent use by a large user community. National agencies in the USA, Canada, Europe, and elsewhere have adopted certain standards for such products (e.g. US Geological Survey's quadrangle maps, Geomatics Canada's topographic map series, British Ordnance Survey maps). While no set of universal standards for all map presentation uses is available, the sources cited here provide a foundation for standardization in map design.

5.4 A practical context for adoption and implementation of standards

Users can take an active role in encouraging and accelerating the standards process, first by becoming informed about standards issues that will impact them and their organizations, and then by using formal avenues to incorporate standards into system design and development activities. Education and information can be facilitated through participation in professional organizations, special training seminars on specific standards issues, and regular reading of key scholarly and industry publications.

The most direct way to influence the computer industry is through procurement specifications that require compliance with standards. However, users must be knowledgeable about the current status of product offerings to make realistic demands on vendors and to avoid setting requirements that are simply not feasible.

A wisely crafted set of standards will promote interoperability, portability, and maintainability. All organizations that are players in a GDI will receive benefits from the creation and use of standards that address low-level and high-level areas. There is a price to pay, however. Developing standards and employing them on a continuing basis takes time up front in the design process, as well as discipline and diligence to ensure that standards are adhered to as a GIS matures. There is also a danger of going too far in defining and dictating standards that are inappropriate or counter-productive to a specific user community. It is possible that an organization could create inflated overheads just to maintain and enforce a set of standards, and there is a risk that users will find complying with standards too time-consuming or inflexible. A well-designed set of standards will take into account the specific needs of an organization within a broader community of users. Those responsible for standards development and monitoring will need to take into account users' needs, communicate the long-term purpose and value of standards, create flexible tools to support their use, and put in place workable procedures to ensure compliance.

While information system standards are critical for organizations that need to share geospatial data with other organizations at a local, regional, or international level, standards are equally important within single organizations regardless of their size.

With a view to both the low-level and high-level standards discussed above, organizations should follow the general rules given below in defining a set of practical standards that will make sense in the long term:

1. Evaluate geospatial information needs and articulate these in the form of applications with a definition of system and data requirements.
2. Establish low-level hardware, operating system, and network standards that are compatible with overall information systems in the organizations and consistent with industry products and trends.
3. Develop system procurement rules and standards that comply with the low-level standards. Regularly update procurement specifications as industry advances occur.
4. With a firm understanding of data requirements, spend adequate time on a database design process and involve users and potential users of the system. Design data formats and classification systems that comply with accepted standards in all appropriate cases.
5. Look closely at the current and future requirements for exchange of data among users within and outside the organization. Assume that the interest in outside data exchange and access will grow in the future and take into account software, data format, and metadata standards that will facilitate this.
6. Make decisions about the application development environment and tools that will be de facto standards in the organization.
7. Select geospatial software products that comply fully with accepted operating system standards and support application development standards adopted by the organization.
8. Set up a sensible and ongoing programme for standards-compliant metadata capture and maintenance, and use these standards in a programme for data access and data maintenance by users.

5.5 Outline for a geospatial data standards manual

This outline is provided as a starting point for an organization interested in preparing a standards manual to guide GDI implementation and ongoing operations. Such a manual should be a living document. While the standards it defines should not be subject to frequent change, a process should be established for formal revision that is appropriate given changes in the organizations and the industry in general. Not all of the suggested sections below will apply in all cases, but for large organizations most of them should be strongly considered for inclusion.

I. Introduction

The introduction provides organizational and project context for standards, rules for standards compliance, and a process for standards revision.

(a) Organizational context
(b) Need for standards
(c) Standards compliance
(d) Process for standards revision.

II. Hardware and network standards

Contains standards and procurement requirements that impact the physical network and hardware to support the GIS. It provides workable specifications to guide procurement and installation of computers and networks. These standards should take into account the standards and the systems already in place with the goal of not duplicating physical networks or services unless necessary. The standards should be strict enough to ensure a sufficient level of interoperability.

(a) Computer operating system standards
(b) Computer hardware standards and specifications
(c) Physical network and protocol standards
(d) Network interface requirements.

III. System administration standards

Addresses standards and conventions for all system administration and database administration, including file naming, organization, procedures, and controls on user access.

(a) File and directory naming standards
(b) System access and security.

IV. Software and application standards

Covers standards and accepted specifications for all application software. This establishes procurement requirements to ensure a high degree of interoperability. Standards for application development, including specific development tools, user interface standards, and application documentation standards, are established.

(a) GIS software package(s)
(b) GIS-related software packages

(c) RDBMS standards
(d) Application development software standards
(e) Application design and documentation standards.

V. Data format standards

Provides all standards that pertain to the storage, management, database design, and exchange of geospatial data and metadata.

(a) Geospatial data format
(b) Data content/data dictionary standards
(c) Data coding and classification standards
(d) Data exchange format standards
(e) Metadata standards.

VI. Data compilation and update standards

Addresses all standards and formal policies that pertain to data compilation and ongoing update. Data compilation standards define acceptable techniques for capturing and compiling data and accuracy and quality requirements. Update standards deal with rules, responsibilities, and procedures for routine data update.

(a) Geospatial data collection and compilation specifications and standards
(b) Map accuracy standards
(c) Database update procedures and responsibilities and schedules.

VII. Product presentation standards

Addresses organization standards in place for the production of standard maps and other output products. Mapping standards include accepted point and line symbology standards, as well as overall layout requirements (sheet format, scale, margin, and legend designs) for standard maps. Design standards for other products (standard reports) may be included here also.

(a) Map symbology standards
(b) Map design and layout standards
(c) Other product output standards.

VIII. System access and data/product distribution standards

Where an organization is establishing a program for external distribution or sale of geospatial data and products, the standards and procedures associated with this program should be documented. This will cover policies for distributing data and products to the outside parties and any restrictions on use that may apply, liability statements, and official fee schedules for product sale.

(a) Policies on external data access
(b) Catalogue and fee schedule for products/services
(c) Procedures for responding to and tracking requests.

Bibliography

ANNITTO, R. and PATTERSON, B. L. (1995). 'Sharing Graphic Files in an Open System Environment', *Journal of the Urban and Regional Information Systems Association*, 7/1: p. 64.

BAMBERGER, W. (1995). 'Sharing geographic information among local governments in the San Diego region', in *Sharing Geographic Information*. New Brunswick: Rutgers, The State University of New Jersey, pp. 119–37.

CARGILL, C. F. (1989). *Information Technology Standardization: Theory, Process, and Organizations*. Houston, USA: Digital Equipment Corporation Press.

COLEMAN, D. J. and MCLAUGHLIN, J. D. (1992). 'Standards for Spatial Information Interchange: A Management Perspective', *The Canadian Institute of Surveying and Mapping Journal*, 46/2: pp. 133–41.

CROSWELL, P. L. (1993). 'Open Systems Mean "Open Sesame" for the GIS Community', *GIS World*, 6/11: August, p. 40.

CROSWELL, P. L. and AHNER, A. (1990). 'Computing standards and GIS: A tutorial', *Proceedings from URISA 1990*, vol. ii, pp. 88–105.

DANIEL, L. (1993). 'Easier Data Exchange Sparks GIS Workplace Evolution', *GIS World*, 6/12: December, p. 48.

DOYLE, A. (1995). 'Software that Plays Together—A System Integrator's Viewpoint in Open GIS', special collection of articles, *Geo Info Systems*, May, pp. 50–1.

DUEKER, K. and VRANA, R. (1995). 'Systems integration: a reason and a means for data sharing', in *Sharing Geographic Information*. New Brunswick: Rutgers, The State University of New Jersey, pp. 149–71.

Federal Geographic Data Committee (1995). *Content Standards for Digital Geospatial Metadata Workbook*, Washington, DC, 24 March.

FEGEAS, R., CASCIO, J., and LAZAR, R. (1992). 'An Overview of FIPS 173, the Spatial Data Transfer Standard', special issue of *Cartography and Geographic Information Systems*, 19/5.

FRIESEN, P. and SONDHEIM, M. (1994). 'Mixing data from diverse sources', *GIS 94 Symposium Proceedings*, Vancouver, BC, Ministry of Supply and Services Canada, Catalogue No. FO 18-16/1994E, 151–6.

GLOVER, J. (1995). 'The Need for Open GIS—Part 1: the Integration Challenge', *Mapping Awareness*, 9/8: p. 30.

JIWANI, Z. (1988). 'Administrative and technical issues of data sharing', *Proceedings of the Ontario Ministry of Natural Resources Geographic Information Systems Seminar: Data Sharing—Myth or Reality*, 3–5 October, pp. 85–97.

KINDRACHUK, E. (1992). 'Dealing with Spatial Data Exchange Issues', *GIS World*, June, pp. 76–9.

KOTTMAN, C. A. (1992). 'Some questions and answers about digital geographic

Information Exchange Standards', Brochure prepared by Intergraph Corporation, 2nd edn., Nov.

LAZAR, B. (1996). 'Understanding SDTS Topological Vector Profile Implementation', *Geo Info Systems*, June, p. 42.

LEE, Y. C. and COLEMAN, D. J. (1990). 'A Framework for Evaluating Interchange Standards', *The Canadian Institute of Surveying and Mapping Journal*, 44/4: pp. 391–402.

LEVINSOHN, A. (1997). 'MERCATOR Promotes GIS Standards and Universal Geospatial Data Access', *GIS World*, 10/2: February, p. 66.

LIBICKI, M. C. (1995). *Information Technology Standards: Quest for the Common Byte.* Newton, Massachusetts: Butterworth-Heinemann.

Microsoft Press (1998). *Networking Essentials*, 2nd edn. Redmond, Washington: Microsoft Press.

MOELLERING, H. and HOGAN, R. (eds.) (1997). *Spatial Database Transfer Standards 2: Characteristics for Assessing Standards and Full Descriptions of the National and International Standards in the World*, Oxford: Pergamon Press/International Cartographic Association.

MOYER, D. and NIEMANN, B. J. Jr. (1993). 'The Why, What, and How of GIS Standards: Issues for Discussion and Resolution', *Journal of the Urban and Regional Information Systems Association*, 5/2: p. 28.

National Institute of Standards and Technology (1993). *Towards SQL Database Language Extensions for Geographic Information Systems*, collection of papers, in ROBINSON, V. B. and TOM, H. (eds.), publication NISTIR 5258.

NEWMAN, I., MEDYCKYJ-SCOTT, D., RUGGLES, C., and WALKER, D. (eds.) (1991). *Metadata in the Geosciences.* Loughborough: Group D.

ONSRUD, H. and RUSHTON, G. (1995). 'Metadata issues in sharing geographic information', in *Sharing Geographic Information.* New Brunswick: Rutgers, The State University of New Jersey, pp. 499–501.

OSTENSEN, O. (1996). 'ISO TC 211', *Geomatics Info Magazine*, 10/3: March, p. 24.

PELED, A. and ADLER, R. (1993). 'A Common Database for Digital Mapping and GIS', *International Journal of Geographical Information Systems*, 7/5: pp. 425–34.

SLONE, J. (1998). *Handbook of Local Area Networks.* Washington, DC: Auerbach Publishers.

TOM, H. (1988). 'Standards: a cardinal direction for geographic information systems', *Proceedings of the 1990 Urban and Regional Information Systems Association Conference*, vol. II.

TRUDEAU, M. (ed.) (1996). 'Open Systems', special collection of articles, *Geo Info Systems*, February, p. 37.

URISA (Urban and Regional Information Systems Association) (1998). *Spatial Information Technology Standards and System Integration*, in P. Croswell (Chief Editor). Park Ridge, Illinois: URISA.

VENTURA, S. (1993*a*). 'Standards Offer LIS/GIS Support', *GIS World*, 6/8: August, p. 48.

——(1993*b*). 'LIS/GIS Standards: an Industry Challenge', *GIS World*, 6/9: September, p. 42.

WELLAR, B. (1972). 'Standardization: issues and directions', *Proceedings of the 1972 Urban and Regional Information Systems Association Conference*, pp. 429–44.

——(1973). 'Information interchange: focus on standards and compatibility', *Proceedings of the 1973 Urban and Regional Information Systems Association Conference*, vol. 1, pp. 234–8.

WIEGAND, N. and ADAMS, T. M. (1994). 'Using Object-Oriented Database Management for Feature-Based Geographic Information Systems', *Journal of the Urban and Regional Information Systems Association*, 6/1: p. 21.

ZAJAC, G. (1994). 'Integration of GIS applications across multiple platform environments', *Proceedings of the 1994 Urban and Regional Information Systems Association*, vol. 1, p. 140.

6

Quality management in GDI

Mark Doucette and Chris Paresi

6.1 Introduction

The quality of a product or service is defined as 'the totality of characteristics of an entity that bear on its ability to satisfy stated and implied needs' (ISO 8402, 1994). Quality is not a value in itself but is always related to the degree of user satisfaction, a result only to be observed when the product or service is used, including its relation to timeliness and price. However, checking the quality of the product at the final stage of the production process is inadequate; complementary to this is the quality control of all elements in the production process. This concept involves the whole organization and is usually referred to as a 'Total Quality' approach. It is defined as 'a management approach of an organization, centred on quality, based on the participation of all its members and aiming at long-term success through customer satisfaction' (ISO 8402, 1994). Realization of this approach requires a Quality Management System (QMS), including definition of the responsibilities, authorities and accountabilities for the interfaces, main processes and procedures necessary to achieve quality objectives of the organization in terms of established quality policy.

The fundamental components of a geospatial data infrastructure are producers of the data, the users of the data, and the technological and institutional environment in which users access and apply the data. It will be evident that the users' perception of quality will depend on the quality management in each of the organizations' processes which comprise the GDI. Inevitably the question arises as to how much confidence we can have in the services within the GDI. How can producers be sure for example that the data protection is functioning well, and that they will be paid reliably for use of their data? Conversely, how can users be sure that what is in the metadata is in fact what they receive? Certification, for example in terms of ISO (9000), series of standards of the QMS in the data-supplying organizations and in the Geospatial Data Service Centres (GDSC) will contribute to increasing the confidence levels. In fact GDSCs supporting a particular application domain or enterprise have a leading and controlling role to play in establishing the QMS for their service provision and the data production within their GDI. An important aspect of this is managing and directing the development of standards manuals.

The traditional data suppliers in the commercial environment face strong motivation to implement a well-considered QMS. An example is CERCO's initiative to develop QMS guidelines for its thirty-five member national mapping agencies in Europe (see for example, CERCO, 1999; Dassonville *et al.*, 1999). But the geomatics or geoinformatics industry is also following the path of certification as can be seen in Geomatics Canada (1996). Transparent QMS in the data-producing agencies is essential for dealing with issues of liability if the data prove to be defective or not what they are supposed to be.

Against this general introduction and in view of the purpose of this book, the discussion on quality management will be brief and limited to some aspects of system performance and data quality.

6.2 The evolving GDI environment

In the past, the primary user issue was the availability of the data. The fact that data users had little choice of where to get the data they required was used by suppliers as a control mechanism. Data suppliers were usually experts on the data they were dealing with, and they used the data in applications over which they had complete control. Today, geospatial data is used in applications that the original data creators could never envisage. The users have varying degrees of knowledge and ability with respect to the original purpose for which the data was collected. For example, hydrographic data collected by the Canadian Hydrographic Service was originally processed and displayed to show the 'minimum' water depth for ships, but it is now being used by environmentalists and engineers to determine how much water there is. The hydrographer was interested in quality data measured against a serious threat to shipping, whereas planners have different interests. Similarly, the data collected in national mapping agencies was used to produce standard, multi-purpose topographical maps. Today the demand is more diverse and includes new products such as the core topographic data or template needed by GIS users as a geometric reference for their thematic data. However, in each example the user community expects up-to-date and quality data sets, especially if they have paid a price for the product or service. This increase in data usage and interest in quality management issues is a direct outcome of the changing environment in which data users now participate. Digital data has made the collecting and sharing of data much easier, and the widespread use of desktop computers and their associated software makes applications requiring geospatial data available to a much broader range of users.

While the proliferation of users and the changing data application environment are impacting the overall concept of quality management in GDI, the integrity of the data itself continues to be the primary quality issue.

6.3 Integrity of GDI

The integrity of GDI is a broad subject worthy of a separate book. Broadly speaking, the subject can be divided into the effective functioning of the associated systems and networks, and the integrity of the data. In Chapter 5, most of the standards issues presented concerning the systems and networks are beyond the control of the GDI user or data supplier. However, some of the systems and network integrity must be taken in hand by the GDSC of an application domain that may function through an Intranet. In this situation the GDSC plays a directing and controlling role in the quality management of the GDI (including aspects of system management and the reliable use of the systems for integrity, continuity, efficiency, and effectiveness). Some examples are:

- assuring that input is processed correctly and completely by the system according to the specifications;
- delivering the required data in a timely manner;
- assuring that the data and applications can be accessed or operated only by authorized persons;
- ensuring that the same data, if resident in different databases, will have the same status, the 'concurrency control';
- guaranteeing the reliability of all transactions (including financial) through the GDSC;
- developing and acting upon performance reports and ensuring continuity of the systems;
- user friendliness in help functions, ease of learning operations, etc.;
- clarity of user manuals;
- controlling the degree of availability of systems functions when the user needs them.

The GDSC is in charge of the quality management programme for the GDI. Consequently it must manage the development of the quality manual for all the players in the GDI. It must further ensure that the quality management is in sympathy with standards' legislation and regulations imposed externally to the GDI, such as national metadata standards and legislation concerning privacy or copyright.

This rather limited sketch of the system integrity shows that data integrity is the single, most important concept in GDI quality management. The nature and degree of data integrity is usually a function of the application in which it is collected or used, which of course involves the data users. While data utility for its original application (or environment) is usually satisfied, the degree to which the data can be successfully integrated with other data involves issues relating to quality. Data quality is usually defined by factors such as data lineage, consistency, completeness, semantic accuracy, temporal accuracy, positional accuracy, and attribute accuracy.

6.3.1 Lineage

Lineage includes information such as descriptions of source materials, methods of derivation, and all data transformations. It also includes dates for all relevant data sources and processing steps. (This is the single most readily available measure of *quality* for many data sets.) While not as 'quantitative' as other measures being espoused by the standards, it does represent the only quality measure for a significant number of non-technical data users. Sometimes the only measure of data quality is that 'the national survey agency collected the information, so it must be good'.

6.3.2 Consistency

Logical consistency describes the fidelity of relationships in a data set, the logical rules of structure, and the attribute rules for spatial data. It describes the compatibility of a datum to other data in the data set. Typically, logical consistency involves testing for attribute database consistency, metric and incidence tests, and topological- and order-based tests. Consistency checks generally involve standard topological tests to build networks and polygons, and checking networks and polygons to determine the logical consistency parameters, such as Euler Counts, Non-node Intersections and Inconsistent Polygon Chains. These tests reflect the 'internal' structure of the data file and are standalone.

6.3.3 Completeness

Data set completeness varies significantly with the intended application and is usually defined as including information about the selection criteria, definitions used, and other relevant mapping rules. It also includes a description of deviations from standard definitions and interpretations and/or a statement on the relationships of the objects represented within the data set. Completeness measures of quality must typically be made against some standard or data set of higher accuracy, i.e. this data set is complete when compared to a data set of equal or greater accuracy (greater accuracy normally refers to larger scale.) For example, a data set collected and displayed at 1 : 50,000 would typically not have all of the lake features that a 1 : 10,000 data set has for the same area. Smaller scale data sets are often generated from larger scale data sets, and on occasion, if a set of standards for generalization are not enforced, features are missed on the smaller scale product that should have been included. The smaller scale data set may not be complete when measured against what should have been included according to the generalization specifications, but it may be 'complete enough' for its intended application (fit-for-use).

6.3.4 Semantic accuracy

Semantic accuracy refers to the quality with which spatial objects are described according to a selected model. It describes the number of features, relationships, or attributes that have been correctly encoded in accordance with a set of feature representation rules. Semantic accuracy is seen as an element of fitness-for-use in the evaluation performed by users. The user and the data provider may have different concepts of semantic accuracy. It typically includes factors such as completeness, consistency, currency, and attribute accuracy. In New Brunswick, Canada, there is a law that a buffer strip along streams must be excluded from harvest when cutting timber. Planning for this activity is completed on 1 : 10,000 maps and at this scale the absence of small streams from the base map is not significant. However, when harvest plans for individual wood lots are developed from the same base map, difficulties arise when harvest volumes have to be adjusted as small streams are discovered in the wood lot. Planning for allowable cut could indicate a profitable harvest based on 'calculated' volumes, whereas the actual cut could prove to be unprofitable because of the buffer requirements. While the 1 : 10,000 data was suitable for the planning of the harvest, it does not prove suitable for individual wood lot management applications.

6.3.5 Temporal accuracy

Temporal information describes the date of observation, type of update, creation, modification, deletion, and the validity periods for geospatial data records. Typically, temporal accuracy is thought of as a time frame or a time of year. However, the temporal accuracy of data can be far reaching, especially when combining data sets. For example, when routing a new highway it is usually necessary to identify all the buildings near the proposed right of way. However, in a particular jurisdiction, the required databases were of a very different vintage. A topographic database, used as a digital mapping base for planning, was compiled from photographs taken during the early 1980s, and was not updated. However, the land use codes used for property assessment and taxation were updated on a continuous basis. The land use codes in the property assessment database indicated a land use, from which some indication of the presence or absence of buildings was made. (For example, if the land use is residential, then it is reasonable to assume that there is a building on the property.) Consequently, the two data sources were not in agreement and their respective measures of data quality were questioned. A temporal accuracy statement would notify the user of this discrepancy. The statement of temporal accuracy could be as follows:

Land use data update frequency: daily
Structures data update frequency: not updated.

6.3.6 *Positional accuracy*

Positional accuracy testifies to the degree of compliance with some preconceived standard of accuracy. This is the measurement of accuracy most readily understood by data users. It is typically reflected in standard deviations, error ellipses, position vectors, etc. This data quality measure most readily adapts to the ISO requirement for the inclusion of Data Collection Specifications. Most spatial digital data is collected to some sort of positional accuracy standard or specification.

6.3.7 *Attribute accuracy*

Attribute accuracy describes how attributes are positioned, and how they are measured and/or determined. Reports on attribute accuracy include the date of the test, and the dates of the materials used in the tests. Accuracy tests may be performed by deductive estimates, tests based on independent samples, or tests based on polygon overlay. These types of accuracy measures are also usually made against a known standard or another data set. For example, a test was conducted in New Brunswick to compare the ownership information in the Property Assessment and Taxation System (PATS) and Property Attribute files. A database table was created using MS ACCESS that contained the property identifier (PID) and the ownership information for the properties. In the Property Attribute file, the ownership information was contained in two fields, the surname field and the first/second name field. In the PATS table, the ownership information was contained in one field with the surname being separated from the first/second name field by a comma. Four different types of comparisons were performed in order to test the accuracy of the data. These were:

1. Exact comparison of the surname fields;
2. Case sensitive 'like' comparison of the PATS owner field using the Property Attribute surname field;
3. Case insensitive 'like' comparison of the PATS owner field using the Property Attribute surname field;
4. Case insensitive comparison of the PATS surname owner field with the Property Attribute surname field and of the PATS first name owner field with the Property Attribute first name field.

The test was performed on the combined database containing 19,872 records and produced the following results:

Test	Number of errors
Exact Surname	3,453
Surname Case Sensitive Like	3,248
Surname Case Insensitive Like	3,248
First and last name Like	14,479

6.4 Some issues of standards

Regardless of the quality of any of these data components, data must be collected, stored, exchanged, and presented according to certain standards. These standards address the fundamental elements of the data and are very much driven, in form and format, by the integrity of the data. Without standards, it becomes impossible to meaningfully, effectively, and efficiently integrate data sets or to exchange data between organizations. Standards may be formal or informal.

Formal standards tend to be long, complex, scientific in nature, and focused on future requirements. Formal standards are usually embraced by larger bodies, such as governments, universities, and large private companies that need to manage or use large data sets. These standards can be entity-wide, or embrace a particular sector or industry. Examples of formal standards are a standard for data formatting used within an organization, or standards such as those currently under consideration by groups such as the International Organization for Standardization (ISO) (see Section 6.7).

Informal standards tend to be only as long and involved as required to address an immediate issue and to solve a particular problem. They are usually used by industry to address production issues driven by customer needs. Informal standards are usually adopted by data users and accepted as norms because of some practical issue that they address. That practical issue can be availability, ease of use, understandability, etc. An example of an informal standard would be the use of AUTOCAD dxf as an exchange format for digital data. While not formally accepted as a standard, its wide availability, ease of use, and ability to address some or most of the fundamental issues required by practitioners makes it the best exchange format to use to get the job done.

When it becomes necessary to integrate or exchange data, the need for widely accepted standards increases, and the degree to which the data exchange or integration can succeed is a function of data compatibility. With the increasing usage of digital geospatial data by users unfamiliar with the purpose and scope of the original data collection, formal spatial data infrastructure is becoming more important. Infrastructure pertaining to the collection, dissemination and maintenance of the data sets becomes critical to the wider acceptance and productive usage of digital data. However, creators of such infrastructure must develop processes that both facilitate the development of formal standards more quickly, and result in the broader use and acceptance of the standard.

6.5 Users of geospatial data infrastructure

Data creation and application has, in the past, been practised by knowledgeable, experienced individuals who understood the limitations of the instruments they were using and the issues concerning the display of the collected information. Whether it was base map production, thematic overlays, or visual impressions, these practitioners had full control over the application. While experts were creating the maps, users were forced to use the products provided and to live with the limitations. Technology has altered this balance; today, techniques that were previously only available to experts are now available to anyone who can afford them (and prices have decreased significantly). However, the spatial data infrastructure required to ensure data/product/system standards and compatibility has not kept apace of the technical developments. Non-technical users do not have the tools available to guide their actions, reactions, and choices.

Global Positioning System (GPS) satellites can now provide positions accurate to the order of a few hundred metres, anywhere in the world, for only a few hundred US dollars. This capability offers enough positional accuracy for many applications that were previously restricted by a lack of any kind of reliable positional information. By supplementing standard, single frequency GPS with differential capabilities, accuracies in the order of a few metres can be obtained with near 100 per cent repeatability, thereby making applications that require accurate, real-time positioning viable world-wide. In their hardware and software tools, manufacturers are now able to embed the capabilities for collecting, processing, and displaying positional information to a highly detailed level, and to a type of user who was previously uninvolved. These users do not know about the theory of satellite positioning systems, nor are they interested. They simply need the positional information to do their work.

The wider acceptance of geospatial data systems is providing the capability of producing detailed maps to marketers, planners, sales people, and those in other unrelated disciplines, untrained in the art and science of producing maps. Issues of scale, generalization, topological structuring, and attribute completeness are not important nor even considered unless they directly impact the picture being created. While traditional map makers may shudder at these applications of new technology and scoff at the inaccuracies inherent in the products, individuals continue to produce products that make their way into the market place and into wide circulation.

Computer mapping tools are also providing options for using historical data that was previously reserved for institutional mappers and organizations. Today, geospatial data is being used in applications totally unrelated to the original purpose for collecting data. Data collected for the singular purpose of showing the water depth under the keel of a deep sea sailing ship can now be acquired from a government agency, manipulated, and then displayed for environmental planning in sensitive coastal areas. The original data collectors did not make any provision for such an application; in fact they were more involved with the opposite issue (hydrographers are concerned about how little water depth there is under a ship's keel and tend to err on the side of less water, i.e. make the depth shallower, while environmental planners are more often concerned with how much water is available).

While these practices are of concern to data producers, they are characteristic of a trend that is not likely to stop; more data will be acquired by more users who are unfamiliar with the original purpose of the data. As these users gain access to data and they become more sophisticated in their applications, data quality issues will arise. Although the users will not be interested in the theory behind why a polygon will not fill properly, or a network that cannot be built, they will nevertheless want these features of the software to deliver on their promise. This makes it necessary for data producers and providers to develop methods for producing good quality data, at a reasonable cost, whilst not attracting undue risk and liability to themselves or their organizations.

6.6 Presentation issues relating to geospatial data infrastructure

In the past, access to geospatial information was controlled in a number of ways, for example by the presentation medium—paper was the first WYSIWYG display (what you see is what you get). You could have the product that was available, but you were forced to accept it as it was. The source data was available from government agencies but its existence was not widely publicized and for those that were aware of the source, access was restricted by the technology limitations of data storage and transfer. The cost of data collection was very high and usually limited to large agencies who needed the information and could afford to collect it.

Today, computer technology provides interested parties with significant mapping possibilities. Data manipulation is much easier owing to increased processing speeds and capabilities, while data storage capacities and computer software capabilities are expanding. All of these are available at increasingly less cost. Data is becoming much more widely available, with large data producers providing access to their data via the Internet through FTP sites. The current limitations for graphical data and of reduced line speed are only temporary and will be overcome.

This increase in technological capability and data availability is encouraging an increase in data usage. The widespread availability of digital data has made collecting and sharing data much easier. Desktop computers and their associated software make applications requiring geospatial data available to a much broader range of users. These advances in technology will continue, making it even more important for data suppliers to address the data quality issues of their current databases.

6.7 The International Organization for Standardization

As indicated in Section 6.1, standards are an important consideration in geospatial information infrastructure. The International Standards Organization (ISO) is currently considering standards that relate to data quality, and metadata standards for spatial data. The ISO is a world-wide organization of national standards' bodies that prepares international standards through ISO-sponsored technical committees. Each member body, and any international, government or non-government organization can participate on a technical committee. International standards are drafted by the technical committees and then circulated to the member bodies for approval. A standard requires at least 75 per cent approval before it is put into effect.

ISO 15046 is a multi-part International Standard for Geographic Information/Geomatics. The standard contains twenty-one parts:

15046–1 Reference model
15046–2 Overview
15046–3 Conceptual schema
15046–4 Terminology
15046–5 Conformance and testing
15046–6 Profiles
15046–7 Spatial subschema
15046–8 Temporal subschema
15046–9 Rules for application schema
15046–10 Feature cataloguing methodology
15046–11 Spatial referencing by co-ordinates
15046–12 Spatial referencing by geographic identifiers
15046–13 Quality principles
15046–14 Quality evaluation procedures
15046–15 Metadata
15046–16 Positioning services
15046–17 Portrayal of geographic information
15046–18 Encoding
15046–19 Services
15046–20 Spatial operators
15046–21 Functional standards.

This ISO standard is exhaustive in its consideration of all aspects of data elements and their associated attributes. However, the question remains as to how comprehensive standards, such as these, can be meaningfully incorporated into the everyday practice of data providers. A recent internal assessment made of official digital data sets being issued by data providers found that of the 401 fields covered in the Metadata Standard, only 58 could be filled in with information readily available from the provider. Is the level of detail stipulated by this standard actually required? The cost of identifying the information required for the remaining fields in the metadata record would no doubt be prohibitive. Are all the fields really necessary? What is the impact of not filling in these fields? These are issues that data users and providers will need to understand to ensure that they can address the issue of data quality, and at an appropriate cost.

Bibliography

CERCO (1999). Good reasons for implementing a Quality Management System in European National Mapping Agencies. CERCO Working Group on Quality, *Laure.Dassonville©ign.fr*

DASSONVILLE, L., VAUGLIN, F., JAKOBSSON, A., and LUZET, C. (1999). Quality Management, data quality and users, metadata for geographical information. CERCO Working Group on Quality, *Laure.Dassonville@ign.fr*

Geomatics Canada (1996). *Quality in Geomatics: A Practical Guide to ISO 9000*, Natural Resources Canada.

ISO 8402 (1994). *Quality Management and Quality Assurance-Vocabulary*.

Recommended Reading

AMRHEIN, C. G. and GRIFFITH, D. A. (1991). 'A model for statistical quality control of spatial data in a GIS', *Proceedings of the Canadian Conference on GIS*, Ottawa, 18–21 March 1991, pp. 91–103.

BOLSTAND, P. V., GESSLER, P., and LILLESAND, T. M. (1990). 'Positional Uncertainty in Manually Digitized Map Data', *International Journal of Geographical Information Systems*, 4/4: pp. 399–412.

British Columbia (1995). 'Spatial archive and interchange format: formal definition', Release 3.2, Surveys and Resource Mapping Branch, Ministry of Environment, Lands and Parks, Province of British Columbia, Canada; also Canadian General Standards Board publication CAN/CGSB-171.1–95; available from anonymous ftp site: *ftp.crl.gov.bc.ca*, file *pub/saif3/saif32ps.zip*, Jan. 1995.

GOODCHILD, M. F. (1992). 'Geographical Data Modeling', *Computers & Geosciences*, 18/4: pp. 401–8.

GUPTILL, S. C. and MORRISON, J. L. (eds.) (1995). *Elements of Spatial Data Quality*. International Cartographic Association.

HEUVELINK, G. B., BURROUGH, P. A., and STEIN, A. (1989). 'Propagation of Errors in Spatial Modelling with GIS', *International Journal of Geographical Information Systems*, 3/4: pp. 303–22.

ISO 9001 (1994). *Quality Systems—Model for Quality Assurance in Design, Development, Production, Installation and Servicing*.

LAM, S. (1994). 'Uncertainty propagation in surface interpolation with fuzzy arithmetic', *Proceedings of the GIS '94 Symposium*, Vancouver, BC, pp. 367–72.

LANGAAS, S. (1995*a*). 'Cartographical data and data quality issues', presented at the UNEP/GRID and CGIAR Workshop (Arendal II), Arendal, Norway.

——(1995*b*). 'Completeness of the digital chart of the world (DCW) database', project report no. 2/1995, UNEP/GRID-Arendal, P.O. Box 1602, Myrene, N-4801 Arendal, Norway, http://www.grida.no, 76 pp.

——and TVEITE, H. (1995). 'To characterise and measure completeness of spatial data: a discussion based on the digital chart of the world (DCW)', in BJØRKE, J. T. (ed.), *Proceedings of the 5th Scandinavian Research Conference on Geographical Information Systems*, 12–14 June, Trondheim, Norway, pp. 155–61.

LANTER, D. P. and VEREGIN, H. (1992). 'A Research Paradigm for Propagating Error in Layer-Based GIS', *Photogrammetric Engineering & Remote Sensing*, 58/6: pp. 825–33.

MIKHAIL, E. M. (1976). *Observations and Least Squares*. IEP—A Dun-Donnelley Publisher, New York.

NATO (1989). *Standardization Agreement (STANAG), Evaluation of Land Maps, Aeronautical Charts and Digital Topographic Data*, 5th edn. North Atlantic Treaty Organization.

PAPOULIS, A. (1991). *Probability, Random Variables, and Stochastic Processes*, 3rd edn. New York: McGraw-Hill, Inc.

SMITH, C. G. and LANGAAS, S. (1995). 'A survey of digital chart of the world (DCW) use and data quality', project report no. 3/1995, UNEP/GRID-Arendal, Longum Park, P.O. Box 1602, Myrene, N-4801 Arendal, Norway, *http://www.grida.no*, 25 pp.

THAPA, K. and BOSSLER, J. (1992). 'Accuracy of Spatial Data used in Geographic Information Systems', *Photogrammetric Engineering & Remote Sensing*, June, 58/6: pp. 835–41.

TVEITE, H. and LANGAAS, S. (1995). 'Accuracy assessments of geographical line data sets: the case of the digital chart of the world', in BJØRKE, J. T. (ed.), *Proceedings of the 5th Scandinavian Research Conference on Geographical Information Systems*, 12–14 June, Trondheim, Norway, pp. 145–54.

USGS (1994*a*). Spatial Data Transfer Standard, available from *http://nsdi.usgs.gov/nsdi*

USGS (1994*b*). Content Standards for Digital Geospatial Metadata, available from anonymous ftp site *fgdc.er.usgs.gov/gdc/metadata/meta.6894.ps*

WaterMark Industries Inc. (1995*a*). 'Managing uncertainty in digital geographic information: survey of technology', Phase I Report for Centre de recherches pour la défense Valcartier, BAT 15, BFC Valcartier, Courcelette, Québec, 16 Feb.

WaterMark Industries Inc. (1995*b*). 'Managing uncertainty in digital geographic information: survey of technology', Phase 2 Final Report for Centre de Recherches pour la défense, Valcartier, BAT 15, BFC Valcartier, Courcelette, Québec, March.

WaterMark Industries Inc. (1997*a*). 'Managing uncertainty in digital geographic information: proof of concept', Final Report for Centre de Recherches pour la défense, Valcartier, BAT 15, BFC Valcartier, Courcelette, Québec, March.

WaterMark Industries Inc. (1997*b*). 'Data integration issues related to data quality', Final Report for New Brunswick Geographic Information Corporation, Fredericton, New Brunswick, August.

WaterMark Industries Inc. (1998). 'Data Quality Visualization', Final Report for Centre de Recherches pour la défense, Valcartier, BAT 15, BFC Valcartier, Courcelette, Québec, March.

WILLIAMS, R. J. (1992). 'Data quality statements for spatial databases', Research Note ERL-0632-RN, Defence Science and Technology Organization (DSTO) Australia, Electronics Research Laboratory, Salisbury, Australia, July.

7

Anticipating cultural factors of GDI

Willem van den Toorn and Erik de Man

7.1 Introduction

Expectations of benefits to society from the use of GIS have been high from their introduction in the 1970s. Subsequently, the notion of the GDI within which GIS applications will be operating has emerged as a precondition for sharing data and information. The Clinton administration and the European Commission described these expectations in high level policy statements in the late 1980s and early 1990s in the context of an emerging information economy. National GDI or even global GDI were described as concepts facilitating access to and responsible use of geospatial data at affordable prices.

An important premise in the realization of these expectations is that universal access to government data would prevail, thus providing the 'raw material' for the potentially huge information market. As Jan Kabel and David Rhind have discussed earlier, different countries are reacting differently to these new concepts. The reasons are at least in part rooted in national culture rather than technical constraints, which may be a barrier to achieving the openness required for full development of the information market across national or cultural boundaries.

Although there have been a growing number of successes in GIS applications, there is also a considerable record of failures. In some cases, the reasons can be traced to an under-estimation of the cultural factors in the organization(s) which affect the adoption of the geospatial data technology. The growing recognition of these 'non-technical' factors in recent literature has not resulted in practical guidelines for taking these factors into account in the planning processes. This chapter discusses the practical issues, with Hofstede's (1980, 1997) research providing our point of departure.

A growing body of literature reflects an increasing interest in the management, organizational and usability issues of modern information and communication technologies. Geospatial data technologies have built, in turn, on advances in these technologies, over the past two decades. There is also considerable literature on how individuals accept and adopt innovations (Rogers, 1983). There are various degrees of social acceptance, ranging from 'simple' compatibility to a ready-to-hand or 'embodied' technology, i.e. to obdurancy and or even institutionalization of that technology within a group or society (Bijker and Law, 1992; De Man, 1996; Zwart, 1993). The matter is further complicated because geospatial data technologies having several specific properties, for example, government organizations rather than individuals, generally provide the focus for decisions on their adoption and diffusion (Campbell and Masser, 1995; Rogers, 1993).

Hence, our knowledge of the factors that determine diffusion and adoption of GIS and IT is still in its infancy. Although this continues to be problematic, the growing literature on the 'non-technical' issues of GIS implementation is most encouraging—see, for example, Obermeyer and Pinto (1994) for its extensive bibliography. Nevertheless, comprehensive and systematic knowledge is still limited as much of the literature is inconclusive as to what can be done in advance to assess the social setting and its implications for the development and implementation planning processes. This chapter explores one assessment model developed by Hofstede's analysis of culture (Hofstede, 1980).

7.1.1 The role of culture

Our thematic issue here is the social acceptability of GIS and we also include the social acceptability of its GDI context. We believe this question needs to be answered from an understanding of both the culture of the society and the interaction between the incoming technology and the recipient society.

We understand culture to mean the learned responses to a group's problems that have worked well enough to be considered valid, and therefore shared by the members of that group and taught to new members and generations. Hence, the way in which human beings perceive problems is largely culturally bound. In this sense, culture is pervasive and basically taken for granted. Different cultures will deal differently with information *vis-à-vis* the problems they face. For example, active societies seek opportunities in their environment for improving their conditions and display a desire for attainment and to be in charge. Passive societies, on the contrary, seek to maintain their status quo and display a tendency to be under the control of natural processes, of social waves and developments, or of active others (Etzioni, 1968). In other words, societies differ in the way they deal with uncertainties: do they perceive them as opportunities or as threats? But active-versus-passive also relates to the different values placed on performing and achieving certain rewards. It is self-evident that seeking new opportunities requires fundamentally different sets of information than the maintenance of a status quo.

Notwithstanding its merits, the active–passive dimension will not be sufficient to describe and explain the interaction between culture and the applications of GIS technology. For example, it does not explain how different bureaucratic ways of management deal with information. To this end, the manner in which a society deals with differences in power (its hierarchy) has to be accounted for as well. A more elaborate view on 'culture' will therefore be necessary.

This is a vast field of study and to keep our 'exploration' manageable we have adopted the four composite cultural indicators identified by Hofstede (1980) to describe national culture differences, namely the ways in which a society accommodates:

- human inequality in power and wealth, including the relationship with authority;
- ways of dealing with uncertainty;
- division of roles between the men and women in society; and
- relationships between individuals and groups (individualism versus collectivism).

At this stage, we need to make three remarks. First, in many countries we observe important cultural differences between the traditional and rural areas, and the large cities with their modern institutions and sophisticated technologies. Such differences are typical for transitional and developing nations. Hence countries cannot be seen as cultural monoliths. Secondly, within the same national culture, there may be organizations with very different cultures. There appear to be at least two interrelated elements of organizational cultures which are important in relation to the use of geospatial data technologies. These are styles of bureaucracy and the approach to decision-making (Campbell, 1996). It goes without saying that strategic decision-making requires different sets of data and information than does operational decision-making. These observations lead to the third point. We think questions about the acceptability of geospatial data technologies revolve around two fundamentally different stages: their *desirability* within a specific cultural connotation, and their *feasibility* with respect to organizational and practical ways for implementation.

7.2 Cultural typology of recipient conditions

7.2.1 Hofstede's 4D model

We have already suggested that culture is a good starting point in understanding social acceptance of information and communication technologies. Specifically, we view differences in culture as an important factor in explaining differences in social acceptance. In order to be able to observe and describe differences in culture, we need operational variables (or dimensions) of culture.

In this section we will adopt those dimensions identified by Hofstede (1980, 1997). From his analysis of 116,000 questionnaires and over 20,000 interviews with employees of the multinational company IBM, in over fifty countries, he found national cultures differed in: *power distance*, *uncertainty avoidance*, *femininity* versus *masculinity*, and *collectivism* versus *individualism*. These terms seem to reflect the basic cultural problems each society faces. These dimensions form a 4D model of differences among national cultures. Each country is characterized by a score in each of the dimensions. This trail-blazing work converts the broad concept of 'culture' into a manageable set of instruments with which culture's consequences on organizational structure, managerial procedures and practices, and motivational patterns can be traced. In addition, it suggests how national culture is a major variable in setting organization and management patterns.

The four dimensions of national culture in Hofstede's model can be described as follows:

1. Power distance (PD). Society's way of accommodating human inequality. High power distance cultures are hierarchical, authoritarian, elitist in the sense of accumulation of the good things in life at the higher levels in the hierarchy, and of the bad things in life at the lower levels. Low power distance cultures demonstrate flat organizations and participation, and significant distribution of the good and bad things in life.
2. Uncertainty avoidance (UA). Society's way of accommodating uncertainty. Strong uncertainty–avoiding cultures show little risk-taking, minimal innovation, extensive institutions to bring security and stability, a conservative nature, and thorough planning. Weak uncertainty–avoidance cultures are innovative and creative, and tolerate different views and behaviour. For such cultures, risk and excitement are greater values than security and stability; these people play life as it comes, adopt incremental planning, and have few contingency scenarios.
3. Masculinity versus femininity (MAS). Society's way of accommodating masculine and feminine values. Masculine cultures focus on achievements and success, are aggressive, winners win/losers lose, and encourage visible success. Feminine cultures are caring cultures emphasizing the quality of life, networking, and relationships as social values, are egalitarian and show compassion.
4. Individualism versus collectivism (IDV). Society's way of accommodating the individual and the group. Individualist cultures comprise calculating citizens: what is in it for me, one's life is one's own, individual views matter, group views are necessary unavoidables to be challenged whenever one feels like it. In collectivist cultures group values dominate: one's sense of life is derived from one's contribution to the common good, and closely knit in-crowds, out-crowds and outside individuals are peripheral.

Hofstede's analysis reveals that countries with a generally large power distance are also likely to be more collectivist, and small power distance countries to be more individualist. In other words, across the full set of countries in his study IDV and PD are strongly and inversely correlated, whereas PD/UA and UA/MAS are not.

The analysis further suggests the relevance of both power distance and uncertainty avoidance for our thinking about organizations. Organizations always evolve around two fundamental questions: (1) who has the power to decide what? and (2) what rules or procedures will be followed to attain the desired ends? The answer to the first question is influenced by cultural norms of power distance; the answer to the second question by cultural norms about uncertainty avoidance. The remaining dimensions, individualism and masculinity, affect our thinking about people in organizations, and particularly impact on motivational patterns rather than on organizations (Hofstede, 1997: p. 140).

The relevance of power distance and uncertainty avoidance for organizational structure, procedures and managerial practices is widely documented. For example, a full (Weberian) bureaucracy is distinguished by these two dimensions (Weber, 1947). The five structural configurations of organizations suggested by Mintzberg (1979), namely simple structure, machine bureaucracy, professional bureaucracy, divisional form, and adhocracy, can effectively be viewed from the perspective of power distance and uncertainty avoidance (Hofstede, 1997; Mintzberg, 1979).

7.2.2 Potential influence of culture on practices of geospatial data technologies

Before we turn to the question of how culture may influence and shape deployment of geospatial data technologies, we will sketch its potential influence on information technologies in general. In the literature, we find the following potential influence of national culture on the management and use of information technologies (Grover *et al.*, 1994).

1. Large or small power distance. Large power distance implies little participation in leadership and a greater centralization of authority. Centralization of decision-making, in turn, implies more centralized computer environments. User involvement and existence of end-user computing may also be lower in centralized environments. As well, there may be less involvement of the information system's staff and management in strategic planning.

2. Strong versus weak uncertainty avoidance. Strong uncertainty avoidance may exhibit centralized tendencies and the creation of formalized policies and operating procedures. Hence, this factor may influence the role of information technologies within the organization. Strong uncertainty avoidance is typically associated with a conservative strategy. The role of information technologies may be more operational than strategic. More traditional technologies will be preferred and more formal justification of expenditures on information technologies will be required.

3. Masculinity versus femininity. Performance, desire to achieve, and visible

achievement motivates masculine societies. Individual accomplishments are emphasized. Therefore this factor may impact the performance appraisal of information technologies. Masculine cultures may focus on more tangible market-based impacts. Such organizations may also exhibit more aggressive, competitive uses of technologies. This factor may also influence resource allocation methods as well as staff reward systems.

4. Individualism versus collectivism. This factor has a strong impact on leadership and management style within the organization and, hence, on the functioning of information technologies. The level of participation in the development and planning of information systems and the level of information sharing may be low in individualistic cultures. This factor also influences the perceived measures of the information system's success (group versus individual achievement).

7.2.3 GIS properties

Which properties of GIS are relevant to these four dimensions of national culture? They may affect many sectors and many levels of decision-making, thus the number of properties to consider could be large, or even infinite. However, we have noted that organizations and their management and staff, rather than individuals in isolation, provide the focus for the application and acceptance of these technologies. It would therefore seem logical to focus acceptance of geospatial data technologies on their organizational properties. We have selected the following properties for further exploration and discussion:

1. Functionality. There is wide understanding in the literature that IT (and GIS) can serve four main organizational functions:

- communication: general information purposes, enhancement of staff corporate identification, 'family' feeling;
- strategic planning: scenario development, simulation and scenario testing, strategy selection;
- operational planning and operational management: goal setting, how to achieve them, routes, responsibilities;
- monitoring and evaluation: management supervisory instrument connected to both the organization's strategic and operational performance.

2. Presentation. GIS is swift and precise, and can provide attractive visual presentations.

3. Organizational impacts. In addition to functionality, GIS may have other organizational impacts since it must:

- interact with an organization's structure and procedures;
- impact the distribution and structure of power in the organization;
- involve tasks and skills not required and/or not available in an organization without GIS.

Of these properties, we believe that functionality in particular is governed by cultural factors. The functionality of a technology is at the heart of its cultural desirability and is concerned with whether a technology is desirable within a particular group or society. The other properties of geospatial data technologies (namely presentation and organizational impacts) are related to organizational structure and management practices. We believe these latter properties are more relevant to the feasibility of implementing technologies. This does not mean that the latter is culturally neutral. We stress that social acceptance of these technologies consists of two stages: desirability and feasibility, both of which are culturally influenced.

7.2.4 Cultural typology of regional groups of countries

How do differences in national culture influence the social acceptance of geospatial data technologies, and specifically their desirability in terms of functionality? Does each country portray its own, different picture in this respect? Theoretically, Hofstede's model characterizes each country by a score on each of the four dimensions. To explore the social acceptance of geospatial data technologies, it might be sufficient to identify cultural profiles of regional groups of countries rather than national cultures of individual countries. Regional differences in the social acceptance of geospatial data technologies may explain emerging regional associations and publications such as *GIS Europe* and *GIS Asia Pacific*. We believe that the role of culture might provide insight into the factors and circumstances contributing to these differences.

As already discussed, organizations rather than individuals generally provide the focus for decisions concerning the adoption and diffusion of technology. It would therefore seem to be logical to focus acceptance of geospatial data technologies on those dimensions of culture that most significantly impact organizational structure and procedures, management practices and staff motivation. The two major indicators of culture from this point of view are power distance and uncertainty avoidance. For motivational aspects we should add the masculinity versus femininity dimension. Table 7.1 identifies seven regions with more or less similar cultural characteristics. Key indicators are power distance and uncertainty avoidance, thus leaving four major regional groups. Some of these regional groups can be further subdivided according to differences in the other two cultural dimensions.

We believe that cultural conditions determine the propensity to socially accept information technologies such as GIS, i.e. cultural conditions primarily determine the desirability. The cultural desirability specifically relates to the functionality of GIS: communication and information sharing, strategic planning, operational planning and management, and monitoring and evaluation.

7.3 Matching recipient culture and GIS properties—an exploration of cultural desirability

Here we specifically consider possible cultural responses to the other properties of geospatial data technologies (namely the organizational impacts). However,

TABLE 7.1 Country cultural typology of recipient conditions (based on Hofstede 1980, 1997)

Type & countries	Power distance (PDI)	Uncertainty avoidance (UAI)	Masculinity/ Femininity	Individualism/ Collectivism
1. Low PDI/low UAI				
Anglo-Saxon	Low	Low	Masculine	Individual
Nordic	Low	Low	Feminine	Individual
2. Low PDI/high UAI				
German	Low	High	Masculine	Individual
3. High PDI/low UAI				
SE Asian	High	Low	Masculine	Collective
4. High PDI/high UAI				
Latin American	High	High	Masculine*	Collective
Mediterranean	High	High	Average†	Diverse
Japan	Average	High	Masculine	Average

* Exception: Chile is Feminine; † Exception: Italy is Masculine

it is important to remember that this is explorative and deductive. The available literature generally reports on more modestly scaled research topics.

7.3.1 Culture's impact on the functionality of GIS; social desirability of the technology

We look at the possible interplay between culture and the functional properties of GIS or of geospatial data technologies in general. We have identified the following functional properties of GIS:

- communication and data/information sharing;
- strategic planning;
- operational planning and management; and
- monitoring and evaluation.

We believe that three dimensions of culture, i.e. power distance, uncertainty avoidance, and masculinity, impact on organizational structure, procedures and managerial practices, as well as the motivational patterns of organizational members. Across the full set of countries, individualism and power distance are strongly and inversely correlated.

1. Power distance

Large power distance cultures are likely to employ GIS mainly to reinforce the instructive hand of management, i.e. operational planning and monitoring. Other functional properties of GIS may trigger profound resistance and opposition on the part of these cultures' authoritarian management. Large power distance cultures are likely to operate on a need-to-know basis, rather than to see sharing and visibility as intrinsically virtuous factors.

Small power distance cultures would welcome the sharing, visibility, and accountability properties of GIS. They would seek to incorporate these properties into the organization's structure and its managerial practices. There would be a culture where individuals are seen to play a role, where there would be a feeling of 'we are in this together, so let us share'. In small power distance cultures, broad sharing of information would be welcomed for its capacity to create and make visible organization-wide accountability.

2. Uncertainty avoidance

Strong uncertainty avoidance cultures would generally be positive with respect to the acceptance of the 'controlling' and conservative functional properties of GIS. These cultures would feel a strong need to adopt those technologies and methodologies that would contribute to security and stability. The higher the score on this indicator, the more desperate the need, and indeed the more influential uncertainty avoidance will be in shaping organizational structure and management practices as well as motivational patterns.

By the same token, weak uncertainty avoidance cultures would generally play events as they come (compared to large uncertainty avoidance cultures, which would seek to be prepared). Weak uncertainty avoidance cultures would be quite prepared to take a risk and apply the new technology in novel situations or if GIS would appear sufficiently attractive for other reasons.

3. Masculinity versus femininity

Masculine cultures would seek GIS technology for its capacity to contribute to the visibility of success. We may expect this feature to show up eclectically in 'operational planning and management' and 'monitoring and evaluation'. Masculine cultures would probably shy away from the wider applications of GIS which may make failure and joint responsibilities more visible.

Feminine cultures would appreciate the networking and relationship-building capacity of GIS. Such cultures would be less interested in the strategic functionality of GIS but not oppose it. In terms of planning, monitoring, and evaluation, feminine cultures would like to be informed and be able to adjust operations. In the latter, 'caring' would probably be an important consideration.

7.3.2 Desirability of GIS

What conclusions can be drawn regarding the desirability of geospatial data technologies and of GIS in particular, i.e. in matching Table 7.1 (regional cultural typologies) and Table 7.2 (impact of culture on GIS functionality)? First of all, whether such technologies are desired or not depends on the cultural context of the host society. We also see that desirability refers to different (functional) properties of these technologies. Consequently, it seems to be impossible to speak in general terms of the desirability of GIS for a particular country or region. Some of the functional properties (namely communication and data/information sharing, strategic planning, operational planning and management, or monitoring and evaluation) may be culturally desirable. And, instead of examining the country as one cultural monolith, individual relevant factors would need to be examined.

7.4 Feasibility of the introduction and implementation of geospatial data technologies

The functional properties of geospatial data technologies discussed above determine their cultural desirability. The other properties—swiftness, precision, visibility, interaction with organization structure and procedures, and with power relationships, new tasks, and skills—can be viewed as influencing the second stage of feasibility in the implementation of these technologies. Issues regarding feasibility may emerge under various degrees of desirability.

The mutual relationship between feasibility and desirability is a complex one. Different possible combinations exist. The implementation of GIS may be:

- desirable and readily feasible;
- desirable but not readily feasible;
- readily feasible but not desirable; or
- neither readily feasible nor desirable.

What determines the feasibility of introduction and implementation of geospatial data technologies? Obvious factors in this respect are resources, skills, commitment, etc. If we see the introduction and implementation of technologies as processes of social change, it will be evident that feasibility also depends on compatibility with the old tasks, i.e. the compatibility of the current and the proposed methods and procedures. If everything remains the same, the resources and skills are already present and no new commitment is necessary. In the literature, we find task congruency as an important variable in the introduction and implementation of information technologies (Shore and Venkatachalam, 1996: pp. 19–35). Low task congruency then occurs if considerable change is required in task content, procedures, organizational structure or power relationships, or any combination of these factors.

An example can illustrate the impact of task congruency: the replacement of an old ledger system by improved technology, while continuing to use similar methods and procedures as in the past, or the upgrading of software already in use for much the same purposes. In both cases, the task congruency

TABLE 7.2 Connotations of cultural indicators *vis-à-vis* functional role*

GIS functional property	Power distance (PD)		Uncertainty avoidance (UA)		Masculinity versus Femininity	
	Large	Small	Strong	Weak	Masculine	Feminine
1. Communication information sharing	L—keep them dumb, top knows	H—we are in this together so let us share	H—thoughtfully structured information creates feeling of security	L—GIS information may be interesting, but who cares	L—who wants to communicate	H—networking, relationship, caring
2. Strategic management	L—top knows, no need for further visibility	H—yes, let us discuss this, united we stand	H—considerable assistance to perceived security and stability	L—strategic management is a red herring, focus on operations	L—provided likely is well advertised in strategy	L—may often be useful away from operations—that is where it matters
3. Operations planning and monitoring	L—reinforcement of management instructive land	H—proper accountability, transparency	H—control, contingency sciences	L—maybe useful, but no need for full-fledged expensive output	L&H: L—only in so far as required to shine success, H—masculinity is defensive at the same time	H—ability to adjust, protection of the privileged

* L, H = low, high desirability of IT/GIS

factor would be high. But, if the introduction of a new accounting system substantially changes the methods and procedures previously used and introduces more complex technology, the congruency factor would be low (Shore and Venkatachalam, 1996: p. 27). In many cases organizations are confronted with significant changes when they discover the need to move to computerization.

However, the feasibility of introducing and implementing geospatial data technologies depends not only on task congruency. A competitive environment may compel individual organizations to adopt innovative technologies as well. The literature suggests that the impact of task congruency on feasibility is somewhat greater than the impact of a competitive environment (Shore and Venkatachalam, 1996: pp. 25–7).

We see feasibility as supporting or constraining, or in some cases even substituting for lack of, desirability of the introduction and implementation of geospatial data technologies. Remember though that cultural desirability relates to individual functional properties of the technology. In Table 7.3 we explore the combined influence of cultural desirability and feasibility on the introduction and implementation of GDI.

This exploration leads to the following conclusions:

- Fully positive, i.e. culturally desirable and feasible. GDI would in this case be culturally desirable in any of the functional properties (namely communication and information sharing, strategic planning, operational planning and management, or monitoring and evaluation). Cultural desirability would suggest that successful implementation could still be effected even when the task congruency and/or competitive environment is relatively low. Strong uncertainty avoidance in this situation (see Table 7.2) would probably demand the actual and full implementation of GDI to be carefully prepared. A great deal of participation would result in a clear script according to which the actual implementation would take place, in terms of the contents of the GDI, the steps to be taken, the time schedule, and the allocation of responsibilities.

TABLE 7.3 Combined influence of desirability and feasibility on the introduction and implementation of GIS/GDI

Task congruency	Competitive environment	Culturally desirable or undesirable	Ease of introduction implementation
High	High	Desirable	+
		Undesirable	+/–
High	Low	Desirable	+/–
		Undesirable	–
Low	High	Desirable	+/–
		Undesirable	–
Low	Low	Desirable	–
		Undesirable	–

At higher levels of task congruency, less care would be needed in planning the implementation. Yet, in this case too, strong uncertainty avoidance tends to caution against a too casual approach to the actual implementation trajectory.

- Fully negative, i.e. culturally undesirable and not feasible. With low task congruency, the successful introduction and implementation would seem to be an uphill struggle. Minimal cultural desirability coupled with considerable discongruence of tasks would be hard to overcome, even if hardline and large power distance management would still see good reasons to try and introduce GDI. These parameters suggest that even if the introduction were to get started, it is highly doubtful whether the technology could be successfully sustained.

  At rather higher levels of task congruency, implementation would be easier. Nevertheless, the negative strength related particularly to adverse motivational patterns would probably lead to a GDI project requiring long-term nurturing and managerial pressure.
- Intermediates. The parameters would suggest that chances for the successful introduction and sustained operation of the GDI, though never great, would be positive only at high levels of task congruency and competitive environments. In other cases, the negative observations made in the previous paragraph would probably hold true.

7.5 Conclusion

We have explored the various dimensions of social acceptance of GDI and other geospatial data technologies. We have tried to create an awareness that acceptance, utilization, and the success of such technologies should not be taken for granted. We have identified some cultural conditions that impact these processes.

Similarly, if the development of the information economy depends on universal access to governments' geospatial and other data, differences in national culture must be taken into account. The way each country or culture perceives this concept and how it believes its citizens will be affected may seriously restrict the acceptance, desirability and feasibility of international economic traffic in government-owned information.

Our exploration of the social acceptability of geospatial data technologies has yielded the following preliminary results:

- Social acceptance of IT and communication technologies depends on the particular cultural conditions of the society. The pervasiveness of culture explains the resistance met when introducing technologies such as GDI. Another consequence of this cultural perspective is that there are no general or globally applicable conditions for the introduction and implementation of geospatial data technologies. We cannot simply copy GIS or GDI applications from elsewhere. Within our own geospatial region we may find similar cultural conditions where experiences with geospatial information technologies can be fruitfully shared and where mutual learning processes can be initiated.

- Social acceptance of geospatial information technologies results from a two-stage process: the cultural desirability and feasibility of their introduction and implementation. The cultural desirability pertains specifically to their functional properties (namely communication and information sharing, strategic planning, operational planning and management, or monitoring and evaluation), whereas feasibility deals with other properties (e.g. presentation and organizational impacts). Consequently, social acceptance depends on the combined influence of desirability and feasibility.
- Social acceptance and cultural desirability of geospatial data technologies are not simple, monolithic measures, rather, they are multi-dimensional in nature. Some but not all of the functional properties may be culturally desirable. Consequently, the process of assessing social acceptance of geospatial data technologies is multi-dimensional as well.
- These cultural dimensions may explain why the implementation of GIS is often preceded by a 'playing with the technology' stage. In this period limitations are discovered, and previous notions of benefits may be replaced by better-informed ones. It is a period in which the local culture has time to adapt to something new before adopting it wholeheartedly but with a larger sense of realism. Overall the communication among the staff about the technology takes on more substance and may lead to the formulation of one or more projects involving major stakeholders possibly from other organizations. The results will probably lead to far more realistic functional specifications, expectations of the time and other resources necessary to develop and implement the applications, but also to the definition of the necessary organizational and institutional supporting environment between the stakeholders.

We hope that we have provided some operational 'tools' to assist you in your specific quest for acceptance and utilization of concrete geospatial data technologies. At the same time, our exploration may serve as an outline for research to acquire in-depth as well as operational knowledge in this challenging field.

Bibliography

BIJKER, W. E. (1995). *Of Bicycles, Bakelites, and Bulbs: Towards a Theory of Sociotechnical Change*. Cambridge, Massachusetts: The MIT Press.

——and LAW, J. (eds.) (1992). *Shaping Technology/Building Society: Studies in Sociotechnical Change*. Cambridge, Massachusetts: The MIT Press.

CAMPBELL, H. (1996). 'Theoretical perspectives on the diffusion of GIS technologies', in MASSER, I., CAMPBELL, H., and CRAGLIA, M. (eds.), *GIS Diffusion: the adoption and use of geographical information systems in local government in Europe*. GIS DATA III. London: Taylor and Francis, pp. 23–45.

—— and MASSER, I. (1995). *GIS and Organizations: How effective are GIS in practice?* London: Taylor & Francis.

DE MAN, W. E. (1996). 'Establishing geographic information systems as institution building', in VAN DER MEULEN, G. G. and ERKELENS, P. A. (eds.), *Urban Habitat: The Environment of Tomorrow*. Eindhoven: Faculty of Architecture, University of Technology.

ETZIONI, A. (1968). *The Active Society*. New York: The Free Press.

GROVER, V., SEGARS, A. H., and DURAND, D. (1994). 'Organizational Practice, Information Resource Deployment and System Success: a Cross Cultural Survey', *Journal of Strategic Information Systems*, 3/2: pp. 85–106.

HOFSTEDE, G. (1980). *Culture's Consequences: International Differences in Work-related Values*. Beverly Hills, California: Sage Publications.

—— (1997). *Cultures and Organizations: Software of the Mind*. London: McGraw-Hill.

MINTZBERG, H. (1979). *The Structuring of Organizations*. Englewood Cliffs, New Jersey: Prentice-Hall.

OBERMEYER, N. J. and PINTO, J. K. (1994) *Managing geographic information systems*. New York: The Guildford Press.

ROGERS, E. M. (1983). *Diffusion of Innovations*, 3rd edn. New York: The Free Press.

—— (1993). 'The diffusion of innovations model', in MASSER, I. and ONSRUD, H. J. (eds.), *Diffusion and Use of Geographic Information Technologies*. Dordrecht: Kluwer.

SCHEIN, E. H. (1969). *Process Consultation*. Reading, Massachusetts: Addison-Wesley.

SHORE, B. and VENKATACHALAM, A. R. (1996). 'Role of National Culture in the Transfer of Information Technology', *Journal of Strategic Information Systems*, 5: pp. 19–35.

WEBER, M. (1947). *The Theory of Social and Economic Organisation* (translated). New York: The Free Press.

ZWART, P. R. (1993). 'Embodied GIS; a concept for GIS diffusion', in MASSER, I. and ONSRUD H. J. (eds.), *Diffusion and Use of Geographic Information Technologies*. Dordrecht: Kluwer, pp. 44–56.

8

The foundation technologies

Wolfgang Kainz

8.1 Introduction

Geospatial infrastructures only function effectively when reliable and efficient computing and communication technologies are in place. Increasingly large databases have to be linked for the transfer of data or to provide a basis for inter-operability of heterogeneous software and hardware systems. Many of these technologies have resulted from military research and applications, with the Internet as perhaps the best known example. Together with the World Wide Web they currently form the most widely used vehicles for geospatial infrastructures.

We start with a discussion of system architectures, paying special attention to the client–server model, and then explore the technical aspects of computer networks and the World Wide Web. Issues of geospatial database architecture and design, their distribution over different physical locations, and the need to acquire information about the availability of geospatial data lead to the introduction of the geospatial data clearinghouse concept. The information technological aspects will be handled in a separate section. A discussion of the role of the Internet and the World Wide Web concludes this chapter.

8.2 System architecture

There are several ways to organize computers so that they can perform their tasks properly. Computer systems may be arranged as individual stand-alone machines, or co-operate in computer networks. Today, computer networks are an indispensable component of any GDI. Special attention is given to the client–server architecture, which is of great importance not only in geospatial data handling but also in general data processing.

Two important basic features of system architectures are open systems architectures and inter-operability. Open systems co-operate with each other using standards for access, processing, and transfer of data. They are not hindered by system-specific structures. Inter-operability is the ability of software and hardware on multiple computers from multiple manufacturers to communicate. Inter-operable databases play an important role in GDI.

8.2.1 Stand-alone computers

Individual computers have been used since the beginning of electronic data processing. They are usually dedicated to specific tasks or applications. However, they are not linked to anything else, and data can only be transferred by physically transporting files on magnetic or optical media such as tapes, disks, or CDs (Figure 8.1).

Advantages of stand-alone computers include knowing precisely who is responsible for them and that they are 'isolated' from each other, which makes unwanted transfers difficult. However, the isolation is also a drawback that causes overheads, redundancy, and frequent duplication. This can be overcome by co-operative computers in various kinds of networks or distributed systems.

8.2.2 Client–server architecture

A client–server architecture is a common form of distributed system where software is split between client and server tasks. In a distributed system, a user does not see which computer is performing which functions. It essentially behaves as a virtual single processor. The characteristic feature of a distributed system is that it is a software system running on top of a computer network.

A client is a process or computer that requests the services of another computer or process (server) using some rules (protocol). The server processes the request and sends the result back to the client. The exchange of information is done through message exchange according to the protocol. A server program (daemon) may run continuously, waiting for requests to arrive, or be involved by some higher level program.

A server computer provides services to client machines. A common example is a file server, which is a computer with a local disk servicing requests from

FIGURE 8.1 Stand-alone computers without co-operation.

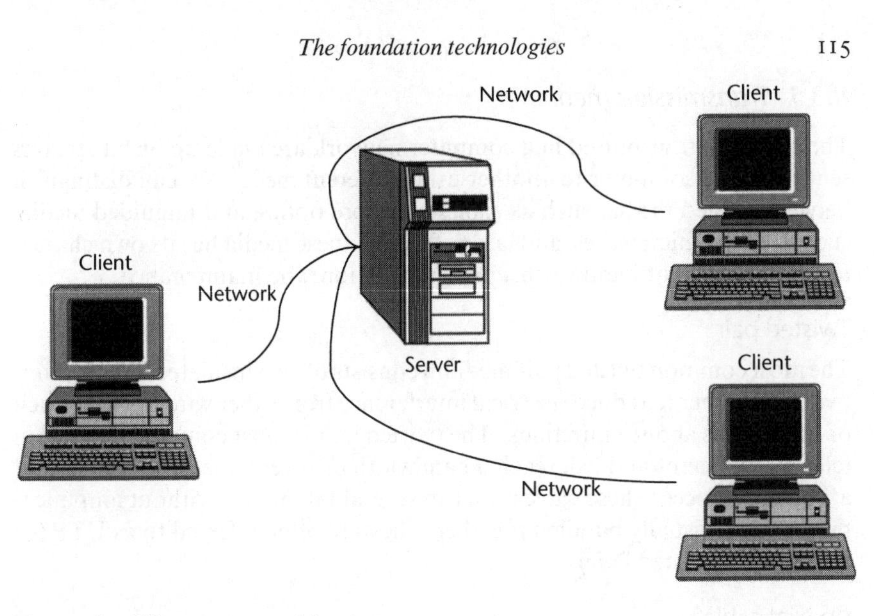

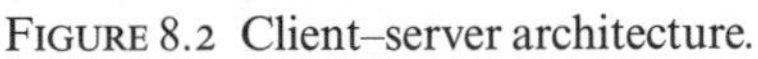
FIGURE 8.2 Client–server architecture.

remote clients (workstations or PCs) to read and write files on that disk. File servers often use the Network File System (NFS) protocol for workstations or Novell Netware for personal computers. In a client–server set-up, we usually find many clients using a few servers. Clients and servers may even run different operating systems, and servers can be clients of other servers. Figure 8.2 illustrates the client–server architecture.

8.3 Computer networks

The exchange of data among computers in a GDI is a necessity. A computer network is an interconnected collection of autonomous computers, i.e. there is no hierarchical relationship between the computers in the network. There are various ways of connecting computers into a network, including the use of wires, fibre optics, microwaves, and communication satellites.

The communication units that are sent between computers in a network are called messages. They can be transmitted one chunk at a time from a sending to a receiving computer (message switching). Often it is necessary to split the message into several data packets to avoid transmission problems (packet switching). One of the principles of packet switching is that individual packets may travel distinct routes to their destination. The speed of transmission is measured in bits per second (or bps). The orders of magnitude used in this discussion are k (kilo or 10^3), M (mega or 10^6) and G (giga or 10^9). Today, transmission speeds of several hundred megabits per second are common.

8.3.1 Transmission media

The messages transmitted in a computer network are made up of bit streams sent from one computer to another using different media. We can distinguish between guided media, such as cables and fibre optics, and unguided media, such as radio, microwaves, and lasers. Each of these media has its own characteristics in terms of bandwidth, costs, installation, and maintenance.

Twisted pair

The most common twisted pair medium consists of two insulated copper wires twisted together to reduce electrical interference from other wires nearby. Each of the wires is about 1 mm thick. The twisted pair is most commonly found in telephone systems and can reach a bandwidth of several megabits per second at short distances. These cables can run several kilometres without amplification, and are usually bundled together. They are often referred to as UTP (or Unshielded Twisted Pair).

Coaxial cable

The coaxial cable is a common medium used for computer and cable television networks. It is better shielded than the twisted pair and allows for higher transmission speeds (up to one or two gigabits per second). These cables consist of a stiff copper wire surrounded by insulating material. This is surrounded by a cylindrical conductor in the form of a braided mesh, which in turn is covered by a protective plastic sheath (Figure 8.3). Coaxial cables are distinguished into baseband coaxial cables for digital transmission and broadband coaxial cables for analogue cable television.

Fibre optics

Fibre cables are an optical transmission system. Such a system consists of three components: a light source, a transmission medium, and a detector. The light source is either an LED (light emitting diode) or a semiconductor laser. The transmission medium is an ultra-thin fibre of glass; the detector is a photodiode generating an electrical pulse when hit by light. Light emitted at a source can travel within the fibre for many kilometres with virtually no loss. A fibre

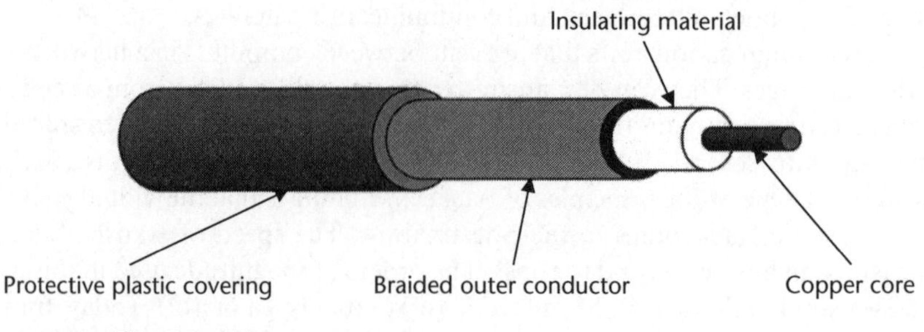

FIGURE 8.3 Coaxial cable.

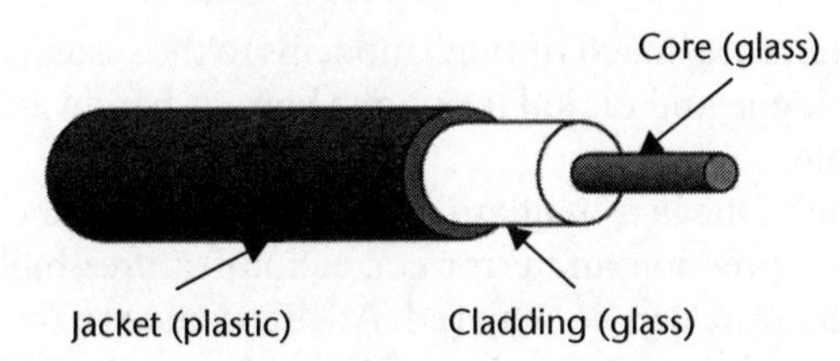

FIGURE 8.4 Fibre cable.

cable is similar to a coaxial cable. It has a glass core about the diameter of a human hair, surrounded by a glass cladding which serves to keep the light in the core. An outer plastic jacket covers the cladding and the core (Figure 8.4). As with twisted pairs, optical fibre cables are usually bundled together and protected by an outer sheath.

With current technology, data transmission through fibre cables reaches a practical bandwidth of one gigabit per second, although theoretically a much higher transmission rate is possible. Fibre optics is used in local as well as long-haul transmissions. Apart from providing a larger bandwidth, fibre cables have other advantages over copper wires. They provide data transmission over longer distances without repeaters, cannot be affected by power surges or electromagnetic inference, are thin, and do not leak light. Moreover, fibre cables are more difficult to tap than copper wires.

Wireless transmission

Electromagnetic waves are used for wireless transmissions. Their advantage is that once broadcast by a transmitting antenna, they travel at the speed of light and can be received by a receiver some distance away. Depending on the range used within the electromagnetic spectrum, we speak of radiowaves, microwaves, infrared, visible light, ultraviolet light, X-rays, and gamma rays. Only radio- and microwaves, as well as infrared and visible light, are in fact used for transmitting information.

The advantage of wireless media is that cables do not have to be installed; one transmitter can, in principle, reach many receivers. Depending on the wavelength used, a repeater must be installed at certain distances. For the frequently used microwave links, a repeater is required about every 50 km. This medium is used mainly for data and voice transmission in the telecommunication business. Infrared transmission is well known for its use in remote control of video and audio equipment within a room. It is also used for linking printers to notebook computers. The nature of infrared requires, however, that the transmitter is in sight of the receiver and that the distance between them is not too large. In the visible light range, lasers are used as a transmission medium. Communication satellites receive a microwave signal, amplify it and rebroadcast it back to the surface of the Earth.

Wireless transmission continues to be a booming market. Notebook and hand-held computers are linked through modems to the systems in the office or base by using analogue and digital telephone lines. A bandwidth of 2 Mbps is typically achievable.

A modem connection via a standard analogue telephone line can reach 28.8 kbps, and with compression and error correction features built in, a slightly higher effective data rate can be achieved. An alternative to the analogue telephone network is the Integrated Services Digital Network (ISDN). Its goal is to provide voice and data transfer simultaneously on a digital network. ISDN comes in two versions, narrowband ISDN (N-ISDN) and broadband ISDN (B-ISDN). The former provides only 64 kbps, whereas the latter is designed for a bandwidth of 155 Mbps. N-ISDN is virtually useless for modern high speed networks, whereas B-ISDN based on Asynchronous Transfer Mode (ATM) technology meets current requirements. The slower N-ISDN can, nevertheless, be very useful for the home user who wishes to download web pages at a higher speed than possible with a 28.8 kbps analogue modem.

8.3.2 Network hardware

We can classify computer networks according to the transmission technology employed, the scale at which they operate, and the network topology, i.e. the shape of the graph of pairwise connections between computers.

There are two types of transmission technology: (1) broadcast networks, and (2) point-to-point networks. In broadcast networks short messages, called packets, are sent from one computer to all other computers. These packets contain address information identifying the computer for which they are intended. Only this computer will process the packets; all others will ignore them. This technique is called broadcasting. If a subset of computers in the network is addressed, this is called multi-casting. In a point-to-point network, each computer is linked to one or more computers and communication is between pairs of machines. If, for instance, a message is to be sent from computer A to computer B, it may travel through intermediate machines before reaching the destination computer. Routing algorithms searching the optimal path through the network are of particular importance. Larger networks are mainly point-to-point, whereas smaller, geographically limited networks use broadcasting.

Another way of classifying networks is according to their scale. Networks of up to a size of about 1 km are known as Local Area Networks (LANs), while Metropolitan Area Networks (MANs) roughly span the area of a city (up to 10 km). Larger areas such as countries or continents are networked through Wide Area Networks (WANs), and at a global scale they are called internetworks, e.g. the Internet.

The structure of nodes and links between them is called the network topology. Figure 8.5 shows the most common network topologies: star, ring, tree, complete graph, and bus.

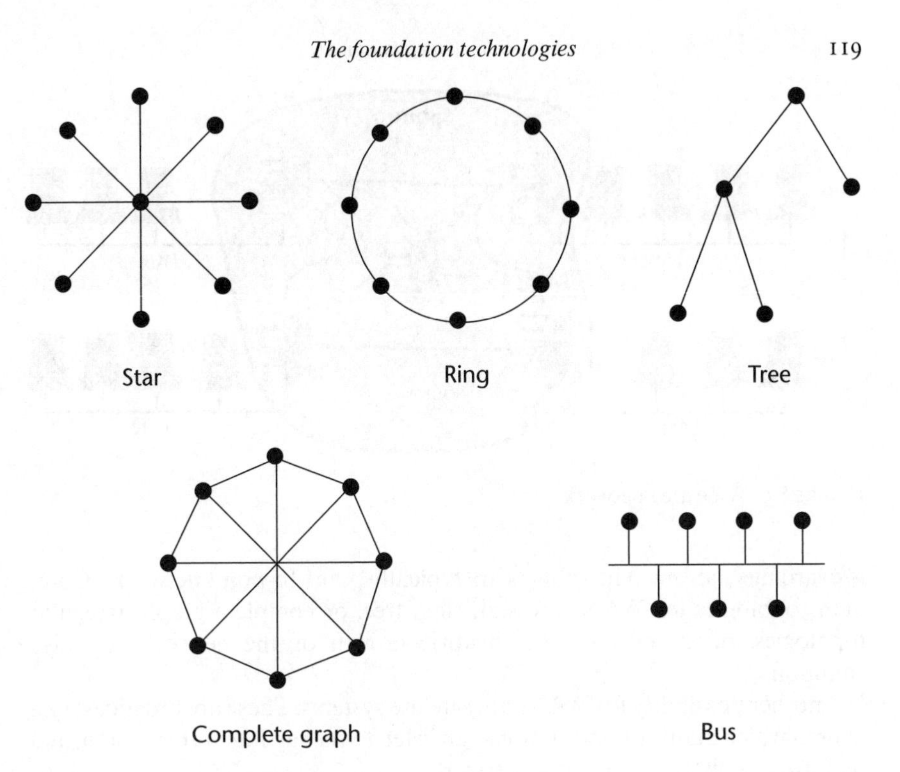

FIGURE 8.5 Network topologies.

Local area networks (LANs)

Local area networks usually link computers within a building or on a campus. They are mainly used to connect PCs, workstations, and other devices such as printers. LANs are broadcast networks, using a single cable to which all computers and devices are linked (bus or ring topology). In a bus topology, only one machine is allowed to send at a time. If two or more machines want to transmit simultaneously, an arbitration mechanism has to resolve the conflict. This mechanism can be centralized or distributed. A typical example is the standard IEEE 802.3 or Ethernet. It uses a decentralized arbitration mechanism and operates at a bandwidth of 10 to 100 Mbps. Another type of topology for an LAN is the ring. The IEEE 802.5 (or IBM token ring) operates at bandwidths of 4 and 16 Mbps.

Wide area networks (WANs)

Wide area networks span a large geographical area and contain a collection of machines called hosts or end systems. The hosts usually belong to an LAN in which there is a special computer, called a router, that connects the LAN to a communication subnet which, in turn, connects to other LANs (Figure 8.6). The routers are also called packet switching nodes. Routers that are not directly connected communicate indirectly via other routers through a store-and-

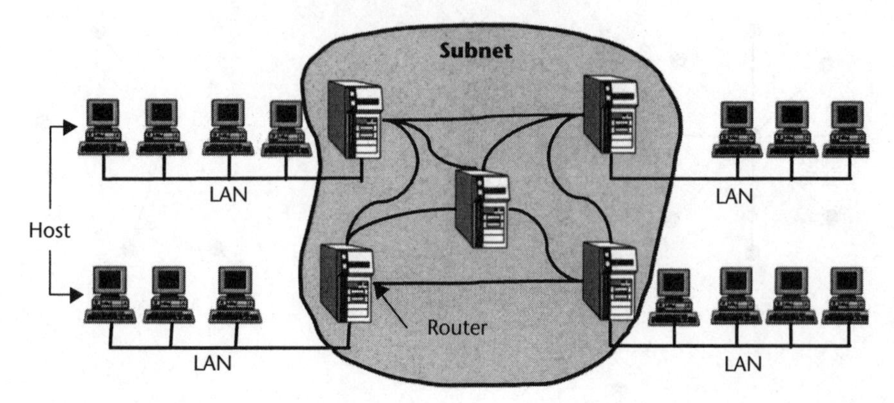

FIGURE 8.6 Wide area network.

forward mechanism. The subnets are typically point-to-point networks. Common topologies for WANs are star, ring, tree, or complete graph. Irregular topologies, often reflecting the historic growth of the network, are also common.

Another possibility for WANs are satellite systems. These are broadcast systems in which many ground stations can listen to the satellite transmission, and they are usually part of a bigger network.

Internetworks

Several heterogeneous networks connected through gateways are called an internetwork. Analogous to WANs, where a router connects LANs, within an internetwork the gateways connect WANs. The Internet is the best known, but not the only example because there is a global internetwork whose origin dates back to a project by the American Department of Defense when the Defense Advanced Research Projects Agency (or DARPA) launched the ARPANET in 1969. By the early 1980s, the ARPANET had grown steadily to include hundreds of host computers and several of today's key protocols were stable and had already been adopted.

When the supercomputer network of the American National Science Foundation NSFNET was connected to the ARPANET the growth became exponential. Many other networks joined such as BITNET, HEPNET, and EARN. By 1990, the ARPANET was shut down and dismantled. Nowadays there are millions of computers connected to the Internet and high-speed links running from 64 kbps to 2 Mbps connect continents.

The Internet has many applications that are indispensable in today's information infrastructure: electronic mail, news, remote login, file transfer, and the World Wide Web. The application with the strongest impact on society is the WWW. Since its invention in the early 1990s and the introduction of browsers (such as Mosaic, Netscape Navigator, and Internet Explorer) websites have been created at an enormous rate. Almost every company, organization, government, and university, as well as private individuals, now have their own home page. And their number is increasing dramatically. A computer is said to be linked to the Internet when it has an IP address, runs the TCP/IP protocol stack, and can send IP packets to all other machines on the Internet.

8.3.3 Network software

When computers communicate with each other they must 'speak' the same language, and implement correct procedures to establish and maintain a connection. Computer networks and network software are complex and are therefore organized as a series of layers or levels. Each layer offers certain services to higher layers. Interfaces between layers define which primitive operations a layer offers to a higher layer. Layer *n* on one computer (host) communicates with layer *n* on another computer according to certain conventions and rules or protocols. The respective layers communicating with each other are called peers. The peers do not communicate directly, rather information is passed to lower layers through the interfaces to the physical medium (below layer 1) where the actual communication takes place. Figure 8.7 illustrates this concept.

A network architecture is a set of layers and protocols but does not include details of implementation or specifications for interfaces, which may vary among systems. A protocol stack is a set of protocols for a certain system, one for each level.

Layer services

Communication between computers means sending messages from one machine to another. If travel in only one direction is allowed we speak of simplex communication. A transfer in both directions, but not simultaneously, is called half-duplex communication, and if there are no restrictions, we have full-duplex communication.

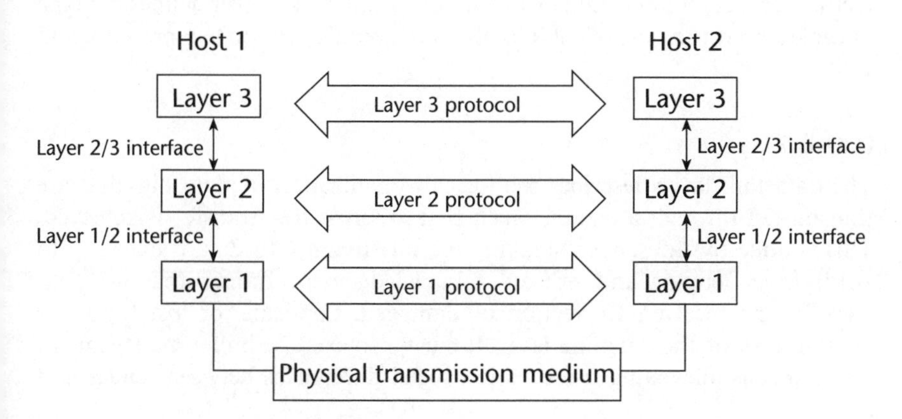

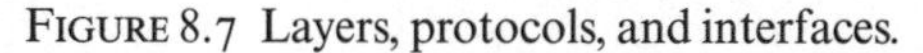

FIGURE 8.7 Layers, protocols, and interfaces.

With so many computers involved, it is extremely important to have an addressing mechanism that allows unique identification of senders and receivers. Rules must be defined for how data can travel between layers. Other services refer to error-detection and error-correction, disassembling and reassembling long messages, avoiding data overflow due to fast transmitters and slow receivers, and routing of messages.

There are two models of significance for modern computer networks: the ISO/OSI and the TCP/IP reference models. The first is an international standard, yet of little practical importance; the latter is less of a model but it is widely used in the Internet.

8.3.4 The OSI reference model

The Open Systems Interconnection (OSI) reference model is a standard approved by the International Standards Organization (ISO). It was developed in the early 1980s and comprises two major components: an abstract model of networking (the seven-layer model) and a set of protocols. The model divides a networking system into layers. Each layer has a certain functionality and communicates with the layer directly under it through an interface. Layers communicate with their corresponding peers in other systems. The ISO/OSI model has a stack of seven layers on top of the physical medium (Figure 8.8): physical layer, data link layer, network layer, transport layer, session layer, presentation layer, and application layer.

This model is important in that it is a means to explain and describe network architectures. Its protocols are of minor importance, having been overtaken by the TCP/IP protocols that play an important role in the Internet.

Physical layer

The physical layer is mainly concerned with the mechanical and electrical interfaces on top of the physical medium guaranteeing that bits are transmitted correctly over a communication channel. For instance, this layer defines the type of cable connectors, size of cable, or the voltage and duration to represent 0 and 1 bits.

Data link layer

The data link layer describes the logical organization of data bits that are transmitted through a channel such that an error-free transfer is achieved. This is done by the sender breaking the bit stream into data frames (up to 10kByte in length) and process acknowledgement frames that are sent back by the receiver. Correction of damaged, duplicate, or lost frames is another task of the data link layer. It must also exercise buffer management, which means alleviating transmission speed differences between sender and receiver.

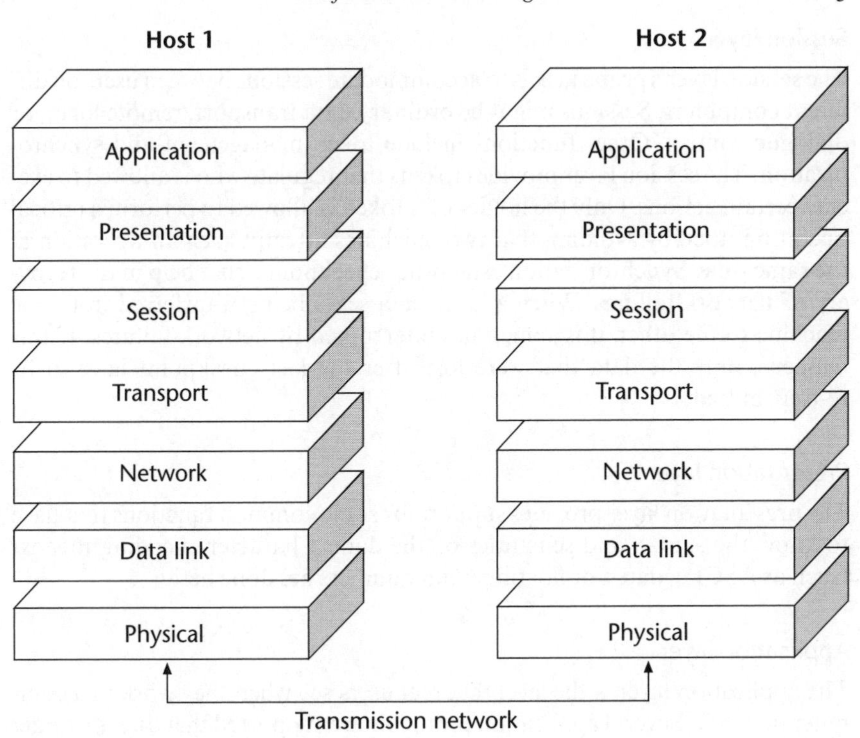

FIGURE 8.8 ISO/OSI reference model.

Network layer

Whereas the data link layer is only responsible for moving frames from one end of a cable to the other, the network link layer must take care of the proper operation of the communication subnet by routing packets from the source to the destination. Routing can become an especially important issue when accounting and billing are involved. Another important task is to resolve naming conflicts (addressing schemes may be different among heterogeneous networks) and differences in protocols.

Transport layer

The transport layer describes the nature and quality of the data delivery. Its major task is to accept data from the session layer (above it), split it into pieces, pass them through to the lower layers, and ensure the correct arrival at the remote end. This layer is the first with a true end-to-end communication, i.e. the transport layer on the source machine conceptually communicates with the transport layer on the destination machine. Another function of the transport layer is flow control, which regulates the flow of information, and establishes and deletes connections across the network.

Session layer

The session layer's prime task is to accommodate sessions between users on different computers. Sessions might be ordinary data transport, remote login, or dialogue control. Other functions include token management and synchronization. The session layer provides tokens that regulate who is allowed to execute certain actions. Only the holder of a token is allowed to perform a critical operation, thereby avoiding that two machines attempt such an operation at the same time. Synchronization establishes checkpoints that help in the recovery of network failures. When a large data set is being transferred from one machine to the other, it is sometimes interrupted by network failures. If this happens, only the data that were lost after the last checkpoint have to be re-transmitted.

Presentation layer

The presentation layer provides support for some common functions that have to know the syntax and semantics of the data. Character encoding for text (such as ASCII), dates, or floating point numbers are done here.

Application layer

The application layer is the level that real users see when they work in a computer network. Several application protocols are supported that directly target the end user. These protocols support file transfer between different operating systems and take care of problems that may arise owing to incompatible file naming conventions (imagine, for instance, the different file naming conventions in DOS and UNIX). Other functions supported are electronic mail, remote job entry, or directory lookup. It must be noted, however, that most of the protocols used on the Internet are not OSI protocols, but TCP/IP model protocols, as explained below.

8.3.5 The TCP/IP reference model

When the Internet evolved from the old ARPANET, and other networks were added, it soon became necessary to design a framework to link these heterogeneous networks. The TCP/IP reference model, named after its two major protocols, the Transmission Control Protocol (TCP) and the Internet Protocol (IP), was defined in 1974.

Figure 8.9 shows a comparison between the ISO/OSI and the TCP/IP reference model layers. The TCP/IP reference model has fewer layers than the ISO/OSI model; in particular, the network access layer (host-to-network) is vaguely defined. The model only requires a few protocols to take care of a correct connection to the network and to guarantee that the data packets from the Internet layer are transmitted correctly. One of these protocols that is widely used to connect (home) computers through telephone lines to host computers of Internet providers is the Point-to-Point Protocol (PPP).

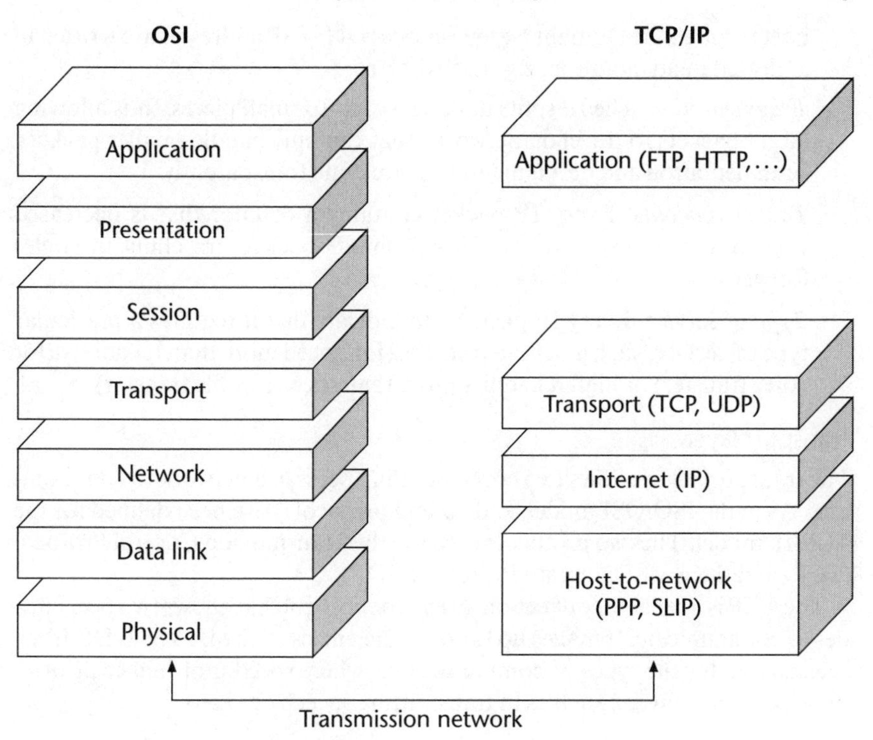

FIGURE 8.9 ISO/OSI and TCP/IP layers.

Internet layer

The Internet layer supports the Internet Protocol (IP) as the most basic means of transporting data packets. IP is a protocol that treats each packet independently. It does not even check whether the packets reach their destination and takes no corrective action if they do not arrive. IP is basically an unreliable protocol because it contains no error detection or correction. However, it does provide the following services:

- *Addressing*. The sending and receiving hosts are identified by 32-bit IP addresses. These addresses are used by intermediate routers to channel the packets through the network. Domain names are alphanumeric strings identifying Internet hosts. Examples of domain names are *www.itc.nl* or *hp18.itc.nl*. In this example, *itc* is a subdomain of the top-level domain *nl*. There are two types of top-level domains: generic and country. Specific generic domains are *com* (commercial), *edu* (educational institutions), *gov* (US government), *org* (non-profit organizations), *int* (international organizations), *mil* (US armed forces), and *net* (network providers). Country domains use the ISO 3166 country codes. Generic domains are generally used in the USA, country domains elsewhere. Domain names are converted into IP addresses, a 32-bit number used to identify an Internet host through the Domain Name Service (DNS). IP addresses are written in a 'dotted quad' notation, e.g. 127.18.53.10.

- *Fragmentation*. The IP splits data packets into small pieces, thus allowing larger packets to travel on networks that can only handle smaller packets. Fragmentation and defragmentation are done transparently.
- *Packet timeouts*. Every IP packet contains a counter that is decreased whenever it passes a router. This prohibits packets travelling in circles forever.
- *Type of service*. Every IP packet can indicate that it requires a particular type of service, such as a high transmission speed more than reliability (for voice transfer) or high reliability more than speed (for file transfer).

Transport layer

The transport layer allows two corresponding peers in a network to communicate. As in the ISO/OSI model, end-to-end protocols have been defined for the TCP/IP model. The two protocols used are the Transmission Control Protocol (TCP) and the User Datagram Protocol (UDP).

The TCP is a reliable connection-oriented protocol that allows two peer entities to communicate. It makes up for the deficiencies of the IP. The UDP, however, is used for the types of communication where speed is of higher priority than accurate delivery, such as in transmitting speech or video.

Application layer

There are no session and presentation layers in the TCP/IP model. In the application layer we find user-level protocols that allow remote login, file transfer, electronic mail, and more. The most important of these protocols are:

- *TELNET*. This is a simple terminal-oriented remote login service. It was and still is popular with text-oriented terminal emulations on remote machines.
- *FTP*. The File Transfer Protocol is one of the oldest Internet protocols and is still in widespread use. It is based on the TCP and is used for the transfer of text and binary data files. Normally, users must identify themselves with a username and password on the host machine when they want to transfer data. Anonymous FTP is a variant of FTP that allows users to access host machines without having a personal account on that particular computer.
- *SMTP* is the Simple Mail Transfer Protocol that allows the sending of electronic mail.
- *DNS* is the Domain Name Service that maps domain names to their corresponding IP addresses.
- *NFS*, the Network File System, is a service that allows TCP/IP clients to share files.
- *NNTP* is the Network News Transfer Protocol that is used to handle newsgroups. Electronic mail normally transfers messages from one sender to

one or more receivers. News messages could potentially reach every Internet host and user.

- *HTTP* (or Hypertext Transfer Protocol) is the youngest of the Internet protocols. It allows communication of multimedia and hypermedia documents that are made available mainly through the services of the WWW. These documents are mostly written in HTML (Hypertext Markup Language).

8.3.6 Comparison between the OSI and TCP/IP reference models

Both reference models have much in common, including the concept of a stack of independent protocols and similar layer functionality. TCP/IP, however, is older and was designed with an already operational network (ARPANET) in mind. Subsequently the ISO/OSI model was developed as a reference model with the aim of defining (not necessarily building) a network architecture. The ISO/OSI model is therefore a better vehicle to explain and describe network architectures, although it is of less practical importance. The TCP/IP model reflects a pragmatic approach and can teach us a lot about practical network design.

The OSI model makes a clear conceptual distinction between services, interfaces, and protocols. The protocols of the TCP/IP model are not well hidden and they are less clear in distinguishing different concepts.

8.3.7 The World Wide Web (WWW)

One of the recent Internet developments which has had an enormous impact on the market as well as on society is the World Wide Web, generally called the Web. It was started at the European centre for nuclear research, CERN, out of the need to exchange documents of various kinds among scientists spread over many countries. A first text-based prototype was demonstrated in late 1991, and the first graphical interface, called Mosaic, was released in 1993. Such interfaces became known as web browsers. Today, two browsers dominate the market: Netscape Communicator from Netscape Communications Corporation, and Microsoft's Internet Explorer. The World Wide Web Consortium was established in 1994 to further the development of the Web. Its home page can be found at *http://www.w3.org*.

The Web is a client–server system; on the client side it presents itself to the user as a vast collection of web pages. These pages contain hyperlinks to other pages. Besides text, different media may be used to present the contents of a page, such as sound, video, or images collectively referred to as hypermedia. To view different types of media, the browsers need external viewers, or helper applications. Frequently these helper applications are built into the browser as a plug-in.

Machines accessing the Web through a browser must be directly connected to the Internet, or at least able to establish a PPP connection with an Internet provider. This is necessary because browsers establish TCP connections with

Web servers to access pages. Sites that provide pages on the Web run a server process that listens to requests from clients (browsers). The hypertext transfer protocol handles these requests.

Web pages are addressed through a Uniform Resource Locator (URL). For instance, the home page of the International Institute for Aerospace Survey and Earth Sciences in Enschede, The Netherlands, has the URL *http://www.itc.nl/home.htm*. A URL consists of three parts: the name of the protocol (*http*), the host name where the page resides (*www.itc.nl*), and the name of the hypertext document (*home.htm*). This is what happens when a web page is displayed in a browser window:

1. A user clicks a hyperlink or enters a URL.
2. The browser determines the URL (when a hyperlink was clicked).
3. The browser asks the domain name service (DNS) for the IP address of the host computer.
4. DNS resolves the name and returns the IP address.
5. The browser establishes a TCP connection to the host.
6. The browser requests the document.
7. The host server sends the requested file.
8. The TCP connection is closed.
9. The browser displays all the text of the document.
10. The browser fetches and displays all images in the document.

Web pages are written in HTML or JAVA. HTML is useful for writing static web pages, while JAVA, developed by Sun Microsystems, is an object-oriented programming language that allows a provider to write applets. A hyperlink can point to an applet that is downloaded and run by the browser locally. Applets provide interactivity and flexibility over static HTML pages. However, by executing 'foreign' code on a local machine, they also pose a potential security threat. A new development is XML (Extensible Markup Language) that not only allows describing XML documents but also partially describes the behaviour of the computer programs processing them.

8.4 Database technology

GDI are based on large amounts of data distributed over many agencies. It is important to provide an organization that supports reliable and efficient sharing of the data among many users. A simple collection of data files or a file directory is not sufficient. The solution is a system comprising a database (DB) and a database management system (DBMS). The database is an organized collection of data, while the database management system is a software package for building and maintaining the database.

The advantages of using a database system include: persistency, efficient storage management, data recovery, concurrency control, ad hoc query support, and data security. Transactions are operations that move a database from one consistent state to another consistent state. If, for whatever reason, a transaction cannot be successfully completed, the database is returned to the consistent state before the transaction (recovery). To avoid inconsistencies caused by concurrent read and write operations to the database by many users accessing the database at the same time, a concurrency control mechanism is employed by the DBMS. This gives every user the impression that he/she is the only one accessing the database.

Not all queries can be known beforehand. A DBMS must provide means for formulating and satisfying ad hoc queries without requiring specialized application programs. Query languages provide this kind of function. Furthermore, a DBMS must allow for multiple users with different rights for accessing the database (data security).

8.4.1 Database architectures

A database architecture is divided into three levels: internal, conceptual, and external level (Figure 8.10). This architecture was proposed by the ANSI/SPARC Study Group on Data Base Management Systems in 1978, and has been widely adopted. The external level is closest to the users; the internal level

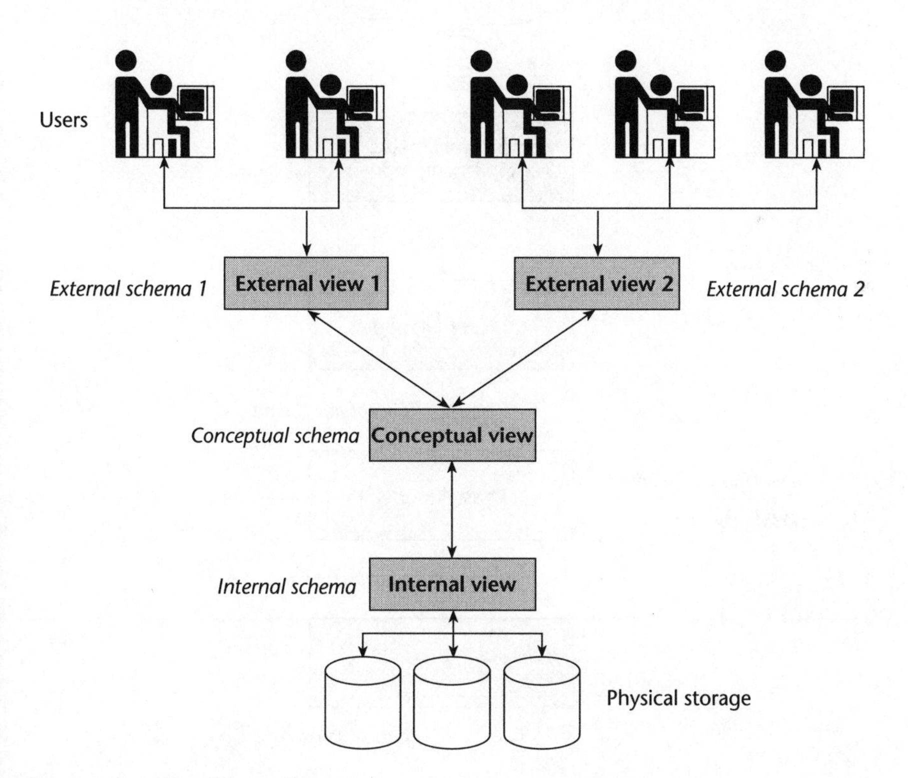

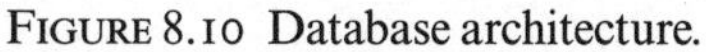
FIGURE 8.10 Database architecture.

is closest to the physical storage. Individual users may have different external views on the database. There is only one conceptual view and one internal view. The views are defined by schemas using a data definition language (DDL). The data manipulation language (DML) is used to describe database objects' processing.

8.4.2 Data models

Databases are organized around models of a perceived reality. In designing a database, several data models are used to describe subsets of the real world at different levels of abstraction. A data model describes the contents, structure, and meaning of data. They can be classified into conceptual-, logical-, and physical data models (Figure 8.11). A database designer starts with an abstract subset of the real world (external model or application model) which is subsequently mapped into a conceptual model. The best known conceptual data

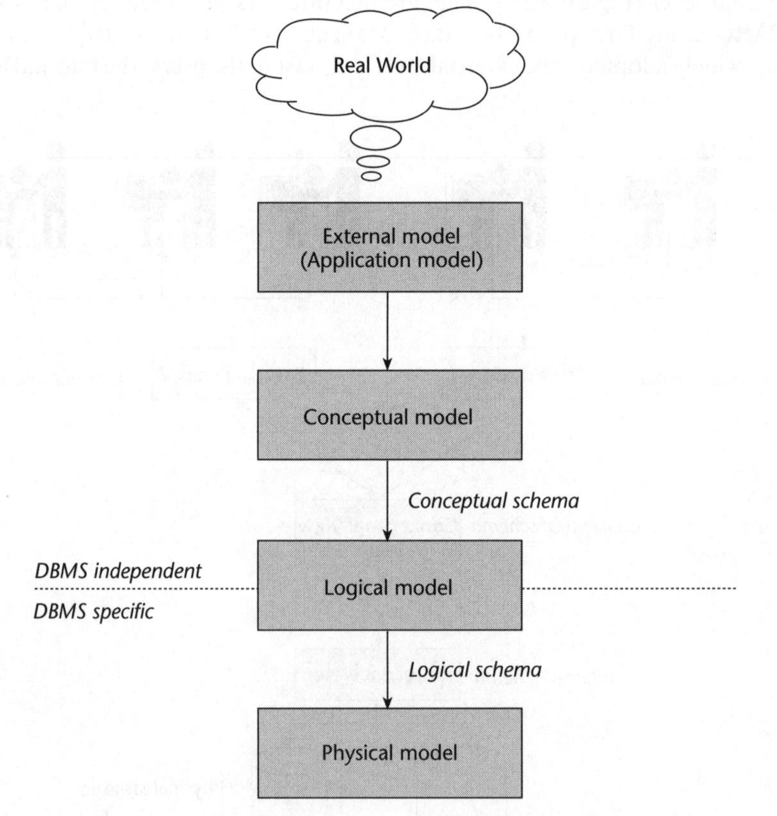

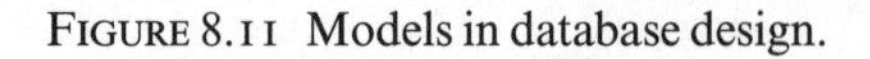

FIGURE 8.11 Models in database design.

model is the entity-relationship model. This, in turn, is mapped into a logical data model, which can be implemented with a commercial database management system. The most popular is the relational data model, introduced in 1970. It uses a simple data structure, the relation. Relations are defined in tables, a concept that is easily understood by human common sense.

The database language that supports the relational data model is SQL (or Structured Query Language). It became an ANSI standard as SQL1 in 1986 (ANSI X3.135) and an ISO standard in 1987. A revised and extended version of the standard was approved as SQL2 (also known as SQL92) in 1992, and SQL3 is nearing its completion. SQL is both a data definition and data manipulation language and is supported by all major relational DBMS vendors.

8.4.3 Distributed databases

In a GDI, many databases are distributed over numerous organizations and agencies. A distributed database consists of a number of sites interconnected through a communication network with each site running an autonomous DBMS. Local applications run only on a local database, while global applications span some or all the sites of the distributed database.

A distributed database looks like a local database to the user. Important concepts of distributed databases are distribution transparency, fragmentation, and replication. Distribution transparency means that a user does not see where the data in the database are physically located. Even if the same data are stored in duplicate at different locations for reasons of efficiency (replication) or different columns of a table are physically located at different local databases (fragmentation), the user always perceives a central database.

Whereas traditional database design involves conceptual-, logical-, and physical design, distributed databases also involve distribution design. There are two possible approaches to the design of distributed databases: top–down and bottom–up. The top–down design is applied to new databases designed from scratch (Figure 8.12).

Very often, however, individual databases already exist, and have been in use for a long time. When the need arises to integrate these databases for a higher level application, a special kind of distribution architecture comes into place, called federated databases. The federated database design approach is shown in Figure 8.13.

8.5 Clearinghouses

The US Federal Geographic Data Committee (FGDC) defines a clearinghouse as a 'system of software and institutions to facilitate the discovery, evaluation, and downloading of digital geospatial data'. Such a clearinghouse usually consists of a number of servers on the Internet that contain information about available digital geospatial data known as metadata. A clearinghouse is an example of a client–server architecture. The server machines hold the

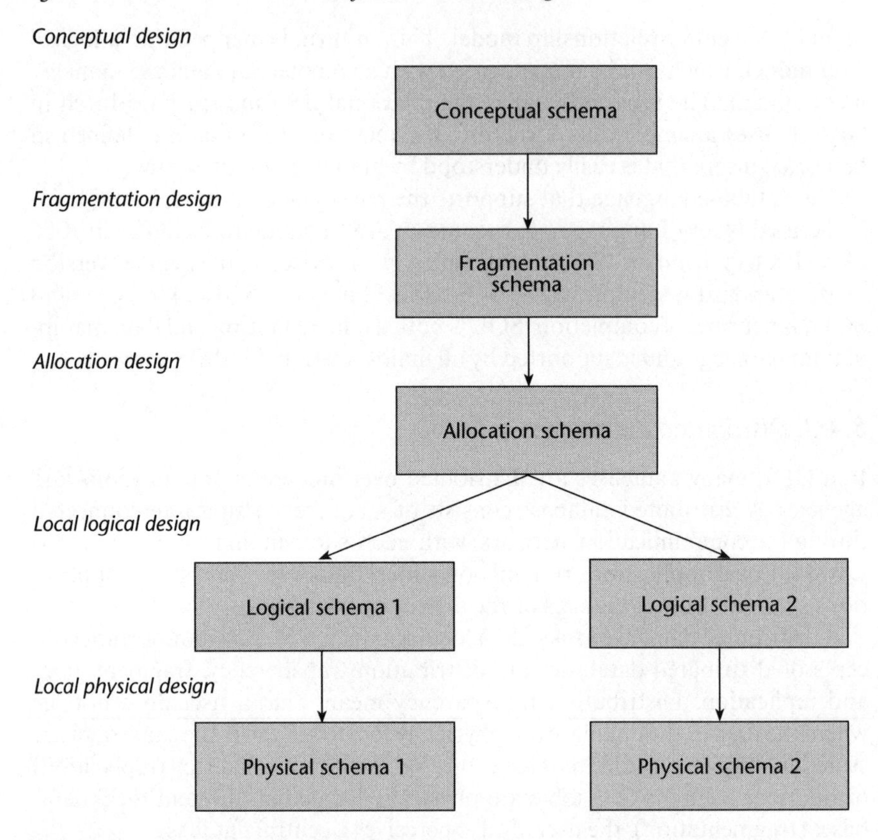

FIGURE 8.12 Top–down distributed database design.

metadata (i.e. data about the content, quality, condition, and other characteristics of data) and the clients request information about the availability of geospatial data by visiting the server nodes, usually through a Web browser. In the ideal case not only the metadata but also the server provides a link to the data itself. There is no hierarchical relationship among the servers. They act as peers within the clearinghouse activities. Servers are usually installed at providers, such as national mapping agencies, at different organizational levels. A typical clearinghouse architecture consists of many inter-operating metadata servers.

Providers of a clearinghouse must have access to a machine directly connected to the Internet. This server holds the metadata and must run server software. One protocol selected to provide search inter-operability among different servers is the ISO 10163-1995 (or ANSI Z39.50-1995) search and retrieve protocol. The Z39.50 was initially developed for the library community. It contains client and server software that establishes a connection, relays a query, returns the query result, and presents retrieved documents in various formats. Figure 8.14 shows the clearinghouse concept. For additional details of the clearinghouse architecture see Chapter 9.

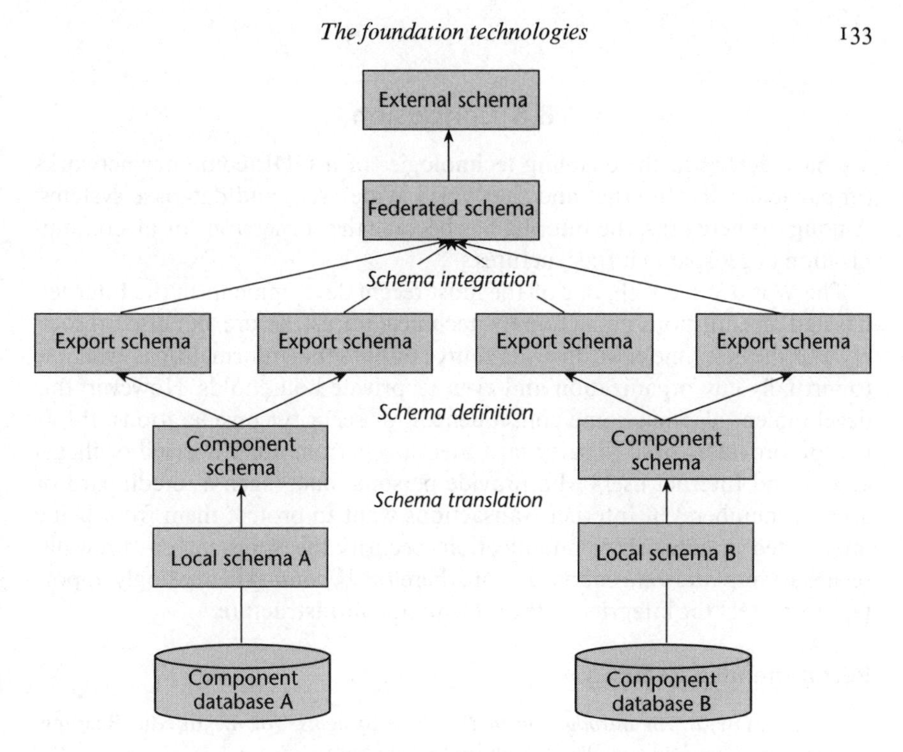

FIGURE 8.13 Federated database design.

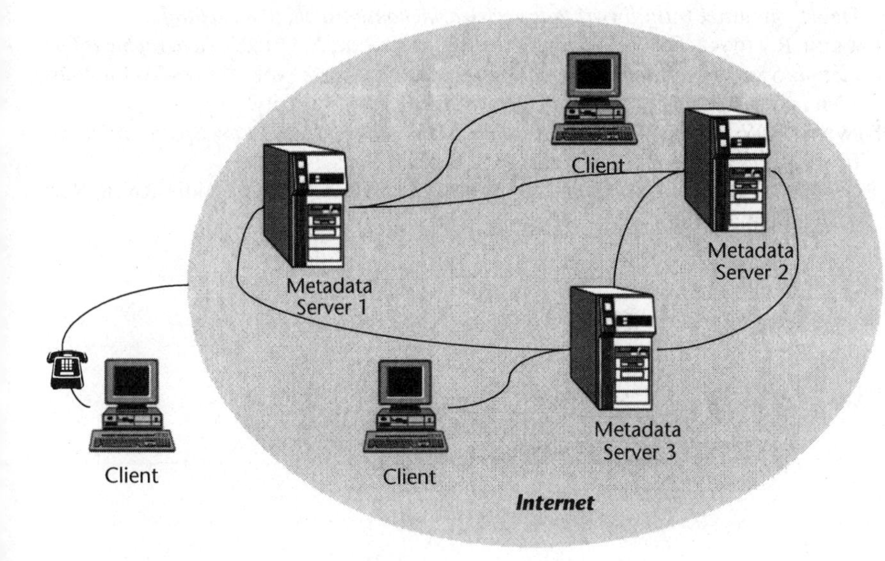

FIGURE 8.14 Clearinghouse concept.

8.6 Conclusion

We have described the enabling technologies of a GDI: computer networks (in particular the Internet and the World Wide Web) and database systems. Among the networks, the Internet has become the cornerstone of all communication in geospatial infrastructures.

The World Wide Web, one of the most recent developments in the Internet, has had an enormous impact on the technical infrastructure, but also on society. For the first time, a world-wide source of potential information is available to virtually any organization and even to private households. However, this development also has some consequences for security considerations. Information providers have to safeguard their assets from unauthorized or illegal access, and Internet users who provide personal data (such as credit card or account numbers) in Internet transactions want to protect them from being intercepted in network communication. Security measures such as firewalls, secure servers, and data encryption are therefore becoming increasingly important to protect the integrity of the information infrastructure.

Recommended Reading

DATE, C. J. (1990). *An Introduction to Database Systems*, vol. I. 5th edn. Reading, Massachusetts: Addison-Wesley Publishing Company.

ELMASRI, R. and NAVATHE, S. B. (1994). *Fundamentals of Database Systems*, 2nd edn. Redwood City, CA: The Benjamin/Cummings Publishing Company.

FGDC (1997). *Metadata to Clearinghouse Hands-on Tutorial*, Federal Geographic Data Committee (*http://www.fgdc.gov/clearinghouse/training/howto.html*).

LAURINI, R. (1995). 'From Electronics to GIS', in FRANK, A. U. (ed.), *Geographic Information Systems—Materials for a Post-Graduate Course*. vol. 2, GIS Technology. Vienna: Department of Geoinformation, Technical University.

NEWTON, P. W., ZWART, P. R., and CAVILL, M. E. (1995). *Networking Spatial Information Systems*, Rev. edn. Chichester: John Wiley & Sons.

TANENBAUM, A. S. (1996). *Computer Networks*, 3rd edn. Upper Saddle River, New Jesey: Prentice-Hall, Inc.

9

GDI architectures

Yaser Bishr and Mostafa Radwan

9.1 Introduction

Recent developments in information technology (IT) are difficult to keep up with. From a general perspective, however, current IT developments are mainly centred around enabling communications between existing (legacy) and newly developed information systems. This is having a profound impact on the GDI industry, including vendors, users, producers, and other service providers. The pressing challenge is not how to collect and store data, nor how to write the most efficient mathematical algorithms, rather it is how to make information known, easily accessible, and understandable to the largest possible groups of stakeholders. The availability of information on the Internet makes it accessible to people well beyond the expert community. This shift is forcing the industry to provide network-enabled software and user-oriented geospatial data services rather than just maps. The availability of information to the general public does not necessarily imply that it should be free of charge. This is a policy issue that differs from country to country, and depends on the local institutional arrangements and limitations, as discussed elsewhere in this book.

Taking Chapter 8 as point of departure, we examine in this chapter possible architectures for GDI in more detail. The architectures discussed here are not exclusive and do not require implementation in their entirety. We provide a schematic description of the potential components and technologies that can be used in designing and implementing GDI.

9.2 Framework of data sharing

Consider a situation where you are employed by a national organization for water management. You are required to produce a damage assessment report for a region that has suffered from a flood. For this you need to produce an inventory of land parcels with their ownership information that were affected by the flood, as well as the magnitude of the damage to the infrastructure. Politicians need this report within two days so that they can decide on a compensation scheme for landowners and on a restoration plan for the infrastructure.

To produce this report will require the latest satellite image of the flooded area. You will also need basic topographic information from the national mapping agency and land ownership information of the parcels affected from the regional cadastre office. You will need a status report of the damage to the infrastructure from the local offices and some statistical information about the social structure of the local municipality in order to support your findings.

In a classic situation, you would contact the relevant offices for this information. After processing your request, the national mapping organization would send you an analogue base map of the area that you would still have to put into digital format. Even if it is already in digital form, you would have to translate it into a format compatible with your own system. The regional cadastre office would send you a database file that you would have to integrate into your base map. The local offices send you long written reports about the damage that you still need to inspect and enter into your database, and then link into your base map and satellite image. It would be impossible to gather all this information and produce the report in two days.

Now consider this scenario. Imagine yourself using a computer connected to a nationwide network that links you to all the other parties. You search the network for the databases containing the required information, you connect to these databases and retrieve what you need in a very short time. You do not have to worry about putting all these data into a format compatible with your computer because the system will do that for you. Now producing such a report becomes possible.

A GDI with its institutional, economic, and technical components in place could provide such services. From a technical perspective, this will require several things:

1. It is clear that the databases of the different parties involved in a GDI serve different purposes and have probably been developed independently. Resolving the resulting heterogeneity problems poses a significant challenge.

2. The software applications installed on the different machines should be able to communicate, i.e. there should be inter-operability between distributed systems.

3. In order to facilitate data sharing between the different communities in these heterogeneous environments, users need to know the characteristics of the data set, especially the syntactic, schematic, and semantic aspects. A network-wide mechanism to share the available data will be required. Data providers should supply precise descriptions of data sets to the user to allow their evaluation, ordering, and transparent access and use. This is, in fact, the role of the clearinghouse.

4. An architecture will eventually be required that brings all the above together in a coherent system. This is the GDI architecture.

9.2.1 Heterogeneity issues

A geospatial database is a computer representation of real world features or phenomena using various abstraction mechanisms. The heterogeneity problem occurs when different communities wanting to share their data with each other have to contend with different views on the real world features, different modelling schemes, and different tools to represent, store, process, and manage geospatial data sets. Bishr (1997) described these heterogeneity issues as syntactic, schematic, and semantic heterogeneity.

Syntactic heterogeneity refers to the differences in software and hardware platforms, database management systems, and the representation of geospatial objects (raster or vector, co-ordinate system, geometric resolution, quality of geometric representation, methods of data acquisition, etc.).

Schematic heterogeneity refers to the differences in database models or schemas, e.g. a particular feature may be classified under different object classes in different databases, or an object in one database may be considered an attribute in another. The classes, attributes, and their relationships can vary within or across disciplines.

Semantic heterogeneity is the way the same real world entity may have several meanings in different databases. This will also influence the geometrical representation of objects, because abstraction of the world is based on the semantics of each discipline. It is intimately tied to the application context or discipline for which the data is collected and used.

9.3 Inter-operability and distributed systems

9.3.1 Distributed systems

The term distributed system refers to a distributed collection of users, data, software, and hardware, whose purpose is to meet some pre-defined objectives (Brenner, 1993; Brunt and Hutt, 1992; Deignan *et al.*, 1993). A comprehensive description of a distributed system design requires three levels of specification: the physical networking, system services, and application software. The overall designed system is known as a distributed computing environment (DCE).

In general, computer systems provide four types of interrelated services, as shown in Figure 9.2. The data storage services provide users with efficient storage media. The data access services provide functions for retrieving data from the storage media. The application services provide users with capabilities to execute specific tasks. Finally, the presentation services provide display facilities and user interfaces to end-users.

There are several design possibilities for distributing any of these services. For example, data can be located in distributed storage media. An application

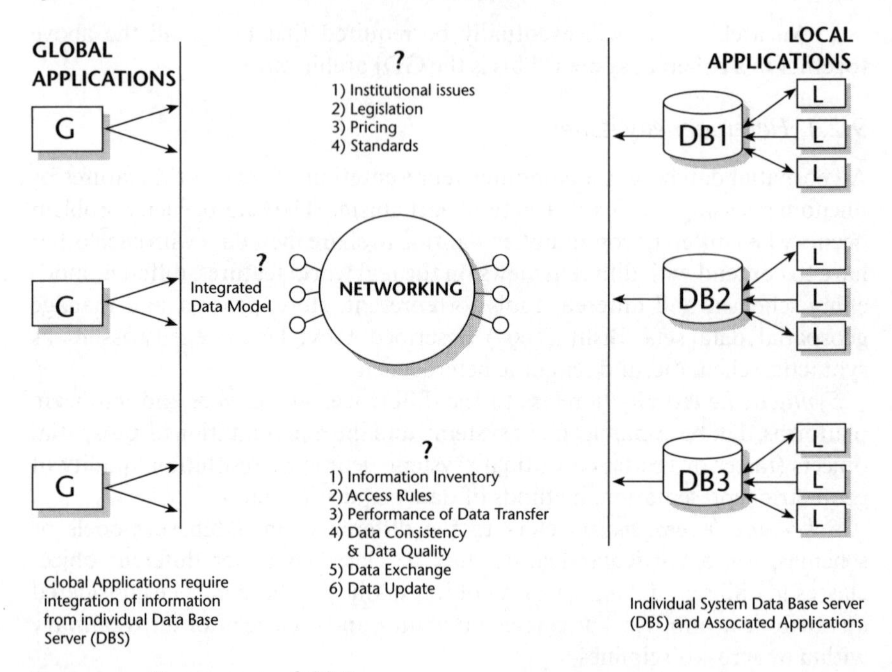

FIGURE 9.1 Components of GDI.

might send requests to several data access services located at different systems. A user might be provided with a single presentation service, which transparently accesses different distributed application services. This particular type of distributed services is known as a distributed database system.

In the context of GDI, the main objective of DCE can be to provide users with the means to share geospatial information, as well as to provide application and representation services in a heterogeneous, distributed environment. The fundamental model with which the DCE is implemented is known as the client–server model.

9.3.2 The client–server model

The client–server model is a computing model, in which system functionality is divided among the components that make requests (the clients) and the components that respond to them (the servers). For example, with regard to Figure 9.2, a presentation service sends requests to the application service to perform some processing. In this case, the presentation service is the client, while the application service is the server. The client and the server components are typically situated in different computers. The main concepts of the model are:

Server: an application component, which performs services in response to requests sent by clients.

Client: an application component, which sends requests to servers and receives the results of the services returned from the server.

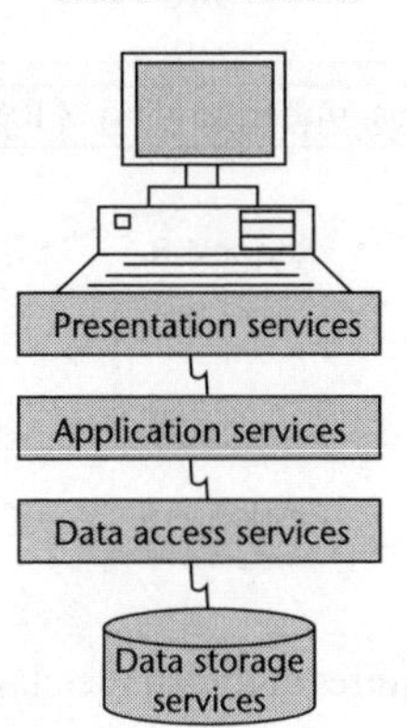

FIGURE 9.2 Distribution possibilities.

Service: data provision, analytical functions, etc., provided by the server to the clients.

Client–server: interaction consisting of one or more service requests and responses.

A major difficulty in the client–server model, and consequently in the DCE, is that many different standards apply. The implementation of the different standards results in heterogeneity among the clients and the servers. The heterogeneity might vary from the cabling system, which links the clients and the servers, on the one hand, to the software applications, on the other.

9.3.3 Inter-operability defined

Inter-operability is the ability of a system or components of a system to provide information sharing and inter-application co-operative process control. As shown in Figure 9.3, two systems *X* and *Y* can inter-operate if *X* can send a request for service *R* to *Y* on the mutual understanding of *R* by *X* and *Y*, and *Y* can return response *S* to *X* based on the mutual understanding of *S*.

9.4 Components of GDI architecture

The design of a GDI should not only focus on the available technologies, but also reflect the organizational and decision-making structure of the jurisdiction. Figure 9.4 shows an example of an architecture for a GDI that provides links between several independent databases and decision-support systems at local, regional, and national levels. These databases can belong to different levels of government or to private organizations. The architecture is based on the client–server model and we can distinguish two types of databases:

- the local databases, which include the basic data required by the different levels;
- the decision-support databases at the client level.

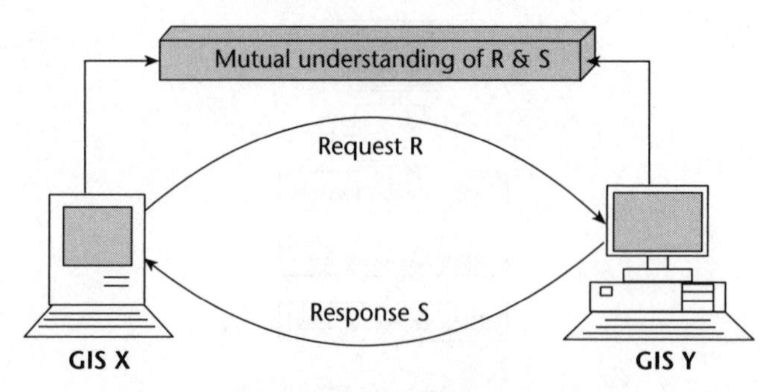

FIGURE 9.3 Inter-operability requires mutual understanding of requests and responses.

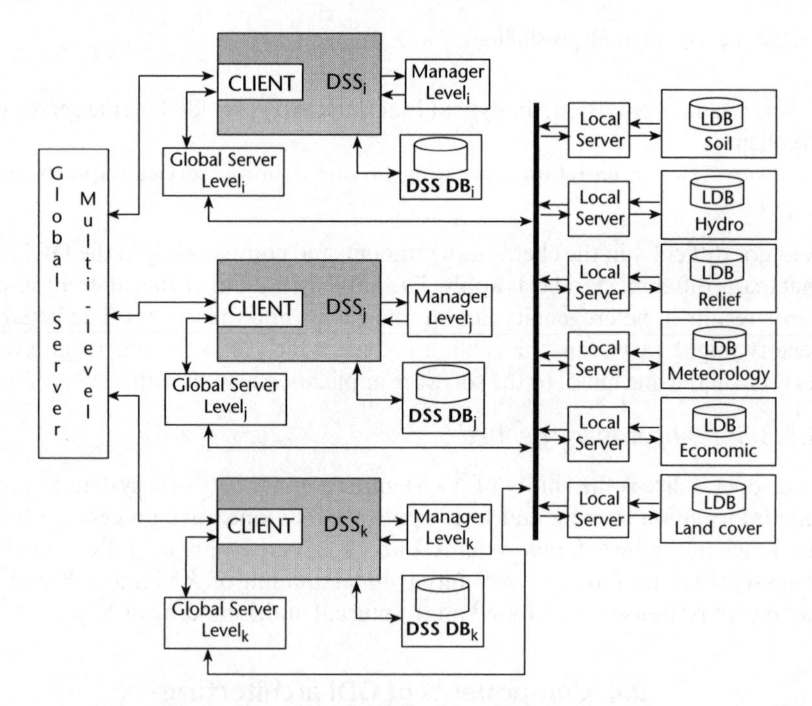

FIGURE 9.4 Example architecture of GDI with a decision-support system at three levels.

Each level has its own database(s) to meet its own objectives and consequently has its own view of the geospatial objects in the basic databases.

It is important to note that this diagram is only conceptual, in the sense that an organization can act at the same time at different levels or it may have different units for different levels. GDI must support actions at different levels and the components of the infrastructure at the different levels must make this possible. The Geospatial Data Service Centre (Figure 1.1) can play a role in this.

9.4.1 Local server

The local server represents the gateway between the local databases and the global server. It contains a description of the information that each participating database is willing to share. The shareable information is described by metadata, which contains information about all the data stored in the database (Bishr, 1997).

In a sense, the database schemas in the local servers are abstractions of the local database. They are responsible for accessing and retrieving information as requested by the users of the individual databases at each level through the global server. The server usually hosts elementary databases from a single application domain, e.g. soil, hydrology, land cover/use, relief, etc. are presented on a detailed geospatial scale. These are also called foundation and framework data. See, for example, MSC (1995).

9.4.2 Global server

The global server has a federated schema, which provides a unified model for external users. To support each level, the corresponding global server links its client with the local servers. On receiving and accepting the client's request, the following operations may be performed:

- analysis of the request to identify and locate the required information and to send the corresponding messages to the appropriate local servers;
- reception of the data sets from the local servers and mapping them to an adequate form understood by the client;
- control of the transactions between the client and data sources;
- maintenance of a global directory, i.e. information about the data available within the federation: location, information on specific data sets, ownership, format, cost, etc. (achieved by storing a comprehensive metadata set in the global server);
- execution of data conversion: units, formats, etc.

9.4.3 Multi-level server

In addition to its functionality being similar to that of the global server, the multi-level server is responsible for linking the different levels, i.e. national, regional, and local to establish the corresponding feedback among them in terms of information, knowledge, and the decisions necessary for their activities. Its tasks include the control of the communication between clients (management levels), and access and retrieval from the corresponding decision-support systems' database at a specific level. Such an architecture has both benefits and drawbacks. The provision of a global data model at the server is not very flexible. This is because the new paradigm of the client–server allows a client in one case to be the server in another. Moreover, global models cannot accommodate system evolution and autonomy requirements very easily. Global schemas cannot always support external users as they must still have an understanding of the elements of the global schema.

9.5 Clearinghouse component

As described earlier, a clearinghouse can be defined as 'a system of software and institutions to facilitate the discovery, evaluation, and downloading of digital geospatial data' (FGDC). However, for our purposes we will extend this definition somewhat. As depicted in Figure 9.5, it is composed of local servers, a clearinghouse server, user interface, global metadata, and local metadata. The local metadata has detailed information about the underlying database, while the global metadata has more generalized information about all the databases connected to the GDI. Each component has its own roles, and functions to fulfil the user's requirement and to accomplish its objective. The clearinghouse shown in Figure 9.4 is a hybrid system which stores high level metadata, i.e. generalized metadata at a central location. It maintains the user interface and performs the search mechanism, while the local metadata, i.e. more detailed metadata, and geospatial data sets are stored in distributed local servers maintaining transactions, data conversions, and the delivery mechanism. See also Chapter 13, pp. 227–9.

In general, the global server of the GDI can also maintain the global metadata service. Similarly, the local servers of the GDI store the local metadata information. This means that the metadata server is also based on the client–server technology that enables clients to request information from the server, while the server performs designated functions to fulfil the user's request. Connection between provider and users can be established through the Internet using TCP/IP and HTTP (see Chapter 8, pp. 126–8).

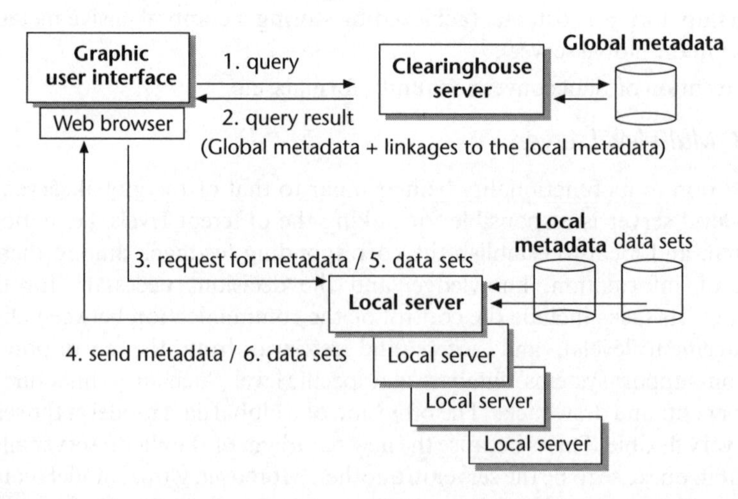

FIGURE 9.5 Conceptual system architecture for the proposed national clearinghouse.

9.5.1 Local server

The local server is composed of several modules to facilitate access, ordering, delivering, and sharing of a data set. The security controller and transaction modules are critical elements for the online transaction in any network-based architecture. It requires specific protocols to prevent unauthorized access and interception of valuable information. Figure 9.6 shows the components of the local server, which are composed of a WWW server, security controller/GUI generator, service controller, service module, local metadata, and geospatial/non-geospatial database. This local server provides three main services to the users: access to the local metadata, access to the geospatial/non-geospatial database, and online transmission of data sets using FTP.

The user can access the local metadata through the clearinghouse without specific interaction with the local server because the global metadata contains links (URL of the local metadata) that directly connects the user to the local metadata.

The other services require the security controller/graphic user interface generator and service controller. When a user accesses the local server, information written in HTML is displayed on the user's web browser to show the available services, to allow the user to select a service, and to enter a 'user name' and 'password' as registered with the local server. The service request invokes the security controller function to check the user's authority to the services, since

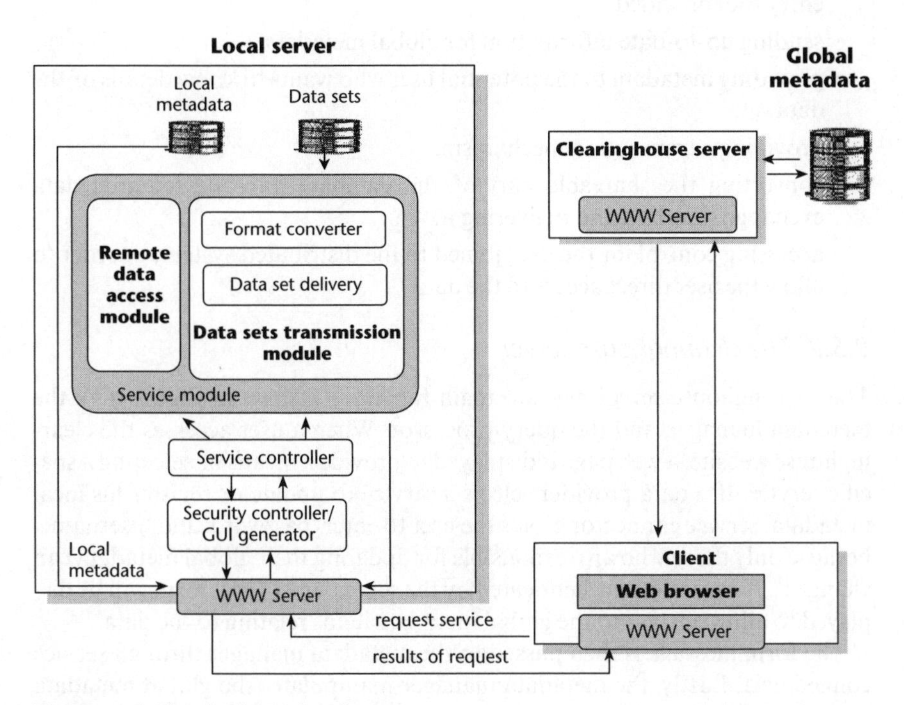

FIGURE 9.6 Clearinghouse local server.

there may be different levels of accessibility to the data. Different GUIs are displayed to the user based on this authority and type of service. The service controller interprets the user's request, invokes the relevant service functions based on the request, and sends the result to the user.

Remote data access service allows the user to access the graphic and non-graphic data set directly by using a specific GUI enabling the remote user to see the graphics as well as to manipulate the graphic and non-graphic data. Current web technologies such as JAVA, ActiveX control, CORBA (Common Object Request Broker Architecture), etc. all provide possible solutions for the remote data access service.

There is a mechanism in the data set's transmission service module to convert the existing geospatial data set format into a standard or specific format based on the user's request. The service controller interprets the request and invokes the data set delivery and format converter functions (if required).

The local server is a node providing metadata and their legacy data sets to the public. The functionality of the local server in terms of data provider includes:

- converting existing metadata information into metadata standard compliant format;
- abstracting metadata into the global metadata;
- registering metadata into the clearinghouse server using the metadata entry tool provided;
- sending up-to-date information for global metadata;
- providing metadata to the potential user who wants to know details of the data set;
- providing a transaction mechanism;
- converting the shareable part of the database into the required data exchange standard and delivering it;
- accessing control for the user joined to the distributed system, in order to allow the user direct access to the data.

9.5.2 The clearinghouse server

The clearinghouse server has two main functions (shown in Figure 9.7): the metadata manager, and the query processor. When a user accesses the clearinghouse website, a web page is displayed to provide a means of selecting a specific service. If a data provider selects a service to update or register his local metadata, service connector 2 asks the user to enter 'password' and 'username' because only those who are responsible for updating their global metadata can change it. If the user is authenticated by the server, an HTML form will be displayed to allow entry into the global metadata items relating to the data.

The form message is then passed to the metadata manager through service connector 2. Lastly, the metadata manager manipulates the global metadata using the bridge or driver (e.g. Open DataBase Connectivity).

If a user selects the query service, the service connector 1 will send a graphic

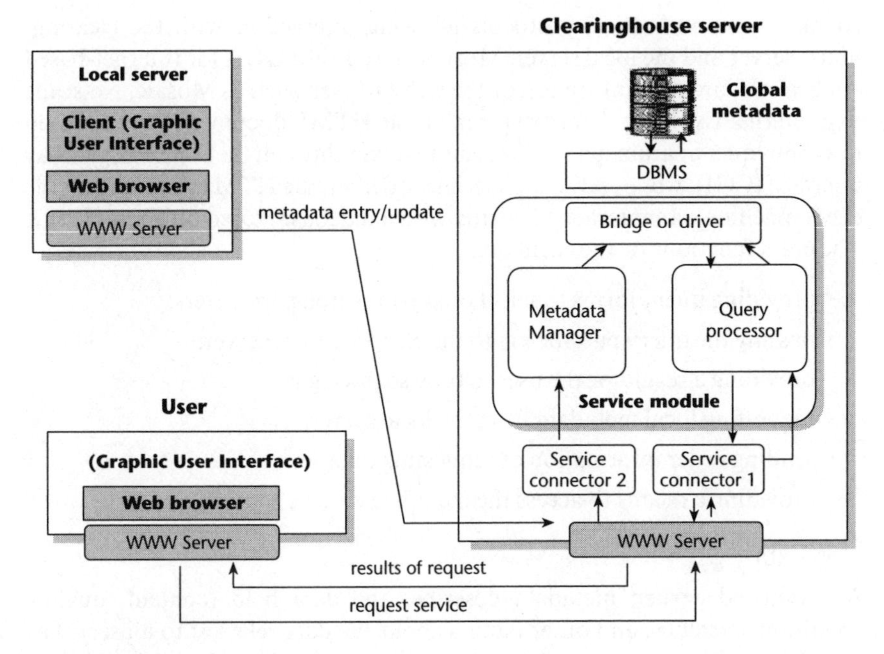

FIGURE 9.7 Clearinghouse server.

user interface designed to enable input of the search parameters. Query parameters are sent to the query processor with the help of the service connector 1 that has the capability to interpret the user's request and to invoke search functions. The search process requires bridges or drivers to retrieve the global metadata managed by specific database management systems; search results are then passed to the user through the service connector 1 which has the mechanism to transform the search result into a web browser-recognizable format (e.g. HTML, TXT, and so on).

The clearinghouse server is a mediator which links the data provider and data user. It provides two main services to the user: searching the global metadata, and providing links to the local metadata to the user. The main functions of the clearinghouse include:

- providing a metadata entry tool to the data provider in order to allow registering and updating of the metadata after checking the user name and password;
- providing a user interface for queries to the user;
- receiving query parameters from the user;
- processing the query against the global metadata;
- sending the query results to the user with links to the local metadata.

Graphic user interface (GUI)

Normally, in a client–server system, the client can request a service from the server through a specific program called a graphic user interface (GUI). A GUI contains communication protocols allowing interaction with the clearinghouse server and the local server. Most of the current GUIs for Internet-based applications are running on top of the web browser, such as Mosaic, Netscape and Internet Explorer. A form written on the HTML document has been used to submit query parameters to the remote server through the common gateway interface (CGI), while JAVA applets embedded on the HTML, running on the client machine, can interact efficiently with a database stored on a remote site. The main functions of the client are:

- providing query forms to select or key-in search parameters;
- passing the query parameters to the clearinghouse server;
- providing a result for the user to browse through;
- requesting local metadata from the local server;
- sending order information or requesting data sets;
- providing a means to access the metadata entry/update tools.

Metadata

As discussed earlier, metadata describes the data held (content, quality, condition, currency, and other factors about the data relevant to a user). The level of details depends on the purpose of the metadata. It can be used by the data provider internally to monitor the status of data sets as well as externally to advertise their data to potential users through the national clearinghouse. For internal use, we refer to 'local' metadata which contains the detailed information about data sets stored on local hardware. For external use, we refer to 'global metadata' which contains a short description of the data sets advertised in the clearinghouse to allow users to find relevant data easily and quickly.

In order to make metadata easily readable and understandable by different disciplines, there should be a standard that provides a common terminology and definitions for the documentation of geospatial data. Key developments in metadata standards have been the ISO Standard 15046-15 Metadata, the FGDC Content Standard for Digital Geospatial Metadata (CSDGM), and the CEN European standard for metadata.

Content of metadata

FGDC specifies the structure and the expected content of more than 220 items that are intended to describe digital geospatial data sets adequately for all purposes. These items are grouped into seven information categories: identification, data quality, geospatial data organization, geospatial reference, entity and attribute, distribution, and metadata reference (FGDC-CSDGM, 1997).

For the European Metadata Standard, the Technical Committee CEN/TC 287 submitted a draft standard for metadata to the Comité Européen de Normalisation (CEN), which is responsible for European standardization in all fields except electro-technology (CENELEC) and telecommunications (ETSI). It includes information about the content, representation, extent (both geometric and temporal), geospatial references, quality, and administration of a geospatial data set. This information is grouped into the following nine categories: data set identification, data set overview, data set quality indicators, geospatial reference system, extent, data definition, classification, administrative metadata, and metadata reference (CEN, 1996).

Both standards well define the items that can be used by a user to judge fitness for use, to order, and to use sharable data sets. The details of the metadata, however, can vary with its purpose. Figures 9.8 and 9.9 show the different levels of details required for different purposes. The sizes of the rectangles in Figure 9.8 represent the amount of information, while the widths represent the number of data sets necessary for each purpose. At the higher level, only a small description is required for the purpose of advertising the data sets, while at the lower level, a more detailed set of metadata is required to judge the fitness for use, to obtain, and to actually use the data sets. The 'advertising' means users can quickly survey the availability of data sets.

Most users require a metadata service that is broad and relatively limited in terms of its data content. There is little demand, in the first instance, for metadata that contains large amounts of technical information or specification details. Once a user has determined that a data set is possibly appropriate, more detailed information about the data sets is needed to make a final judgement.

Figure 9.9 shows the Internet-based meta-metadata system in which global metadata are stored at level 1 in a single site (the clearinghouse global metadata), while the local metadata are stored at each organization or data provider who has the capability to perform online marketing and delivery with high security.

Two different metadata contents are required, one for local metadata and the other for global metadata. In order to develop the global metadata, we need to:

- identify the core data element from the local metadata if a standard exists, otherwise with the agreement of participants in the GDI;
- develop an abstraction mechanism which converts local metadata into global metadata.

Two approaches can be considered in the development strategy. One is to set up the local metadata standard first, based on modifying arbitrary metadata of existing data sets into a standard format, and then establishing the global metadata. The other approach is first to define the global metadata items, establish a global metadata server, and then define the local metadata standard based on the global metadata. The latter approach is preferred in the urgent implementation of a national clearinghouse, because the global system can be established quickly, enabling the GDI to start work even with a limited functionality.

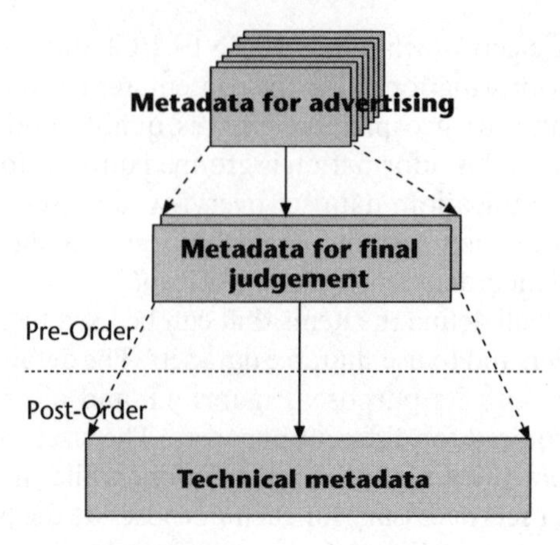

FIGURE 9.8 Levels of detail for metadata.

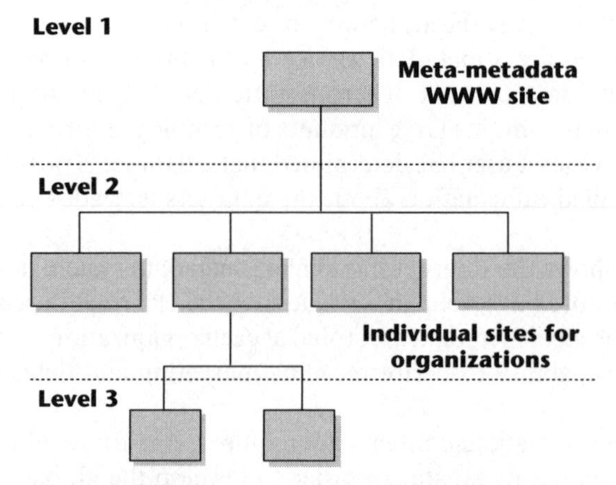

FIGURE 9.9 Meta-metadata system architecture.

9.6 Geospatial Data Service Centre

So far this chapter has presented GDI architecture in functional and relational terms. When examined from an organizational perspective, the functions of global servers, local servers, and clearinghouse are often combined in a data service centre, or Geospatial Data Service Centre (GDSC). See Figure 1.1.

For an application domain (e.g. sustainable land management) integration of a significant number of data sets is required for policy analysis, decision-making processes and the operational management and monitoring processes. Many specialists using GDI require routine and reliable access to data, which are standardized for this domain. This makes data integration and correlation possible but also optimizes data sharing. A large portion of the data will be drawn from existing sources. For each of the GDI users in the domain, to address individual data sources would be unmanageable in terms of the multiple use of the data acquired for the domain as a whole. Also, any form of quality control in the standardization process would be difficult if not impossible. In such an application domain there also needs to be some consistency in the development of contracts with data suppliers, or adherence to possible NGDI standards, to ensure that the data within the domain is accessible to others through the NGDI. In the latter case the domain GSDC needs to develop its own access, use, and pricing policies in sympathy with higher level legislation. Hence the role of the GDSC is to provide the technical administrative architecture of the application domain GDI. On behalf of the application domain, it also co-ordinates the access to other data sources including compliance with the information policy of the owners, and payment, as well as assuring that the data will fit within the domain standards. In other words, the domain GDSC is responsible for the integrity of the service which matches data demand and supply in two directions. It manages the relationship between outside data suppliers and domain applications according to information policy of the application domain. It also manages the external demand for data within the domain according to the same information policy (see, for example, Groot, 1997).

9.7 Conclusion

In this chapter we have reviewed the essential technical components of GDI. The concepts of client–server architecture and distributed databases play a major role in GDI design, and current technology provides great advantages for implementing such a system, for example, the Common Object Request Broker (CORBA), JAVA, and Extensible Markup Language (EML) all provide distributed computing platforms for implementing GDI. Furthermore, the OpenGIS™ aims to provide the essential technology that allows us to implement the geospatial dimension of the GDI architecture.

Bibliography

BISHR, Y. (1997). 'Semantic Aspects of Interoperable GIS', *ITC Publications*, no. 50.

BRENNER, J. (ed.) (1993). *Distributed Application Services*, Open Framework, Prentice Hall International Ltd, Campus 400, Maylands Avenue, Hemel Hempstead, Hertfordshire HP2 7EZ, UK.

BRUNT, R. and HUTT, A. (eds.) (1992). *The Systems Architecture: an Introduction*, Open Framework, Prentice Hall International Ltd, Campus 400, Maylands Avenue, Hemel Hempstead, Hertfordshire HP2 7EZ, UK.

Comité Européen de Normalisation (CEN) (1996). *Doc: prEN 287009*.

DEIGNAN, F. and HOLLINGSWORTH, D. (eds.) (1993). *Networking Services*, Open Framework, Prentice Hall International Ltd, Campus 400, Maylands Avenue, Hemel Hempstead, Hertfordshire HP2 7EZ, UK.

FGDC-CSDGM (1997). *Content Standard For Digital Geospatial Metadata*. http://www.fgde.gov/

Groot, R. (1997). 'Spatial Data infrastructure (SDI) for Sustainable Land Management', *ITC Journal*, 3/4: pp. 287–93.

Mapping Sciences Committee (MSC) (1995). *A Data Foundation for the National Spatial Data Infrastructure*. Washington, DC: National Academy Press.

10

Conceptual tools for specifying geospatial descriptions

Martien Molenaar

10.1 Introduction

The pre-eminent objectives of GDI are to achieve efficiencies in geospatial data handling by means of data sharing and to open economic opportunities in a broader sense through the emerging 'information market'. A major tool in achieving this aim is to facilitate access to government-owned information at prices that are set in a consistent and transparent manner.

As we have seen in previous chapters, geospatial data can be classified into broad categories: foundation data, framework data, and application-specific data. The first type is collected with broad data sharing in mind and serves as a topographic/geometric framework within many domains of different specialists, while the second type is usually the broadly sharable data within a specialist domain. The third type is collected paying little attention to data sharing owing to its highly specialized context. In all cases, however, the question arises as to what extent data produced in one context can be shared for application in another. The question is opportune because a decision to use data already available from another source may be far less expensive than surveying or developing the data separately. Hence there is a financial element to consider, but also the element of consistency, especially if the data is to be used in multiple levels of abstraction in classification and aggregation hierarchies.

This chapter describes the conceptual tools for defining geospatial descriptions and specifying the semantics of the content of geospatial databases. These are important tools for dealing with the above questions, and for determining the fitness-for-use of data for a particular application within an economic and timeliness context.

10.2 Geographic phenomena and geospatial data modelling

The Earth's surface can be considered as a spatio-temporal continuum in which many different processes take place. These processes include the development of vegetation covers, geological processes, demographic processes, the development of land use, etc. Each process results from interacting forces, which are caused and affected by internal and external factors, and these forces lead to geospatial patterns of field characteristics that will change with time. There are two major classes of processes:

1. the first class refers to the processes with a field characteristic, i.e. the strength of the interacting forces is a function of the position within the field, and the resulting pattern can also be expressed in terms of position-dependent field values;
2. the second class of processes is based on the behaviour of (spatially interacting) objects; the pattern resulting from such a process can be expressed by the spatial distribution and the state of these objects.

The geospatial data models discussed here were developed to represent field situations in a geospatial database environment. Each field situation is then considered as the state of such a process at a specified moment, i.e. it can be considered as a time slice of the spatio-temporal continuum in which such a process has been defined. Time, therefore, will not be dealt with explicitly in this section and the emphasis will be on the relationships between the geometric and non-geometric (thematic) aspects of the field descriptions.

This section aims to explain how such data models can be formulated and to raise the reader's awareness of some important decisions that have to be made in this modelling step. The semantic aspects of field descriptions will therefore be explained emphasizing the field approaches and object approaches, i.e. describing the state of field- and object-structured processes, respectively (Goodchild, 1992). We will study the relationships between the thematic and geometric information for both approaches.

10.2.1 Fields, objects, and geometry

For most applications the thematic aspects of a field description are of prime importance; this means that the data querying and processing will be organized and formulated primarily from a thematic perspective. The structuring and formulation of the analysis of the geometric aspects of the data will be secondary, i.e. this formulation will depend on the thematic problem formulation.

We will consider two principal structures for linking thematic and geometric data (Burrough and McDonnell, 1998; Chrisman, 1997; Laurini and Thompson, 1993; Peuquet, 1990):

1. The first structure is the field approach, which considers the Earth's surface as a spatio-temporal continuum. Thematic aspects of the terrain can be

given in the form of attributes, the values of which are considered to depend on their position. The representation of such a field in a geospatial database requires the continuum to be descretized in the form of points or finite cells, often in a regular grid or raster format. The attribute values are then evaluated for each point or cell. This structure is shown in Figure 10.1*A*.

2. The second structure assumes that terrain features or objects can be defined by their geometric position and shape and they may have several non-geometric characteristics. These objects are represented in an information system by means of an identifier, to which the thematic data and the geometric data are linked, as in Figure 10.1*B*. This is called the 'terrain feature-oriented approach' or '(terrain) object-oriented approach'. Here we prefer to use '*object-structured approach*', to avoid confusion with modern computer science techniques.

Both types of representation will be discussed in the following sections. Section 10.3 explains the field approach and how field data can be handled in raster-structured data models. Thereafter, object representation in geospatial information systems is discussed, largely based on the presentation of the subject matter in Molenaar (1995, 1998). Section 10.4 discusses how objects can be represented in a raster and a vector geometry, and how topological relationships between objects can be dealt with. Section 10.5 discusses several semantic aspects of object modelling, the role of thematic object classes and classification hierarchies for organizing thematic object descriptions, how aggregation hierarchies can be used to link objects at different levels of complexity, and lastly, it explains the similarities and differences between object aggregations and object associations. Section 10.6 briefly discusses the different types of object dynamics, and Section 10.7 finally explains how object

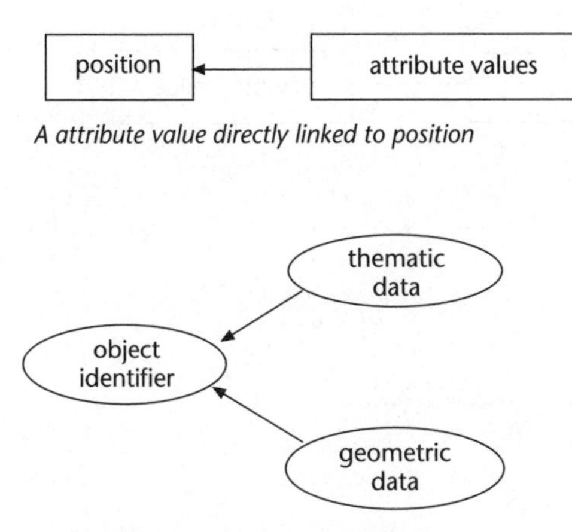

FIGURE 10.1 Two basic structures for spatial data.

definitions and object descriptions are related to the context in which a geospatial database will be used.

10.3 Fields and rasters

The name 'field approach' refers to the fact that the values of the terrain attributes are treated as field functions that take a value at any position in a two-dimensional space. A terrain description of this type can be based either on randomly distributed positions, or on positions that are regularly organized in a grid, see Figure 10.2 (Molenaar, 1998).

An irregular point pattern often occurs when attribute data are collected in field surveys; these may be soil surveys, vegetation surveys, the collection of terrain height data, etc. A surveyor will choose sample points depending on the type of attributes to be evaluated, and also on the expected spatial interrelationship between these values and the structure of the terrain. The resulting point pattern will be irregular, but it is not necessarily random (see Figure 10.2). The actual choice of the point positions contains information about the thematic field to be mapped. The required accuracy of the point position in the terrain depends on the aim and contents of the mapping. For a soil or vegetation survey, a position accuracy of 1–2 m or worse is often sufficient, but the sampling of soil pollution near an industrial plant may require an accuracy in the order of 1–2 dm. In the latter case, accurate interpolation techniques or geostatistical methods will be applied to identify the pollution patterns in the soil exactly. The 'triangular irregular networks' or TINs form an example of commonly used interpolation methods based on finite element techniques.

The application of remote sensing techniques or the use of scanned aerial photography will result in a raster-structured terrain description that follows

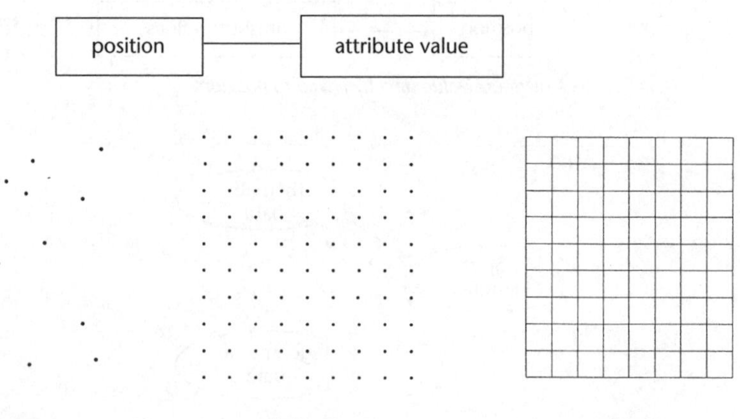

FIGURE 10.2 Several configurations for linking attribute values to position.

from the data acquisition technique. The raster structure might also be applied because it is convenient for the processing of the data (Molenaar, 1998; Tomlin, 1990). This grid or raster structure can be interpreted in two ways (see Figure 10.2):

1. if the observations are considered to refer to observational points with an exact location, then the raster is a point raster; or
2. if the observations are considered to refer to area segments, which are represented by raster cells, then the raster is a cell raster.

A height raster will often be interpreted as a point raster because height observations refer, in general, to specific positions. Land use data or data about population density will generally be represented in a cell raster because these data refer to area segments rather than to specific points.

10.3.1 The geometry of rasters

A raster is a collection of points or cells that cover the terrain in a regular grid. In a point raster, each raster element contains thematic data that refer to the position of the terrain point represented by that element. In a cell raster, the thematic information refers to an area segment represented by each element.

According to Figure 10.3, a raster can be defined geometrically by:

- The choice of the origin of the axes along which the rows and columns will be counted. This origin has the co-ordinates (X_0, Y_0).
- The choice of the direction of the X-axis and Y-axis, which are mutually orthogonal. By convention, the columns of the raster will be counted along the X-axis and the rows along the Y-axis.
- The choice of the step sizes ΔX and ΔY. In a point raster, these are the distances between the successive columns and the successive rows, respectively; in a cell raster, these step sizes define the size of the cell sides.

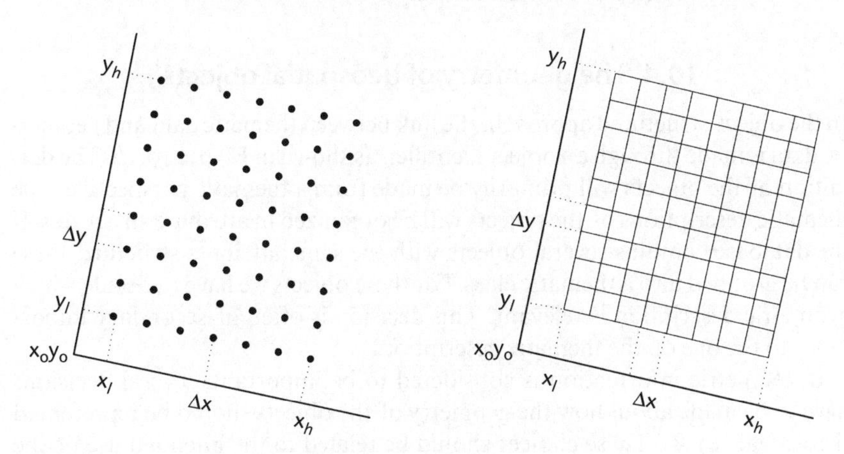

FIGURE 10.3 The geometric definition of rasters.

- The size of a point raster, which will be defined by the elements with the lowest and the highest co-ordinate values, (X_l, Y_l) and (X_h, Y_h) respectively. In a cell raster, these are the co-ordinate values of the lower left corner of the lower left cell and the upper right corner of the upper right cell, respectively.

The position of the raster elements can also be indicated by means of indices (i, j) instead of co-ordinates (X, Y). These indices can be computed from the co-ordinates by the following rules:

$$X_1 \rightarrow i=0,$$
$$Y_1 \rightarrow j=0,$$
$$X = X_1 + p^*\Delta X \rightarrow i=p,$$
$$Y = Y_1 + q^*\Delta Y \rightarrow j=q.$$

Co-ordinates can be computed from the indices by applying these rules in reverse.

The step sizes ΔX and ΔY define the resolution of the raster, so that the smaller the steps, the greater the resolution. This means that at a higher resolution, the terrain will be described in more geometric detail. A high resolution means that many raster elements are required to cover the terrain, whereas a lower resolution will reduce the number of elements. Hence, a high resolution requires that the raster processing handles a large amount of data. In many cases data compressing techniques can be applied, such as run length coding and quadtrees, which might reduce the actual data storage to some 20–30 per cent of the original number of raster elements (see for example, Samet, 1989). The processing time for many of the important operations on rasters increases more or less linearly in proportion to the amount of data. So we should find a balance between the detail required for the terrain description and the amount of data to be acquired and processed. The resolution should be adjusted to the size of the terrain details we want to analyse.

10.4 The geometry of geospatial objects

In the object-structured approach, the link between thematic data and geometric data is made through an object identifier, as shown in Figure 10.1*b*. The definition of the objects will primarily be made from a thematic perspective. The thematic descriptions of the objects will be organized in attribute structures. If the database contains several objects with the same attribute structure, these can be grouped into a thematic class. For these objects we have to decide which geometric description is relevant. This decision is often of secondary importance to the one on the thematic description.

If geometric information is considered to be important, several decisions have to be made about how the geometry of the objects should be represented (Molenaar, 1998). These choices should be related to the intended use of the information.

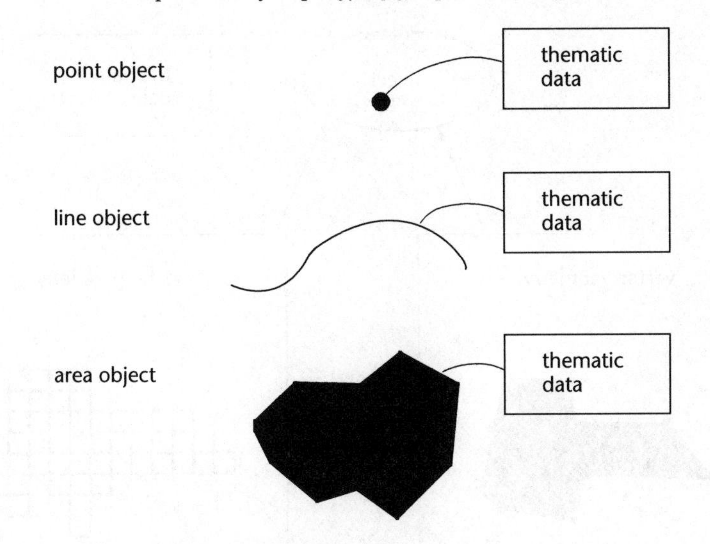

FIGURE 10.4 Three geometric object types.

- For each class of objects, we have to decide whether the objects will be represented as point, line, or area objects (see Figure 10.4):
  - *point objects*: only the position is stored;
  - *line objects*: the position and shape are stored, 'length' is the only size measure given;
  - *area objects*: position and shape will be given, the length of the perimeter and the area are the size measures.

This choice depends primarily on the role that the objects play in the terrain description and analysis rather than on the actual appearance of the objects. In turn, their role depends largely on the thematic aspects of the objects and on the scale of the terrain description.

- We should decide the accuracy required for the geometric description of the terrain objects. This decision involves both the locational accuracy and the precision of the shape description or the required geometric detail.
- There is also the choice of the geometric structure of the description. This can be a raster structure or a vector structure (see Figure 10.5). In addition, we should decide whether the topological relationships of the objects need to be shown explicitly in the data structure. Examples of topological object relationships are given in Figure 10.9.

10.4.1 Objects represented in rasters

The representation of objects in a raster can best be done in a cell raster. The cells represent area segments, so this geometry is most suitable for representing

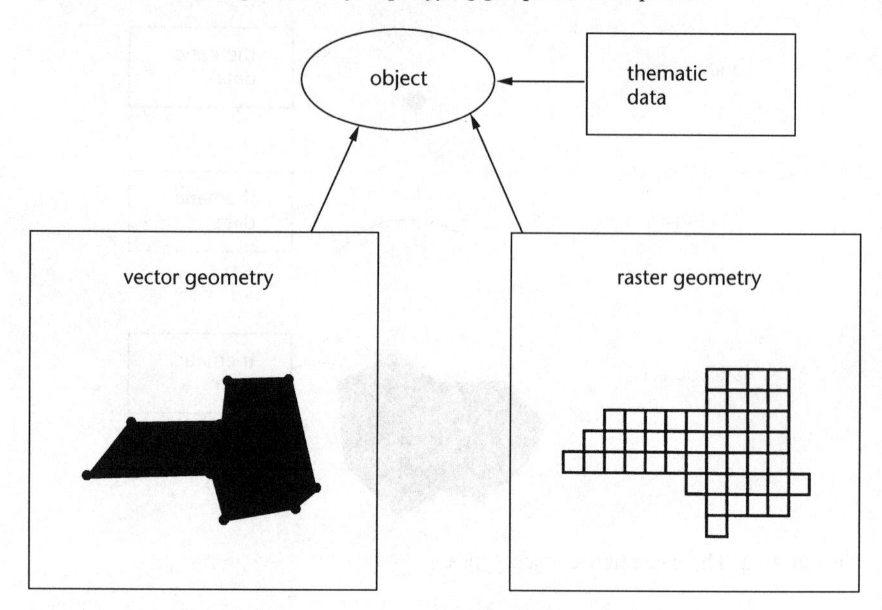

FIGURE 10.5 Two geometric structures for terrain objects.

area objects. Each cell is then labelled to indicate to which object it belongs. There are several possibilities for assigning labels to cells that are on the border between two area objects. Two possibilities that are used often assign a label to the cell indicating that it is on the border, or assign a label referring to the object that has the largest overlap with the cell.

Rasters are less suitable for representing point or line objects. It is possible, however, to indicate for each cell which point objects fall inside the cell and which line objects intersect the cell. The position of the objects within the cells is not known exactly, but it can be approximated with an accuracy determined by the resolution of the raster.

The topology of the raster helps to find the geometric structure of the objects. This can be done by checking for each raster element which neighbour (or full neighbour) has the same label; objects can then be identified as contiguous sets of cells with the same label, as in Figure 10.6.

When objects are represented in a raster format, the cells of the raster give the geometric object description. The resolution of the raster determines how precise (i.e. in how much geometric detail) the object geometry is represented. The spatial (topological) relationships of the objects can be analysed by algorithms based on the raster topology.

10.4.2 The vector structure

The description of a terrain situation in a vector structure represents the geometry of each terrain object by its linear characteristics. This means that the lin-

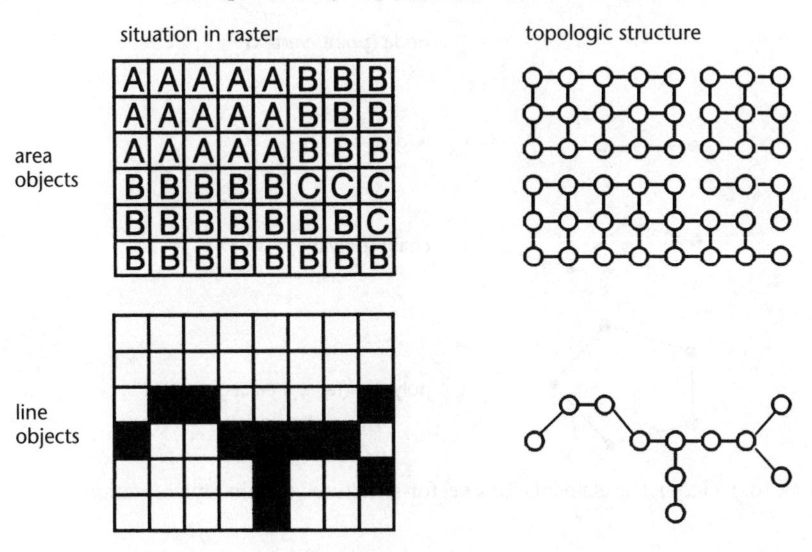

FIGURE 10.6 Objects represented in a raster.

ear structure of line objects, the boundaries of area objects, and the position of point objects are represented. The geometric elements shown in Figure 10.7 are used. The point objects are represented by points or nodes, the line objects by a chain of edges, and the area objects by polygons.

This can be done in several ways, as shown in Figure 10.8. Figure 10.8*A* shows a terrain situation that has been represented in different ways in Figure 10.8*B*, *C*, and *D*. Figure 10.8*B* represents the geometry correctly but it is not possible to link the objects to the geometric elements in an unambiguous way. It is therefore difficult to derive geometric information about the objects from this structure. In Figure 10.8*C*, it is possible to link the objects and the geometric elements in an unambiguous way and to derive information about the position, shape, and size of the individual objects. It is rather difficult, however, to analyse the topological relationships between the objects, because these are not expressed explicitly in this structure but must be computed from the polygons and chains representing the objects.

The representation of Figure 10.8*D* allows an unambiguous linking of the objects and the geometric elements so that geometric information can be easily obtained about the individual objects, but the topological relationships between the objects can also easily be derived from this structure. The edges in this figure can be linked to the area objects to their left and right sides, and a link can be made to the line objects if they are part of them. These links support the analysis of topological relationships between objects like those between adjacent area objects, and between line objects and the areas they intersect, and many others (Molenaar, 1998).

If a geospatial database has been designed with the topological structure of

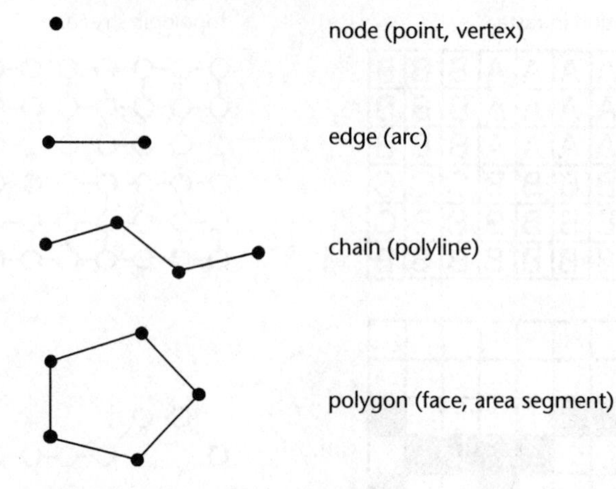

FIGURE 10.7 Geometric elements for a vector-structured terrain representation.

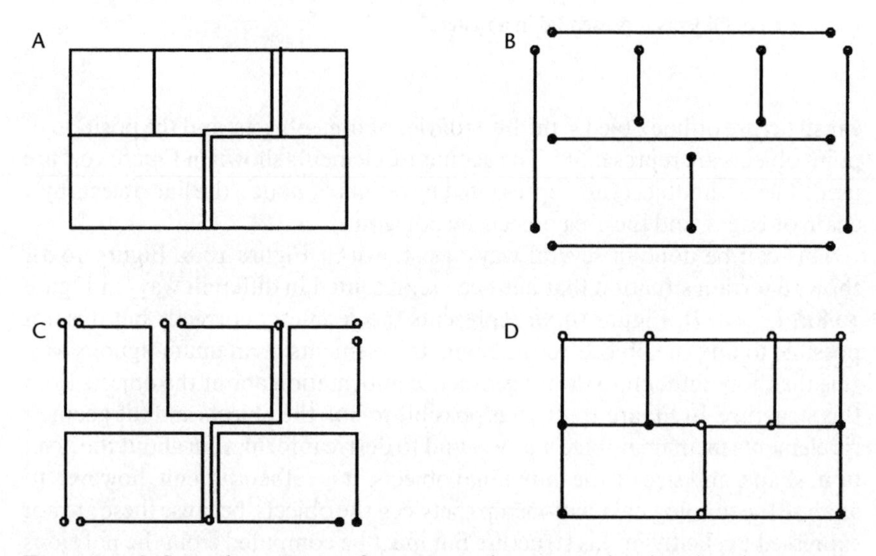

FIGURE 10.8 Several structures for representing a terrain situation in a vector geometry.

Figure 10.8*D*, it is possible to query the database to find a large variety of topological object relationships between the geospatial objects, such as those of Figure 10.9 (Egenhofer and Herring, 1992; Floriani *et al.*, 1993).

10.5 Object hierarchies

In Section 10.4 the concept of terrain objects was introduced; these objects were categorized into classes by their thematic aspects, but they could also be

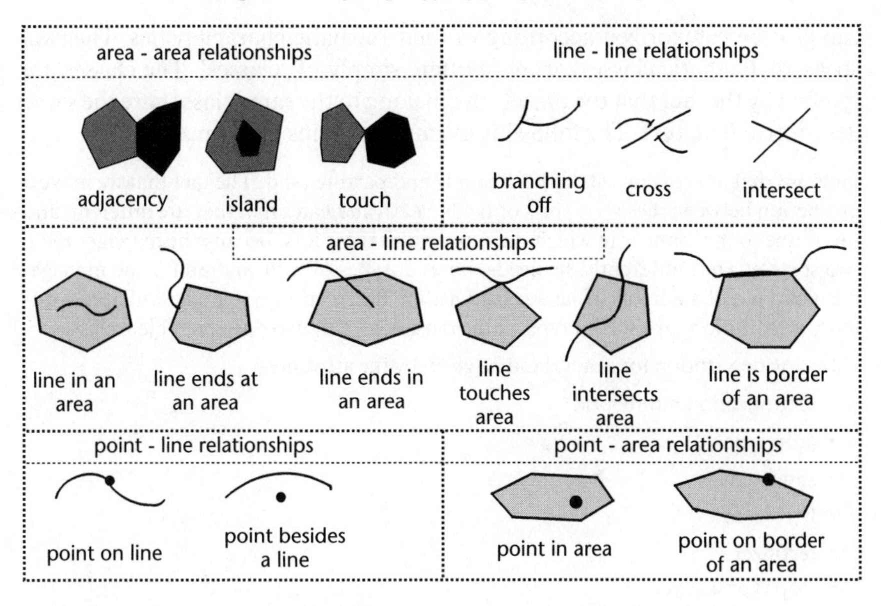

FIGURE 10.9 Several topologic relationships between spatial objects.

classified according to their geometric aspects. Objects classes define the characteristics which their members have in common. This principle has been exploited in computer science in techniques implemented in object-oriented programming languages and object-oriented database management techniques (Cox, 1987; Hughes, 1991; Smith and Smith, 1977). These classes define description structures for their member objects and operations that can be applied to them. There is abundant literature on this field, but some good concise introductions to object-oriented data modelling and database management are given by Brodie (1984) and Brodie and Ridjanovic (1984). The concepts in computer science have been interpreted and modified by several authors for geospatial database applications (Egenhofer and Frank, 1989; Fritsch and Anders, 1996; Kemp, 1990; Molenaar, 1993, 1998; Nyerges, 1991). Many of them have adapted the database techniques to the requirements for operations on geospatial data. Some authors use the concepts for the formulation of semantic data models to represent spatial phenomena, although their emphasis is not on the definition of database operations but rather on the development of conceptual tools for describing the real world. Their activities are considered to be in the domain of geo-information theory rather than in computer science. Molenaar (1998) is an example of this group and the discussions here follow this line.

10.5.1 Terrain object classes and generalization hierarchies

In the previous sections, terrain objects were divided into three types (point, line, and area), based on their geometric characteristics. In addition, objects can also be categorized according to their thematic characteristics. Then we speak of thematic classes, or more often, simply of 'classes'. The classes are typified by the fact that the objects that belong to the same class share the same descriptive structure. The following example explains this concept.

Suppose that a farm consists of both arable and pasture land. The fact that we make a distinction between these two sorts of fields clearly indicates that they are different, and this is due to the manner in which the farmer uses these lots. Because he manages these two sorts of land differently, he needs different types of information for the management and use of such lots. If he wants to use an information system, he will need a different description for each lot type, i.e. he must deal with two different object classes.

Let the description for arable land be given by the attributes:

- identification number (id)
- crop type
- sowing date
- herbicide
- fertilizer
- crop type last year.

Let the description for pasture land be given by:

- identification number (id)
- grass type
- (monthly) biomass production
- fertilizer.

A table can now be generated for each class. In Figure 10.10, columns W_1, W_2, and W_3 represent the thematic attributes of the class of pasture land, and columns A_1–A_5 are the thematic attributes of arable land; attribute names are given separately. Each class has its own attribute structure and a value will be assigned to every attribute for each object.

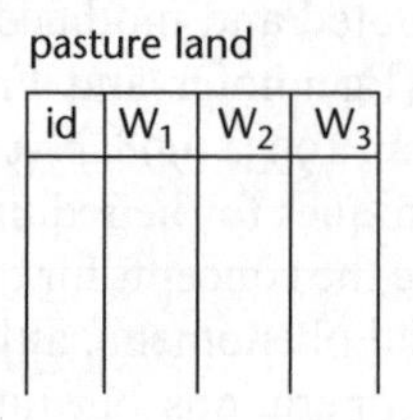

thematic attributes

W_1 = grass type

W_2 = biomass production

W_3 = fertilizer

A_1 = crop type

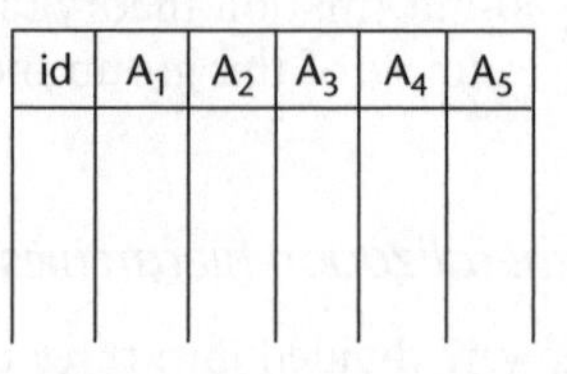

A_2 = sowing date

A_3 = fertilizer

A_4 = herbicide

A_5 = crop type last year

FIGURE 10.10 Tables for two classes of agricultural land use.

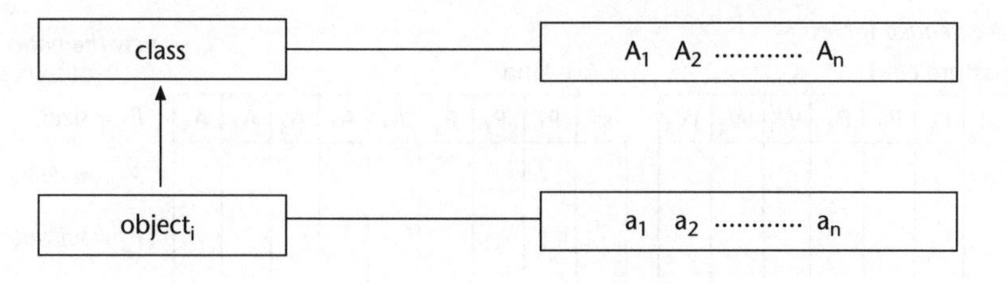

FIGURE 10.11 Diagram representing the relationship between objects, classes, and attributes.

These values must fall within the range of the attribute domain, which must be defined prior to the actual assignment of attribute values. This relationship between objects, classes and attributes is shown in Figure 10.11. Each class has its own unique list of attributes, and each object in a class has a list consisting of one value for every attribute. We will assume that each object belongs to only one class, so that the attribute structure of an object is completely determined by the class to which it belongs. We thus speak of an object inheriting the class attribute structure.

Different classes have different attribute structures, but that is not to say that all of the attributes are different. If we extend the list of attributes of both classes in our example with data concerning the following factors:

- area
- soil water level
- soil type

then we have two options: extending both existing tables with the new attributes (Figure 10.12*A*), or creating a new table with these new attributes (Figure 10.12*B*). Choosing the latter option implies that all the objects that appeared in the two original tables must also appear in the new table. This new table, called farm lot, is a more generalized description of the objects. The distinction between arable land and pasture land is, in fact, a further specification of the objects.

This is apparent in the fact that, per class, a more detailed specification of attributes is added to less specific classes of farm lot, as can be seen clearly in the extended lists of attributes for the tables in Figure 10.12*A*. We speak of 'farm lot' as a generalized class or superclass above the classes 'arable land' and 'pasture land'. An object that belongs to the class 'pasture land' inherits not only the attributes of this class, but also those of the superclass 'farm lot'. Figure 10.13 represents a hierarchy of classes created in this way.

All classes in a hierarchy can be distinguished by their own unique attribute structure. Within a hierarchic line, these structures are handed down, i.e. objects that belong to a specific class inherit not only the attributes of that class but also all those of the superclass or superclasses above it. In a strict hierarchy, the relation between a given level and the level above it is always $n:1$ (many to one), thus a superclass can have many lower classes, but a class on a lower level

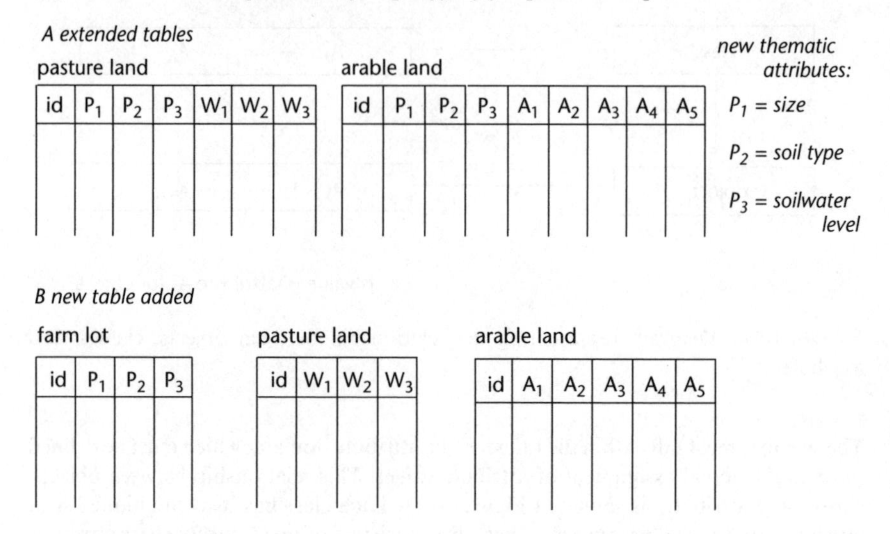

FIGURE 10.12 Table structure extended for new thematic attributes.

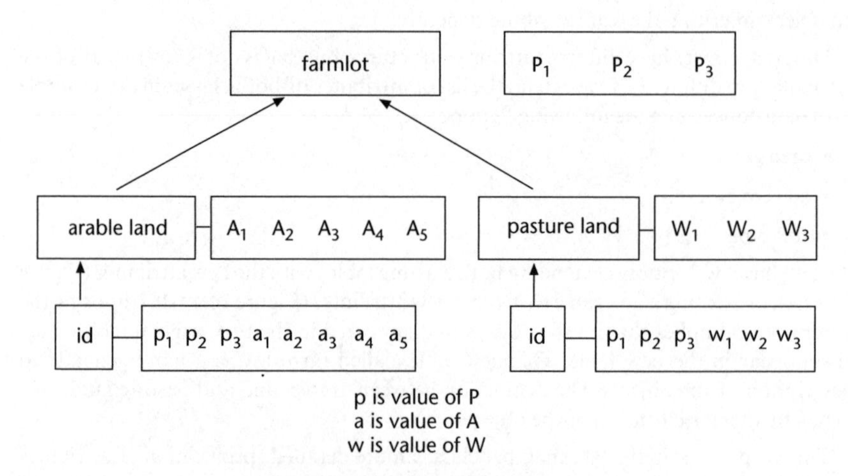

FIGURE 10.13 Class hierarchy for agricultural objects.

can belong to only one superclass on the level directly above it. Objects belong only to one class. Thus, when descending through a hierarchy, we see that at each level a following, increasingly detailed part of the object's attribute structure is defined. We speak in this case of specialization. At the level of the objects, there is no further extension of the attribute list, but the values of attributes are assigned per object. As we ascend in a hierarchy, the description of the objects becomes less specific, and we speak of generalization.

A classification system should be set up so that it is complete and exclusive, i.e. in such a way that all objects identified in a mapping belong to exactly one class. This implies that, within a hierarchy, classes receive their attribute structure via only one inheritance line.

Hierarchical lines, or inheritance lines, as described above, can co-exist in a classification system, forming one or more trees. A classification system structured in this manner is a manifestation of the 'thematic field' concept. It is characterized by a collection of classification hierarchies, i.e. classes and superclasses with their hierarchical relations. Further characteristics include the attribute structure of classes and superclasses, and the attribute domains. A thematic terrain description is the complete set of objects with a list of attribute values per object. In an information system, such a thematic field is always defined in a specific user context. This context is determined by many factors including, but not limited to, the mapping discipline, the point in time or era of the mapping, and the scale or aggregation level on which the work is carried out.

The upwards relation in a classification hierarchy is called an 'ISA' relation. Thus Amsterdam ISA city ISA built-up area. This example shows that an ISA relation assigns objects to a class and its superclasses. Classes and superclasses are typified by their attribute structures.

Terrain objects occur at the lowest level in a classification hierarchy and can therefore be seen as the elementary objects within the thematic field, represented by a given classification system. This implies that the decision of whether to include certain objects as elementary objects must be made within the context of the thematic field into which they are to be classified. This decision is thus referred to as a context-dependent decision. Objects that are elementary within one thematic field are not necessarily elementary in another.

10.5.2 Aggregation hierarchies

The previous section dealt with elementary objects thereby implying that there can also be composite objects, or aggregates. These can be defined within the framework of aggregation hierarchies. An aggregation hierarchy describes the way in which composite objects are built up with elementary objects, and how these composite objects, in turn, can be combined to form even more complex objects.

An aggregation hierarchy is shown in Figure 10.14. In the first step, from level 1 to level 2, farm fields are combined to form lots. Next, these lots are combined with a farmyard to form a farm. In the third step, a number of farms are combined to form an agricultural district. An aggregation hierarchy has a bottom–up character, in the sense that the elementary objects from the lowest level are combined to compose increasingly complex objects as the hierarchy is ascended. The compound objects inherit the thematic data from the objects of which they are composed.

The rules for constructing complex objects consist generally of two sorts. First, there are rules indicating the object classes from which a given compound object can be composed. Secondly, there are rules indicating which lower level objects can be included in a composite object on the next level. In a GDI, these

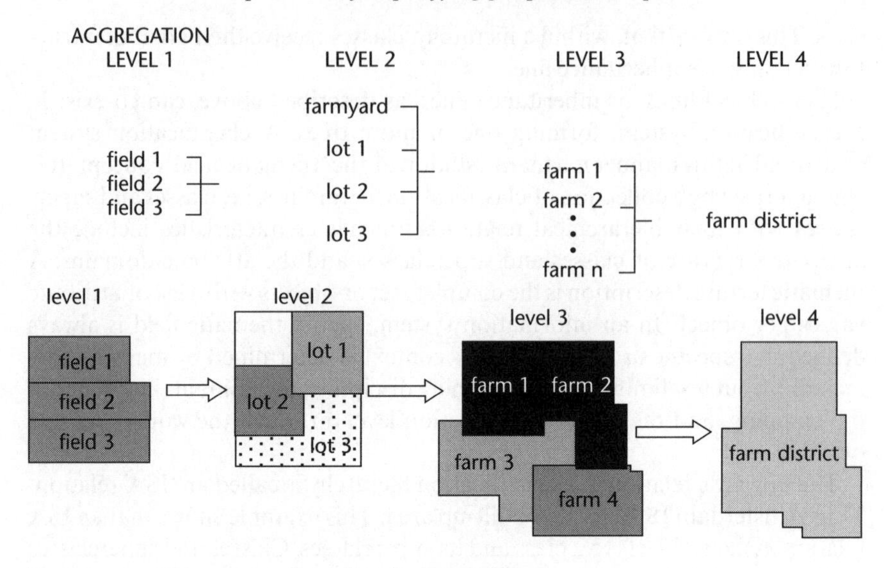

FIGURE 10.14 The aggregation of agricultural objects.

rules are often based on topological relations between objects. The agricultural district in Figure 10.14 is formed from farms that are mutually adjacent and fall within a communal boundary.

This means that aggregation types can be determined by their construction rules (note that these types must not be confused with the object classes of classification hierarchies). If elementary objects are combined to form a composite object, their attribute values are often aggregated too. Farm yield is the sum of field yields, and district yield is the sum of farm yields. We speak of upwards inheritance in aggregation hierarchies. Upwards relations in an aggregate hierarchy are 'PART OF' relations. For example, 'St James Park is PART OF Westminster is PART OF London'. The PART OF relations connect groups of objects with a certain aggregate and possibly on a higher level with another, even more complex aggregate, and so on.

Another characteristic distinguishing an aggregation hierarchy from a classification hierarchy is that within a thematic field an elementary object can belong to only one class, and thus has only one line of inheritance within a classification structure. In an aggregation hierarchy, however, it may be part of several different aggregates. Thus a set of aggregate types need not be either exclusive or complete. This implies that not all elementary objects are necessarily part of an aggregate. For example, hydrological systems and shipping routes are non-exclusive aggregates of waterways. A river can be part of a hydrological system composed of rivers, lakes, and streams. This same river can also belong to the shipping routes made up of rivers, lakes, and canals. It does make sense, however, to define aggregates within a hierarchy in such a way that the aggregates of one type are mutually exclusive within that type, i.e. that elementary objects belong to one aggregate of a given type. In this restricted sense, the relationships between objects and aggregates are many to one, $m:1$. Thus, a house can only be part of one neighbourhood, which can only be part of one municipality. In the same manner, a river can only belong to one hydrological system.

10.5.3 Object associations

The two types of hierarchies discussed so far have a clear description. A classification hierarchy has a top–down, step-by-step introduction of attribute structures for terrain objects. Classes are thus collections of objects with the same attribute structure. An aggregation hierarchy is defined by the construction rules that describe how the objects on a given level are composed of objects from the underlying level. From the bottom up, the levels have an increasing complexity.

A third possibility, object associations, is more loosely defined. The formation of object associations is not bound by a strict set of rules, but objects are grouped on the basis of some common factor. The relationships formed are not necessarily $m:1$ (many to one) relations, but may also consist of $m:n$ (many to many) relations. This implies that the associations of one type need not be exclusive. The following example is clarification of this principle.

The neighbourhoods in Figure 10.15 form associations. The neighbourhood of plot 3 consists of the plots that are adjacent to it. In this respect, the composition of an association resembles the composition of an aggregate. The difference, however, is that these plots also belong to other neighbourhoods. For plot 4 we can also define a neighbourhood which overlaps with the neighbourhood of plot 3.

The road network in Figure 10.15 can be regarded as an aggregate, whereas the routes in this network are another example of associations. The route from A to F consists of different roads or roads segments. The roads (and segments) that are a part of the route from A to F can also be a part of other routes, such as that from D to F. Thus, the routes are not mutually exclusive.

These examples should make it clear that associations consist of $m:n$ relations. This means that a given object can be part of several associations of the same type, i.e. they do not exclude each other as in the case of aggregations. The relation between an object and an association is called a 'MEMBER OF' relation. For example, 'plot 5 is a MEMBER OF the neighbourhood of plot 4'.

10.6 Object dynamics

Terrain objects may change with time so that their representation in a GDI must be changed through an updating of operations. The most obvious changes are that a new object appears and that an old object disappears. During its existence, however, an object may also go through several state transitions (Janssen and Molenaar, 1995). This can affect its representation in GDI in several ways:

ASSOCIATIONS

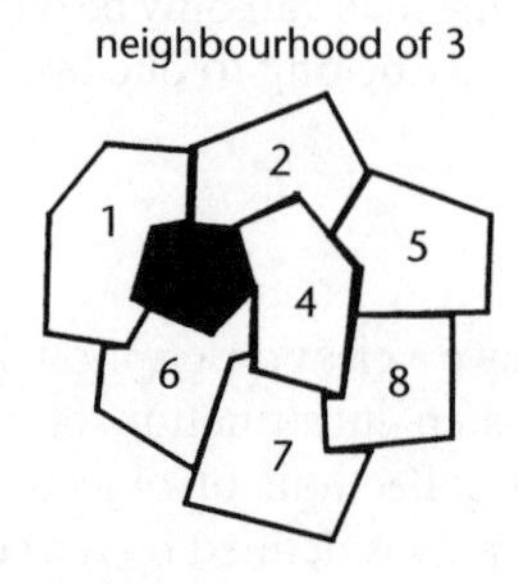

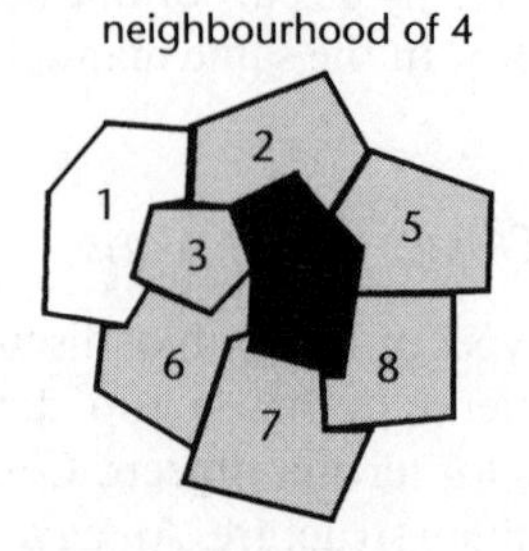

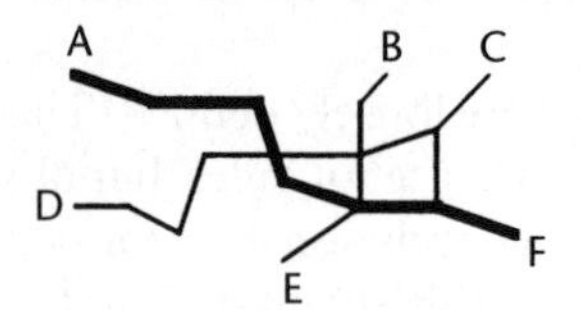

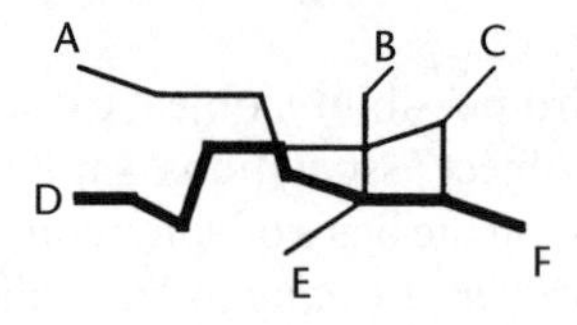

FIGURE 10.15 Examples of object associations.

- *First*, the thematic aspects of an object may change. In the simplest case, the value of one or more attributes change, e.g. a property has a new owner, or the cover type of an agricultural field changes. Another possibility is that an object is reclassified, e.g. the land use class of a field changes from farmland into built-up area. Such a migration from one object class to another may imply a change of attribute structure, and all the new attributes should then be re-evaluated.
- *Secondly*, the geometric aspects of an object may change. This might be a change of position or shape or size, or any combination of these. These changes may lead to changes of topological object relationships (see Figure 10.16).
- *Thirdly*, an object may change its aggregation structure. The aggregation structure indicates how a terrain object can be considered as a composite of smaller objects. Here too several possibilities exist for such a transition (see Figure 10.17).

The fact that only the internal structure of the composite or aggregated object changes implies that its external relationships remain unchanged. These relationships can be transferred through its inner structure to the composing elementary objects, i.e. the relationships will be transferred from the geometric elements of the aggregated objects to the geometric elements of the constituent objects.

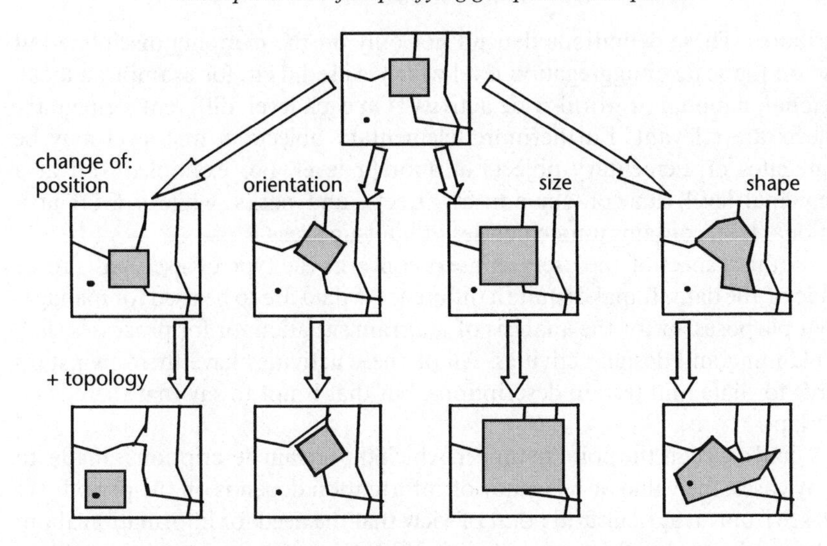

FIGURE 10.16 Geometric object changes.

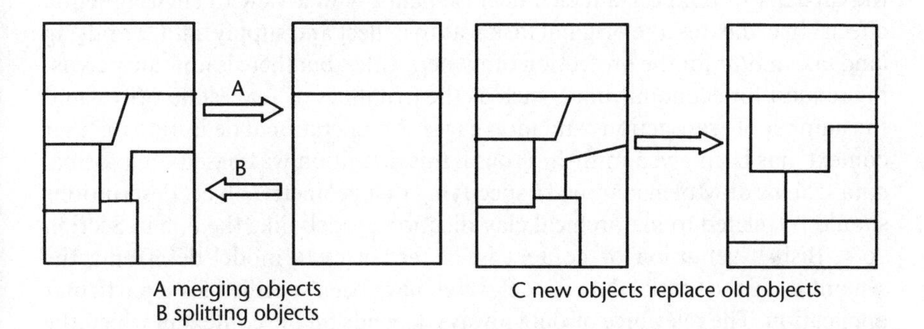

FIGURE 10.17 Changes of the collections of objects: A. elementary objects dissolve into one larger object. B. an elementary object is fragmented into smaller objects. C. change of aggregation structure of a composite object.

10.7 Object definitions and context

In Sections 10.2 and 10.4 we saw, in general, how a terrain object is represented in an information system via an identifier with associated geometric and thematic data. Section 10.5 explained that such objects are meaningful within a certain classification system. Thus, before a database can be built, the classification structure must be chosen. This choice must be made within the users' context and considering the following characteristics.

The first aspect considers the discipline(s) of the users. They could be dealing, for example, with a soil map or a demographic study or a cadastral system. Each discipline has its own definition of terrain objects, with classes and attributes. These definitions depend not only on the mapping discipline but also on the scale or aggregation level which is used. Is it, for example, a local, regional, national or world-wide activity? On each level, different elementary objects are relevant. Furthermore, elementary objects at one level may be aggregates of elementary objects at another level. For example, GDI at a municipal level can contain houses, streets, and parks, while a GDI at a national level contains municipalities or built-up areas.

Another aspect of the users' context concerns the type of use that is to be made of the data. It makes quite a difference if data are to be used for management purposes, or for the analysis of a terrain situation, or for processes such as planning and design activities. All of these activities have their own standards for data and terrain descriptions, but that is not to say that there is no overlap.

A final aspect is the point in time in which the terrain description is made. In many cases, the value or relevancy of information depends on the period. We can see from an agricultural point of view that the need for information about soils has changed in the course of time. Whereas the major interest used to be in the suitability of soils for certain crops, nowadays there is more interest in the capacity to bind certain chemical elements, with a view to environmental effects. In cadastres, the original task was to collect and supply data for raising land tax, and/or for the protection of owners' titles, but there is now an increasing request for economic data, such as the dynamics of real estate prices, and the number of transactions and mortgages. An operational definition of 'user context' has been given in Bishr (1997); this definition was based on a formal data scheme of Molenaar (1998), specifying that geometric object descriptions should be related to hierarchical classification models, like the one in Section 10.5. Bishr's definition of context is, in fact, a meta model describing the semantics of the spatial data model which have been specified for a particular application. The relevance of data always depends on the context in which the data will be used.

If the data are modelled according to the concepts presented in this section, a context will be expressed through the semantic definitions of the objects and the actual descriptive structures that have to be expressed with these formalisms. This means that in such a context the elementary objects are represented with their classes at the different generalization levels. Several class hierarchies may coexist, i.e. the collection of classes may form one or more trees. If the classes of these hierarchies are defined so that, at each level, each object of the database is a member of exactly one class, then they form a thematic partition per level. This implies that each object inherits its attribute structure through exactly one inheritance line of such a system. A classification system structured in this manner is a manifestation of a 'thematic field'. It is characterized by a hierarchical classification system, i.e. classes and a superclass with their hierarchical relations. Further characteristics include the attribute structure of classes and superclasses, and the attribute domains. A thematic terrain description is the complete set of objects with a list of attribute values per object. In an information system, such a thematic field is always defined in a specific user context.

Furthermore, the choice of geometric type must be made for these objects. This again depends on the role the objects are to play in a terrain description. A river can be regarded as a line object in a hydrological database, but the same river can be handled as an area object in the database of an organization that manages waterways. In the same way, a city can be seen as an area object in a database for demographic studies, whereas the same city appears as a point object in a database for continental traffic lines. Thus, the decision on which geometric aspects of a given class of terrain objects are relevant (i.e. the choice of treating objects as points, lines, or areas) always depends on the user context. This implies that the choice of which objects should be regarded as elementary, with their relevant thematic and geometric characteristics, also depends on the user context. Quite often the definition and identification of elementary objects follows their thematic specification as expressed through the classification system. Within such a context, decisions must also be made as to which object aggregates and associations are relevant. These are not necessarily explicitly stored in a database, but may be in the form of generic models, i.e. in the form of rules and procedures to use in generating these aggregates and associations.

10.8 Conclusion

Data modelling tools have been presented here which make it possible to describe data requirements in classification and aggregation hierarchies in a consistent manner. They serve to analyse the data requirements at all levels in the application domain and establish the relationships between those levels. The designer of the data services within the application domain can then determine what data may be available from other sources (Bishr, 1997) and assess the trade-offs that may have to be made to take advantage of available data, as opposed to having to develop the data separately by surveys. These trade-offs will, for example, take into account the timeliness, cost, and reliability of the result.

In order to carry out this assessment, data suppliers must tag their data with a large amount of qualifying information. However, in an 'information economy', suppliers will be motivated to make this investment. We should expect that government data owners will, in due time, be required by regulation to add this information and thus satisfy the public requirement to facilitate access to government geospatial information. However, taking into account David Rhind's observations (Section 4.9) this assumption may be optimistic. Adding this information is a demanding task because it has consequences for the overall performance of the application domain, both in terms of content and economics. It is also a specialized task, to be carried out as an integral part of the activity of a Geospatial Data Service Centre (see Section 9.6).

Bibliography

BISHR, Y. (1997). *Semantic Aspects of Interoperable GIS*, ITC Publication Series, No. 56, Enschede, The Netherlands, 154 pp.

BRODIE, M. L. (1984). 'On the development of data models', in BRODIE, M. L., MYLOPOULUS, J., and SCHMIDT, J. W. (eds.), *On Conceptual Modelling*. Berlin: Springer-Verlag, pp. 19–47.

——and RIDJANOVIC, D. (1984). 'On the design and specification data base transactions', in BRODIE, M. L., MYLOPOULOS, J., and SCHMIDT, J. W. (eds.), *On Conceptual Modelling*. Berlin: Springer-Verlag, pp. 277–306.

BURROUGH, P. A. and McDONNELL, R. A. (1998). *Principles of Geographical Information Systems*. Oxford: Oxford University Press.

CHRISMAN, N. R. (1997). *Exploring Geographic Information Systems*. New York: John Wiley & Sons Inc.

COX, B. J. (1987). *Object Oriented Programming*. Reading, Massachusetts: Addison-Wesley Publishing Company.

DE FLORIANI, L., MARZANO, P., and PUPPO, E. (1993). 'Spatial queries and data models', in FRANK, A. U. and CAMPARI, I. (eds.), *Spatial Information Theory, a Theoretical Basis for GIS*. Berlin: Springer-Verlag, pp. 113–38.

EGENHOFER, M. J. and FRANK, A. U. (1989). 'Object-oriented modelling in GIS: inheritance and propagation', *Proceedings of Auto-Carto*, 9, Bethesda: ACSM & ASPRS, pp. 588–98.

——and HERRING, J. R. (1992). *Categorizing Binary Topological Relationships Between Regions, Lines, and Points in Geographic Databases*, Technical report, Department of Surveying Engineering, Orono: University of Maine.

FRITSCH, D. and ANDERS, K. H. (1996). 'Objectorientierte Konzepte in Geo-Informations systeme', *Geo-Informations-Systeme*, 9.2: pp. 2–14.

GOODCHILD, M. F. (1992). 'Geographical Information Science', *International Journal of Geographical Information Systems*, 6: pp. 31–46.

HUGHES, J. G. (1991). *Object-oriented Databases*. New York: Prentice Hall.

JANSSEN, L. L. F. and MOLENAAR, M. (1995). 'Terrain Objects, their Dynamics and their Monitoring by the Integration of GIS and Remote Sensing', *IEEE Transactions on Geosciences and Remote Sensing*, 33: pp. 749–58.

KEMP, Z. (1990). 'An object-oriented model for spatial data', in BRASSEL, K. and KISHIMOTO, H. (eds.), *Proceedings of the 4th International Symposium on Spatial Data Handling*, University of Zürich, pp. 659–68.

LAURINI, R. and THOMPSON, D. (1993). *Fundamentals of Spatial Information Systems*. London: Academic Press.

MOLENAAR, M. (1993). 'Object Hierarchies and Uncertainty in GIS or Why is Standardisation so Difficult', *Geo-Informations-Systeme*, 6: pp. 22–8.

——(1995). 'Spatial Concepts as Implemented in GIS', in FRANK, A. U. (ed.), *Geographic Information Systems—Materials for a Post-Graduate Course*. Department of Geoinformation, Technical University Vienna, pp. 91–154.

——(1998). *An Introduction to the Theory of Spatial Object Modelling for GIS*. London: Taylor & Francis, 229 pp.

NYERGES, T. L. (1991). 'Representing geographical meaning', in BUTTENFIELD, B. P. and McMASTER, R. B. (eds.), *Map Generalisation: Making Rules for Knowledge Representation*. London: Longman, pp. 59–85.

PEUQUET, D. J. (1990). 'A conceptual framework and comparison of spatial data models', in PEUQUET, D. J. and MARBLE, D. F. (eds.), *Introductory Readings in GIS*. London: Taylor & Francis, pp. 250–85.

SAMET, H. (1989). *The Design and Analysis of Spatial Data Structures*. Reading, Massachusetts: Addison-Wesley Publishing Company.

SMITH, J. M. and SMITH, D. C. P. (1977). 'Database abstractions: aggregation and generalization', *ACM Transactions on Database Systems*, 2: pp. 105–33.

TOMLIN, C. D. (1990). *Geographic Information Systems and Cartographic Modelling*. Englewood Cliffs, New Jersey: Prentice Hall.

11

Spatial referencing

Marco Hofman, Erik de Min, and Ruben Dood

11.1 Introduction

A consistent definition of the position of different geospatial data sets is a necessary condition for their geospatial integration. For a variety of reasons, different data sets may have been spatially referenced using different techniques. For example, in projects crossing international boundaries, the positions of data sets on roads or vegetation cover, etc., are usually related to reference systems which pertain to each country. In a geographic area, different data sets may have to be drawn from maps with different map projections. Data sets referenced by GPS, which works on an international reference system, may have to be merged with data collected on a local reference system.

The quality of the decisions based on data integration in a GIS application therefore depends, in part, on the quality applied to bringing these different data sets spatially under one common denominator. This chapter aims to generate insight amongst developers and users of GDI into the possible reasons for spatially mismatching data sets, and to provide an ability to assess these problems in the light of the accuracy requirements of their applications.

The point of departure for this chapter is that for all spatial referencing the concept of a geodetic datum is used:

A geodetic datum is a set of constants specifying the co-ordinate system used for geodetic control, i.e. for calculating co-ordinates of points on the Earth.

(The term datum is synonymous with the frequently used term 'reference system'.)

Section 11.2 introduces ellipsoids as a mathematical model of the Earth, and describes how the step from this 3D model to a 2D map can be made through a variety of purpose-specific map projections. Section 11.3 describes the realization of a datum, and presents a summary of geospatial data acquisition methods including dynamic positioning and GPS applications in the context of the emerging dynamic spatial referencing infrastructure. Section 11.4 presents an overview of the issues associated with the combination of data sets of different datums and discusses a practical example.

11.2 Datums

11.2.1 The shape of the Earth

The shape of the Earth is the starting point for all spatial referencing. Since the Earth is slightly flattened at the poles, the geometrical figure used to describe the shape of the Earth is the ellipsoid of revolution, a figure obtained by rotating an ellipse around its shorter axis, in this case the rotational axis of the Earth. A reference ellipsoid is defined by two parameters: its major axis (diameter) and the flattening. Determining the size and shape of the earth has a long history, which we will not deal with here. A good reference for this is: 'Geodesy for the Layman' (*http://www.nima.mil/geospatial/geospatial.html*).

The joint efforts of geodesy and navigation have led to the realization of many 'reference ellipsoids' that best matched the geographic areas they described. In most countries, older reference systems, called 'local datums', are still in use; they use different ellipsoids and map projection methods. These are always fixed to 'reality' using benchmarks in the terrain for which the co-ordinates (latitude/longitude, northing/easting, *X*/*Y*) are known and published. This is called the realization of a datum.

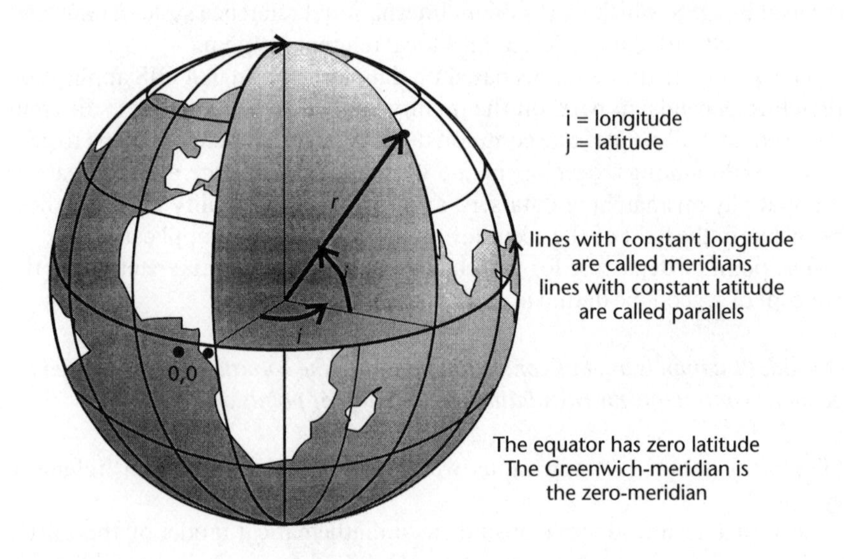
i = longitude
j = latitude
r
j
i
0,0
lines with constant longitude
are called meridians
lines with constant latitude
are called parallels
The equator has zero latitude
The Greenwich-meridian is
the zero-meridian

FIGURE 11.1 Latitude and longitude.

The 1927 North American Datum (NAD, 1927) used in the USA and Canada has been replaced by the World Geoditic System (WGS84) datum. The Hayford ellipsoid is called the international ellipsoid and is often used in maritime environments (e.g. the North Sea area). Other examples are Bessel's (1841) and Clarke's (1880) ellipsoids, both of which are used in Europe. A list of datums can be found at: *http://www.nima.mil/geospatial/products/GandG/historic/hdatums.html.*

The era of ballistic missiles and satellites introduced the need to obtain accurate relative positions and orientations of different national ellipsoids. Long distance travellers as well as future users of the global GDI will want to navigate on a world-wide ellipsoid, and not hop from one national system to the next. The world geodetic reference system, WGS84, introduces the 'ultimate' description of the shape of the Earth, including its gravity field. This world-wide reference system is now replacing many existing reference ellipsoids.

11.2.2 Defining a local datum

Besides the two parameters defining a reference ellipsoid, at least six more constants are needed to describe the local X, Y system: three to specify the point of origin on the ellipsoid, and three more to describe the orientation of the co-ordinate system. The X-axis usually points north, and the Y-axis usually points east. Finally, projections are necessary to transform the ellipsoidal co-ordinates to plane co-ordinates for mapping. A datum therefore consists of:

- a reference ellipsoid with its major axis and flattening;
- an origin of the co-ordinate system;
- an orientation of the co-ordinate system;
- one or more map projections.

11.2.3 Height references

The most common description of altitude is 'height above sea level'. The problem is that sea level cannot be described by a simple mathematical shape like an ellipsoid. The 'real' shape of the Earth is a bit like a potato, even if we replace all the continents with water. This shape is described by the 'geoid', an equipotential[1] surface mostly coinciding with (global) mean sea level. Physical heights are heights with respect to the local geoid. The system is usually set up using benchmarks for which the exact height is known and published. Differences in the height referencing between countries can cause problems for GIS applications in, for example, international watershed management and hydrological projects, e.g. concerning the River Danube or River Rhine.

[1] An equipotential surface is a surface on which the potential energy of gravity is everywhere equal.

11.2.4 Map projections

Introduction

There are many ways of projecting the 3D Earth on to a 2D map, but in all cases this will cause distortions. Each type of projection determines what the distortion will be. Hence, by carefully selecting the projection that best fits the purpose of the map, the effect of the distortion on the use of the map can be minimized.

The basic problem consists of the two steps needed to convert the position on Earth to the position on a map (see Figure 11.2). The Earth is approximated by the geoid. A position on this geoid is projected on to the ellipsoid. A position on the ellipsoid is specified as two angles: latitude and longitude (see Figure 11.1). The second step consists of projecting the ellipsoid on to a plane for mapping purposes (see Figure 11.3 for example). For almost any kind of projection, the type of distortion can be specified.

Distortion

There are three basic forms of distortion: angle (or shape), distance, and area. The absence of a certain type of distortion gives a classification of projections. The type of distortion can be visualized by projecting a circle on to different parts of the map. Usually there will be a point (or a line) on the map where the distortion is zero. The projected circle at this point is the reference circle. Comparing the projected circles to the reference circle gives an idea of the distortion at different parts of the map. Such a circle is called Tissot's indicatrix.

The absence of angle distortion is called conformal. These types of maps are important for navigational purposes. A line with a constant compass course is a straight line on the map (a rhumb line). These types of projections are also used because the form of the map elements is not distorted, whereas the size of the elements can be greatly distorted. Figure 11.4 shows a conformal map (Mercator) with Tissot's indicatrices. Along a parallel, the scale is constant, but along a meridian the scale increases rapidly from the Equator to the poles.

A map with an absence of distance distortion is called an equidistant projection. Normally a projection will only be equidistant along certain lines. Figure 11.5 shows an equidistant projection with Tissot's indicatrices. This projection is equidistant along the meridians. The indicatrices have the same diameter for

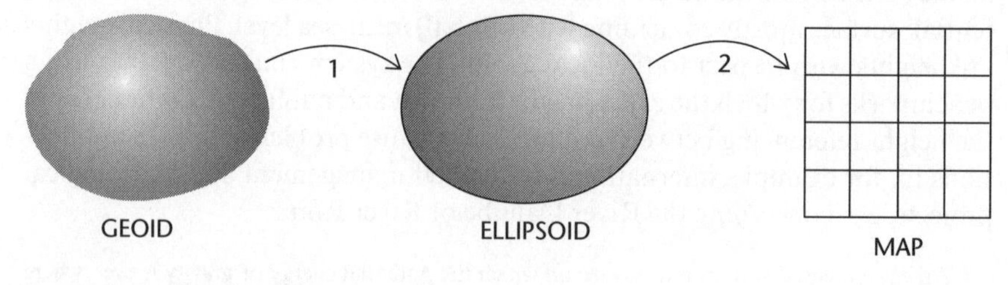

FIGURE 11.2 Geoid—ellipsoid—map.

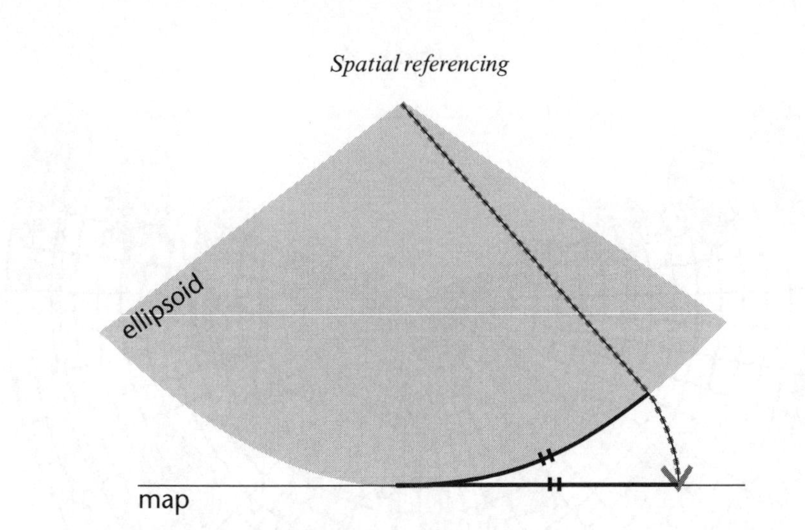

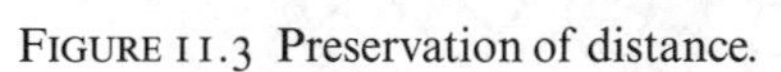
FIGURE 11.3 Preservation of distance.

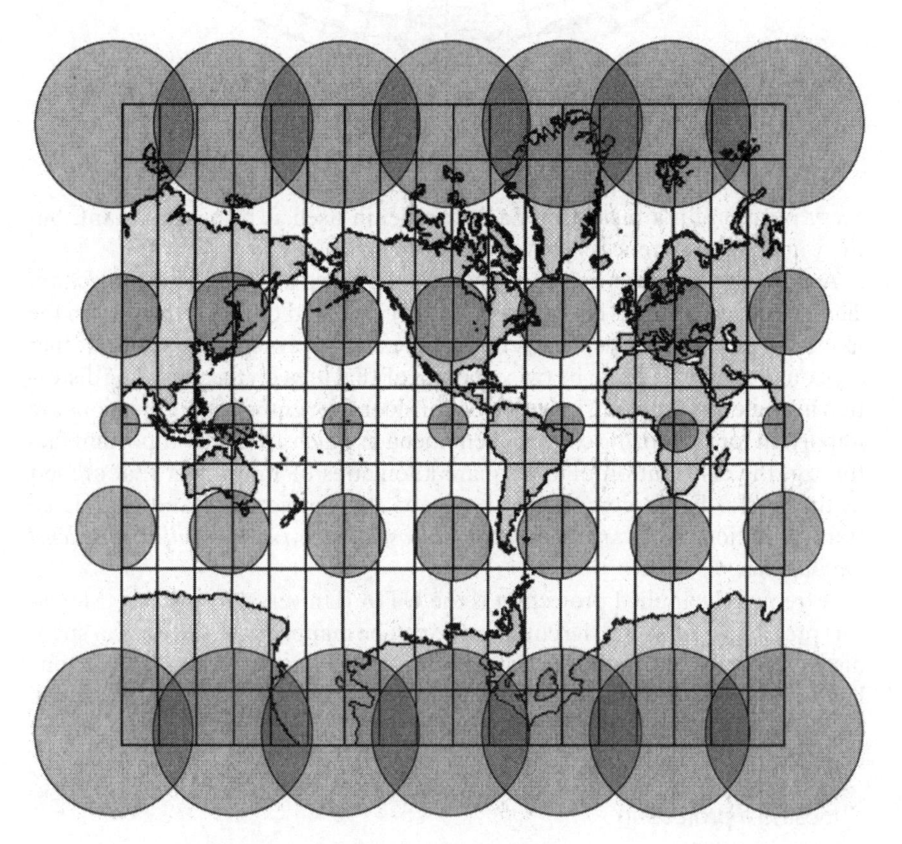

FIGURE 11.4 Conformal projection.

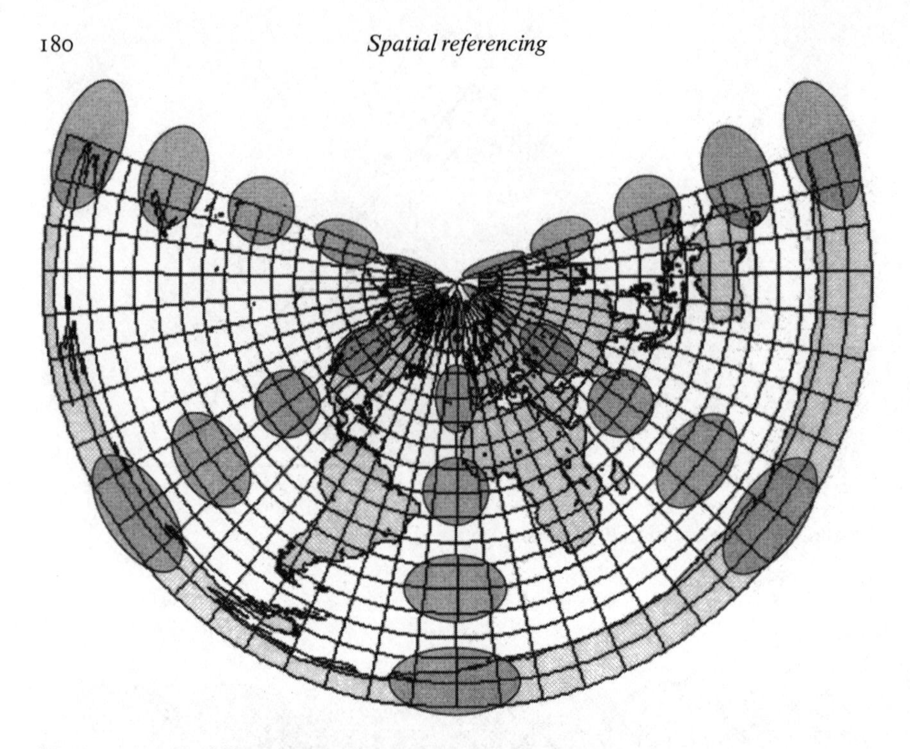

FIGURE 11.5 Equidistant projection.

every position on a meridian. Each parallel in itself is also equidistant, but there are huge differences between the parallels.

A map with an absence of area distortion is called an equal area projection. The objects on the map have the same size (compared to each other) as on the globe (Figure 11.6). This means that an area that is twice the size of another area on the globe will also be twice the size of that area on the map. The distortion in angle or shape can, however, be considerable. Equal area projections are important for visualizing certain phenomena in which area is an important factor, e.g. the distribution of the square kilometres of rain forest that are left in the world. For a comprehensive set of map projections and associated transformation software see *http://kartoserver.frw.ruu.nl/html/staff/oddens/mapsat14.htm*.

A frequently applied projection is the UTM (Universal Transverse Mercator) projection, in which the entire globe can be mapped into sixty zones spanning six degrees of longitude (i.e. the cylinder has sixty positions). Each zone has a central meridian, at which the distortion is zero. Each zone has a unique number so that projections can be easily recognized. Most NATO military mapping has been produced using this projection.

Choosing a projection

Deciding which projection to use is difficult. The choice depends on the location, size, and shape of the area to be mapped, and on the purpose of the map.

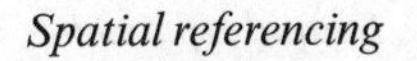

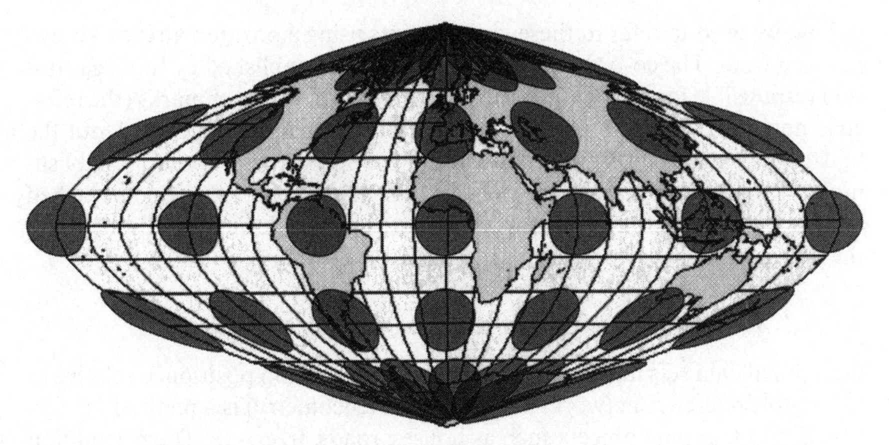

FIGURE 11.6 Equal area projection.

The position of the area to be mapped on the Earth can be decisive. The shape and size of the area to be mapped can be important. For a small area anywhere on the Earth that fits neatly in a circle (such as The Netherlands) an oblique azimuthal projection might be chosen. For north–south elongated areas, such as Chile or Italy, a transversal cylindrical projection is best. Areas such as the USA or Russia are best projected with a conical projection because of their shape, orientation, and location on the Earth's surface.

11.2.5 Other reference systems

Postal codes, civic addresses, and areas defined by administrative boundaries are also important spatial references. Their main application is in social and economic surveys such as the population census, labour force surveys, the agricultural census, etc. For business applications requiring the integration of social and economic survey data with that of physical geography, a transformation to co-ordinates may be required. These reference methods are not dealt with here.

11.3 The realization of a datum and a summary of geospatial data acquisition methods

We should appreciate the great progress made in the last thirty years in the efficiency, consistency, and accuracy of positioning methods and systems. We should also be aware of the differences in geometrical quality of the positioning these different methods have provided. This in turn affects the quality of data integration, and ultimately may adversely affect the effectiveness of a GDI.

11.3.1 Realizing the datum

A datum or a co-ordinate system needs to be realized in the field by means of benchmarks or other recognizable features. To determine a position in a certain datum, we need to refer to these visible points using a position already known in that datum. The co-ordinates of these points are published by the organization responsible for the maintenance of the network of benchmarks (the reference network), which is usually by the national mapping agency. Until the mid-1970s triangulation and trilateration were the basic methods for establishing and maintaining reference networks in many countries, which reached accuracies to the order of several centimetres to some decimetres. In the 1970s the first satellite-based positioning systems came into use.

11.3.2 Summary of geospatial data acquisition methods

Geospatial data sets may be surveyed on the ground and positioned relative to this visible reference network. Tachymetry (tacheometry) is a method for surveying and mapping objects such as houses, roads, trees, etc. The orientation and distance to an object is measured from a fixed observation point. The precision of the relative positions observed using tachymetry can be better than 1 cm. Currently, tacheometry is carried out using a system with automatic registration of measurements called a 'Total Station'. Land surveying will result in data sets of the same datum as the reference points. Geospatial data may also be acquired by means of photogrammetry and remote sensing, and techniques for obtaining information from images from air- or spaceborne platforms. These techniques and their respective performances are presented in Chapter 12.

The last classic method to be mentioned is levelling. Levelling is a way of determining height differences using an instrument in which the line of sight is strictly horizontal (Figure 11.7). Long distances can be bridged by using this method repeatedly, obtaining height differences to an accuracy in the order of some millimetres over distances of several kilometres.

11.3.3 Dynamic positioning

The Second World War brought revolutionary developments in the field of regional navigation and positioning. DECCA (UK) and Loran-A (USA) were the first radio positioning systems to allow navigation over large regions, for example, along the coast of Europe or the Eastern USA, with a precision of up to some 100 m. DECCA is still operational in several areas around the world. Loran-A was superseded by Loran-C, which is operational today in the USA and several other areas around the world. In Europe the system is being modernized and new transmitters are still being built.

11.3.4 Towards GPS and a new positioning infrastructure

By 1956 the USA had launched the first positioning satellite for what became the Transit system. This was the first system to permit world-wide position determinations with homogeneous precision. At the end of the 1970s, the USA launched the first satellite of the NAVSTAR Global Positioning System

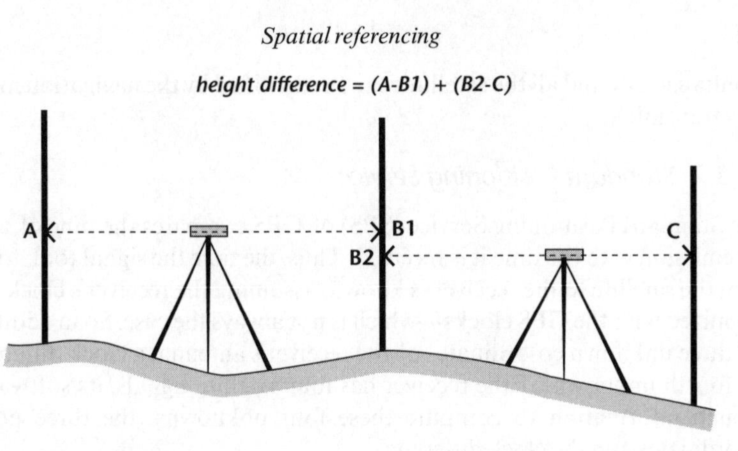

FIGURE 11.7 Levelling.

(GPS). Today, 24 GPS satellites are operational, and the world-wide standard precision is 100 m. GPS has revolutionized all aspects of position referencing. It has become, in Lance McKee's terminology, 'mundane' (Chapter 2, p. 14).

11.3.5 Practical use of GPS

GPS is becoming the basic positioning system. It is used in many applications from the realization and maintenance of datums and the basic networks for mapping projects to dynamic mapping applications, like continuous array echo-sounding for nautical map production and determination of the sensor's position in remote sensing applications.

In this dynamic mode GPS is finding significant applications in the transportation industry. Vehicle location and in-car navigation charts combined with local communication systems open the way to intensive traffic management systems and optimal routing of ambulances, fire service, and police vehicles. Hikers, bikers, and canoeists can use hand-held systems, which cost less than $US500. However, this also carries with it the danger that the equipment will be used inappropriately. If field data is being collected by means of GPS and consistency of accuracy in positioning of the order of a metre is required, then a $US500 receiver will not suffice. Matching newly surveyed data with existing data may not work reliably with this equipment. Yet the advertising for the receivers is sometimes ambiguous in this respect, which may lead prospective customers to have higher expectations than warranted. It is with this in mind that a variety of GPS positioning methods are explained in this chapter.

11.3.6 Global Positioning System

The GPS satellites orbit around the Earth at a height of 20,000 km with each satellite transmitting the time and its position. This is synchronized by stable atomic clocks, and corrections can be given by ground segments from a number of monitoring stations spaced around the world. All the clocks run simultaneously and all the satellites can be identified by the navigation message they transmit.

11.3.7 Standard Positioning Service

The Standard Positioning Service (SPS) of GPS compares the time of the signal emission with the time it is received. Thus, the time the signal took to travel from the satellite to the receiver is known, assuming the receiver's clock is synchronized with the GPS clocks—which is not always the case. So in addition to the three unknown co-ordinates of the receiver's antenna, a clock difference is the fourth unknown. If the receiver has four satellite signals, its software has enough information to compute these four unknowns: the three position co-ordinates and the clock difference.

The accuracy of the standard service of GPS is 100 m horizontal and 150 m in height in 95 per cent of the observations. This accuracy could be better, but for the intentional errors in the satellite's navigation message (time and orbit parameter errors called 'Selective Availability' or SA), which were introduced for military security reasons. A second service, Precise Positioning Service (PPS), is available to the military community only. PPS uses a special code not freely available. If a military P-code receiver is used, position accuracy is in the order of 10 m.

GPS measurements always result in data with WGS84 latitudes and longitudes. Most receivers, however, can transform this to several other datums, even by on-line computations.

11.3.8 Pseudo-range differential GPS, and the new dynamic spatial referencing infrastructure

The errors introduced by SA and by some secondary error sources, like the atmospheric delay of the signal, are the same for each receiver used in a certain area. If there is one receiver with known co-ordinates, the position error in that receiver can be calculated. Such a fixed, known receiver is called a reference station. All other receivers in the neighbourhood (using the same satellites) will have the same error at the same time. Since different receivers could use different satellites, however, it is better to compute error values for each signal and generate a signal 'as it should be'. Corrections needed to change a standard signal to an error-free signal are then transmitted to all receivers in the neighbourhood. These corrections change constantly. The remaining error in the user's receiver is the result of the time lag between the calculation, the transmission of the corrections, and the receipt and application of the corrections. This means that the faster the data link, the better the position accuracy that can be achieved.

Typical values are 1 m for fast local systems (local with respect to the reference station) to 10 m for slower longer-range systems. This principle is called Differential GPS. Several services supply Differential GPS corrections, by determining the corrections in a reference station, and broadcasting them using a radio data link. Several countries have installed a regularly spaced network of these GPS reference stations in support of this technique and of dynamic positioning applications (see below). This results in a facility usually called an Active Control System, which is the new dynamic spatial referencing infrastructure. It is dynamic in the sense that it continually updates the reference system, which stands in sharp contrast to the static geodetic control system represented by benchmarks on the ground (see, for example, Georgiadou 1999).

11.3.9 Carrier wave phase difference

Another, even more precise, method uses not the code of the signal, but the carrier wave itself. The two frequencies of GPS provide wavelengths of approximately 20 cm. The receiver can calculate the difference in distance very accurately by comparing the signals from two satellites. A moment later both satellites are at different positions and the result of the calculation changes. By observing GPS in two receivers, the vectors between both receivers can be computed. If the position of one receiver is known, the position of the other can be computed. In order to determine the vector between the reference station and the unknown station, a certain minimum number of observations is needed. The longer the vector or 'baseline' and the better the accuracy required, the longer the observation time. Using fast and efficient methods, an accuracy of some centimetres is easily achieved, while extra time and effort can give an accuracy of some millimetres.

11.3.10 Dynamic positioning

Ultimately, the preferred method of using GPS is in 'kinematic' mode in real time. This means the determination of positions by observing GPS while on the move, making use of observations done by the reference station and transmitted to the moving receiver in real time. Using a fast data link, and with baselines shorter than 30 km, a precision of some 10 cm can be obtained. This mode finds extensive applications in real time, and dynamic navigation techniques both on land and at sea. More details about GPS can be found in many geodetic text books (e.g. Seeber, 1993).

11.3.11 GLONASS

The Russian Federation operates a second satellite navigation system, called GLONASS (GLObal NAvigation Satellite System). It is much like GPS, but has different time and co-ordinate references, different orbits (though at the same altitude), and a somewhat different signal structure. Only lately has the integrated commercial use of both systems become available to professionals. Since the maintenance of the satellite system is proving to be a problem, the future of GLONASS is uncertain.

11.3.12 Conclusions on GPS

From the user's perspective, the following summarizes the state-of-the-art application of GPS:

- When consistent accuracy is not the most important requirement, a simple, low cost GPS receiver should give good results.
- To determine positions to within 1–10 m, Differential GPS is required, and a provider must be found who broadcasts the differential corrections. The equipment will be more expensive, and the service provider may charge a fee. However, radio beacons installed along the coasts and providing a public broadcasting service may be used.
- Accuracy better than 1 m requires specialized equipment and software, and usually the services of a land or hydrographic surveyor. Access to special local service infrastructure, such as reference stations and data links, is needed. The higher the accuracy required, the more expensive the hardware, software, and professional support.

11.4 Combining data sets

11.4.1 The problem

The discussions on the different geodetic ellipsoids and the different map projections and survey methods makes it clear that when data sets are related to different datums, they cannot be easily combined. The first step must be to transform all the data to one datum. Next the positional quality of the data is determined, and then the data sets can be combined. If one or both data sets are given in a local *X*/*Y* datum, these co-ordinates should first be expressed with respect to its ellipsoid by means of an inverse mapping function. Next the data can be transformed to the other reference ellipsoid, by applying a seven-parameter similarity transformation, after which another mapping function can be applied to produce the results in a new map projection. The result after these three steps is an expression of the co-ordinates of one local datum in the other local datum. These transformations will be shown in this section.

Further information on co-ordinate transformations can be found in specialized publications. Transformation software (MADTRAN) is available from several sites on the Internet, e.g. from
http://www.ntis.gov/fcpc/cpn5198.htm or from
http://www.nima.mil/geospatial/products/GandG/madtran/madtran.html.

There are four basic cases:

Case I. The datum is known for both data sets. Both reference ellipsoids are known, the position of the origin and the orientation of both data sets are known and, if applicable, the map projections are known. The transformations between the ellipsoids are also known.

Case II. The complete datum is known for both data sets. Both reference ellipsoids are known, the position of the origin and the orientation of both co-ordinate systems are known and, if applicable, the map projections are known but the transformation between the two ellipsoids is not known. However, there are corresponding points in both data sets.

Case III. The datum is not known for one or both of the two data sets. However, there are corresponding points in the data sets.

Case IV. The complete datum is not known for one or both data sets. There are no corresponding points in the data sets.

For the first three situations, a possible solution will be given. The fourth case is not dealt with here since the data sets cannot be combined. The three cases will be discussed using Figure 11.8.

11.4.2 Description of the cases

Case I. Both datums are known, ellipsoids can be transformed into each other

In the first case both data sets are given in a co-ordinate system for which the relation between the map co-ordinates and the underlying (local) reference ellipsoid is known. The first data set results in map I, the second in map II. The relation (A) from map co-ordinates to the underlying reference ellipsoid is known, hence the first data set can be 'reverse-projected' on to the ellipsoid. The co-ordinate transformation between the two reference ellipsoids is also known (B), so the data can be transformed to the reference ellipsoid of the second data set. Finally, the map projection of the second data can be applied to the first data set (C). Now both data sets are in the same datum.

Example 1. German (Gauss-Krüger) and Belgian (Lambert72) data sets to WGS84

A GIS user is making a road map of Europe and has collected many national maps with this information. All the maps are given in the local co-ordinate system, which consists of an underlying reference ellipsoid (that optimally fits the local geoid) and a map projection. Our GIS user has chosen WGS84 as the reference system for the new data set.

Two maps with data were obtained from Belgian and German colleagues. Figure 11.9 shows the steps the GIS user has to take to obtain the information in WGS84 so that it can be included in the database. For the German

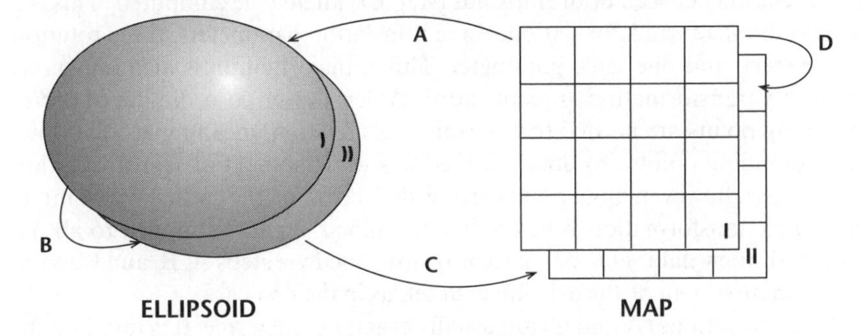

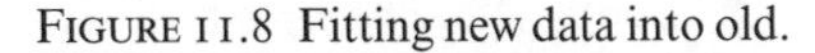
FIGURE 11.8 Fitting new data into old.

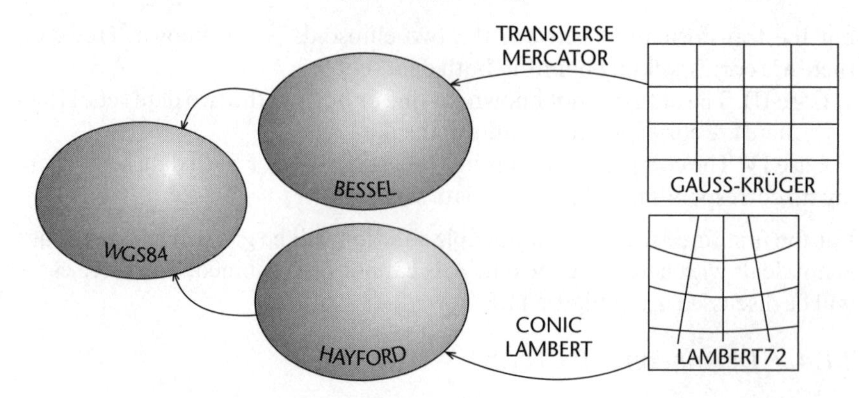

FIGURE 11.9 Lambert and Gauss-Krüger.

map, the data are given in Gauss-Krüger, which is based on a transverse Mercator projection on the Bessel ellipsoid. The data have to be transformed by an inverse map projection to the Bessel ellipsoid. Then the ellipsoidal transformation of the German Bessel ellipsoid to WGS84 has to be known, to compute the co-ordinates in WGS84. A similar procedure is applied to the Belgian data. The inverse Lambert72 projection is applied, giving co-ordinates with respect to the Hayford ellipsoid. The formula of the ellipsoidal co-ordinate transformation can be found in many geodesy textbooks.

Many transformation parameters are published in Hooijberg (1997), while more and more information is also becoming available on the Internet (search for 'datums').

Case II. Both datums are known

This could be the case of matching new GPS-based data with older maps for which the datum is known. In this case step B is not known (Figure 11.8). There are, however, corresponding points that occur in both data sets. We can apply step A and the inverse step C to these points in both data sets to yield the co-ordinates of the points in both ellipsoidal co-ordinate systems. The exact relation between both ellipsoids (step B) can now be computed. This is a seven-parameter transformation: three translation parameters, three rotation parameters, and one scale parameter. This transformation is also known as similarity transformation or rectification. At least seven co-ordinates of corresponding points are needed to determine this relation, meaning at least three corresponding points. To obtain a precise and reliable set of transformation parameters, however, more points are needed, covering the entire area of interest. Once transformation B has been determined, it can be applied to all the data in the new data set needing to be transformed via steps A, B, and C to the co-ordinate system of the existing data set, as in the first case.

Map projections (A and C) are usually exact, i.e. error free. It is just a mathematical formula describing the relation between map co-ordinates *X*, *Y* and ellipsoidal co-ordinates ϕ and λ. The precision of co-ordinate transformation B depends on:

- the number and position of corresponding points that are used to determine the relation;
- the occurrence of systematic disturbances and local measurement errors in the measured network(s) in both local datums;
- the size of the overlapping area.

GPS measurements are much more accurate than the old classical measurements. When GPS is used to verify the co-ordinates of the most important benchmarks, discrepancies may emerge. If the network covers an area up to 100×100 km then the discrepancies may be in the order of 1 dm; but networks up to $1{,}000 \times 1{,}000$ km may well have errors up to 1 m. Because whole co-ordinate systems depend on these benchmarks, they cannot be corrected individually without causing tremendous disruption in the whole spatial referencing system. We simply have to live with them and sometimes make do with local adjustments to meet specific requirements. For example, the transformation parameters to WGS84 are determined for a whole network at once. When considering the complete network they more or less fit, but in certain parts there may be unacceptable systematic errors in terms of their application. Even if step A and the inverse step C project the networks on to the same ellipsoid (e.g. WGS84), the corresponding points could have serious local systematic errors. In this case, even if step B is known, it should be considered as unknown, and local transformation parameters should be computed.

Case III. At least one datum is unknown, and there are corresponding points

This could be the case where new GPS-based data need to be merged with maps of unknown datum. As in case II, a relation between two co-ordinate systems can be determined if corresponding points are available in both data sets. Here we need a transformation from one map directly to the other (step D in Figure 11.8). There are several possibilities for determining such a transformation and the choice depends on the number of points available in both data sets and the quality of the sets. The transformations most commonly used and most often available in GIS systems are:

- 2D similarity transformation;
- rubber sheeting;
- edge matching.

The relation between two *X*–*Y* co-ordinate systems can be described in many ways, with many different degrees of freedom. In geodesy, a similarity transformation is usually applied with four degrees of freedom. The two maps are related by a shift (or translation) in both the *X*- and *Y*-directions, a rotation of the map, and a scale parameter. This transformation is called (2D) rectification or similarity transformation, because it leaves the form of the map unchanged. An extension of this transformation is the affine transformation, where one or both co-ordinate systems have co-ordinate axes that are not perpendicular, and they may have different scales for the *X*- and *Y*-axes. In this case, two more unknowns have to be determined for the transformation.

Example 2. Similarity transformation (see Figure 11.10)

A GIS user has a map for which the reference ellipsoid and projection are not known, but part of the map overlaps an old map for which the datum is known. The data sets can be connected by fitting the objects that occur on both maps with a 2D similarity transformation. First the new map is moved over the old one so that the corresponding objects match up in a global sense. Then the new map is rotated a little. Finally a scale factor correction is applied, i.e. the new map is stretched in such a way that the objects match up as well as possible. These three steps, a translation (or shift), rotation, and scaling form the 2D similarity transformation.

Rubber sheeting

Sometimes a similarity transformation is not sufficient: the corresponding objects that occur in both maps are clearly not in similar positions. It is also possible that a similarity transformation does not result in closely fitting data sets. In this case more than four degrees of freedom can be given, so that the

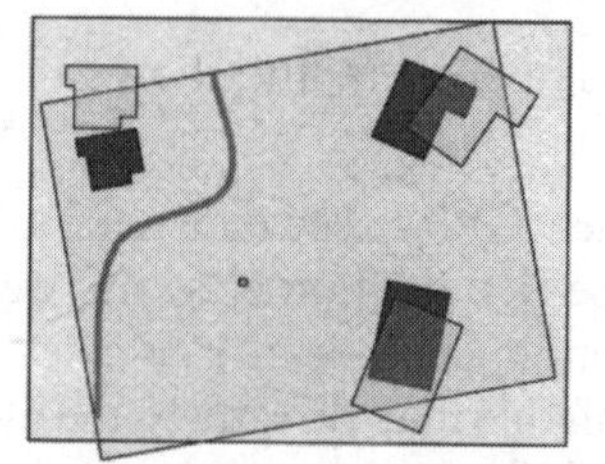

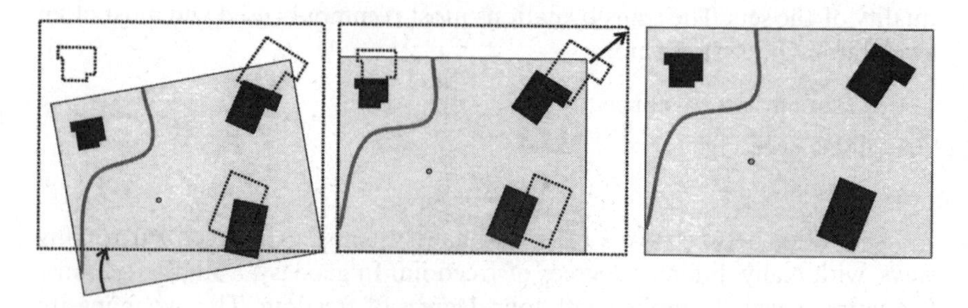

FIGURE 11.10 2D similarity transformation.

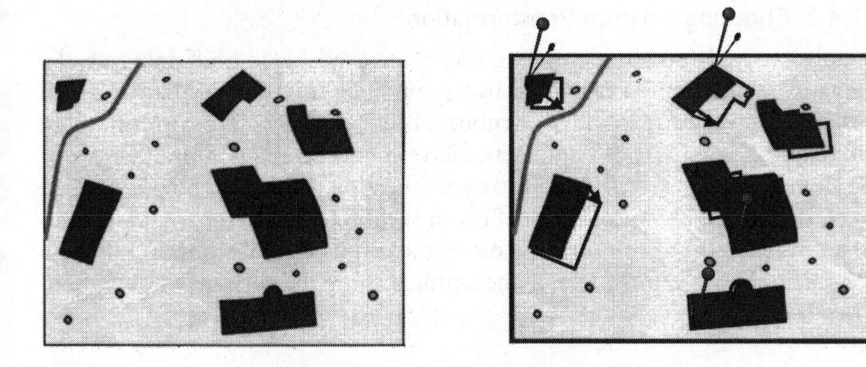

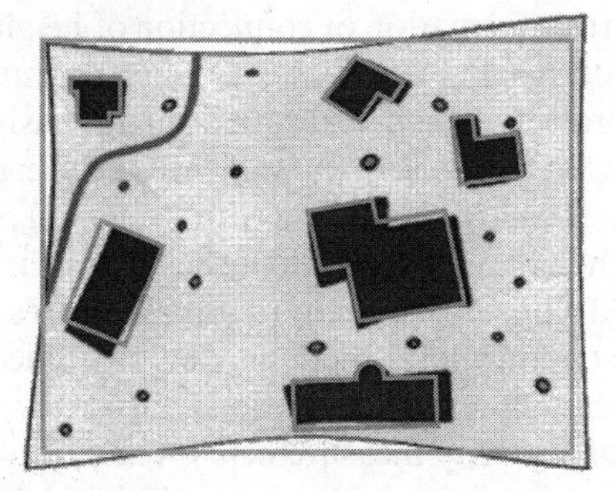

FIGURE 11.11 Rubber sheeting.

two maps will fit better. This can best be understood by considering the areas as several sub-areas, each with a different similarity transformation. This is called rubber sheeting. There are a lot of (mathematical) possibilities to rubber sheet a map or data set. We will not go into details here, but most GIS systems contain this tool for combining data sets (see Figure 11.11).

Combining data sets in this way is very flexible and the data can always fit completely because as many degrees of freedom as are required are allowed. This transformation is purely empirical.

Edge matching

Edge matching is applied when there are two maps that contain common data points, but generally no significant overlap. Here a kind of rubber sheeting in one direction is applied, but how the translations in this along-edge direction affect other points in both maps is a matter of preference. We can choose to move the entire vertical line along with the edge points, or to have a small strip in both maps along the bounding edges affected (by inverse distance weights for example), while the remainder of the maps are left unchanged.

11.4.3 Choosing a datum transformation

If only two or a few corresponding points are available in both data sets, the only way to determine a relation is by the 2D similarity transformation, which has four degrees of freedom. For rubber sheeting, many more corresponding points are needed in both data sets, since in general many more degrees of freedom than four are used to create a connection. Edge matching is applied when there is an obvious stretch effect in the direction of the bounding map edges or when the absolute precision of the results is not so important but a consistent map is wanted (e.g. if the features along the edge may not contain irregularities).

11.4.4 Heights

A special topic is the transformation or connection of heights. Within the GDI context, problems with heights usually occur when crossing national boundaries. Examples are projects, which deal with drainage basins or the regimes of large river systems. From the point of view of the GDI, height referencing consistency becomes important when dealing with the global perspective.

Different kinds of heights can be distinguished: physical heights (e.g. orthometric or normal) and ellipsoidal heights (with respect to a reference ellipsoid, e.g. WGS84). At the country level there may be differences between the two that cannot be neglected in some applications. Physical heights are obtained from spirit levelling and gravity measurements. The surfaces of equal heights roughly follow the gravity field disturbances. Physical heights indicate the direction water will flow and are based on heights above Mean Sea Level (MSL). Apart from effects like wind, temperature, and salinity variations, the mean sea surface is a surface of equal gravity potential, and hence of equal physical height (namely zero). This surface is not as smooth as the surface of a reference ellipsoid due to the gravity variations that are caused by mass density variations in the upper and lower layers of the Earth's surface. Ellipsoidal heights are important because they are used in GPS.

Combining or connecting height data

Five cases of combining or connecting height data are considered below.

1. Both data are physical heights in the same datum

In principle the data sets can simply be combined. Height networks are based on a few benchmarks for which the height above MSL is measured. This is done using tide gauges, of which each country has its own set. If two networks are combined that use different benchmarks, an offset for one data set may be needed. For heights at sea level, the difference between orthometric heights and normal heights are in the order of millimetres and may be neglected for most purposes. At heights of up to a few thousands metres above sea level, the differences between orthometric and normal heights may be a few decimetres up to more than 1 m. In most GDI cases this can also be neglected.

2. Both data are physical heights, but in different datums

These different datums are the different tide gauges that are used to determine the zero level of a height network. If both heights are physical heights, an offset is enough to connect the two data sets.

3. Both data are ellipsoidal heights in the same datum

This is a simple case. The heights can simply be combined into one data set. If one or both datum points of the different data sets contain an error, the heights of the corresponding points may not be equal. Here again, an offset will solve the problem, although it may be difficult to determine which of the two data sets is correct, and which had the error in the datum point.

4. Both data are ellipsoidal heights, but in different datums

Here an ellipsoidal co-ordinate transformation must be applied. A simple offset is usually not good enough. Since the different positions of origins and the different shapes of the reference ellipsoids are not identical, the heights not only differ by a constant but also by a tilt. After the transformation to the same reference ellipsoid the data can simply be combined.

5. One data set contains physical heights and the other ellipsoidal heights

This is the most difficult case. Since the heights are of completely different natures, even height profiles in overlapping areas do not look alike. A geoid model gives the geoid height for every point, as the distance between the geoid and reference ellipsoid. Geoid heights can differ by tens of metres with respect to global ellipsoids (WGS84). With respect to a national ellipsoid (that optimally fits the geoid), the geoid heights are within a few metres of the physical heights.

Thus if physical heights are to be combined with ellipsoid heights, a geoid model is needed. Nowadays, geoid models are available for more and more areas in the world. Since the precision of the geoid is highly dependent on the availability of dense and accurate gravity measurements, the quality of local or regional geoid models may differ greatly per area. In countries with dense gravity networks, a precision of less than centimetre-level can be obtained for the geoid over distances up to 100 km. For countries with poor gravity coverage, only global models are available which lack the local gravity variations. Here a precision of several decimetres can be obtained for a geoid over 100 km.

11.5 Conclusion

In Europe, the USA, and Japan, three initiatives are underway to improve the applicability of GLONASS and GPS by developing GNSS, the Geostationary Navigation Satellite Service. Using geostationary satellites to transmit corrections and additional information to improve the quality of positioning for both GPS and GLONASS, a service will be provided that will permit metre-range positioning on a global scale. The European initiative is called the European Geostationary Navigation Overlay Service (EGNOS), the US version is the Wide Area Augmentation Service (WAAS), and the Japanese are building the Multi-(functional transport satellite) Satellite-based Augmentation Service (MSAS). These three systems should be inter-operable.

In the meantime, the USA is building the next, much improved generation of GPS satellites, and Europe is planning a GNSS-2, which may be a European satellite navigation system, or an upgrade of GLONASS, or an overlay on GPS. Development continues apace and much can be expected from the future.

Whether we consider a GDI for a particular application domain, a municipal, or a national one, we do not expect the end users, i.e. GIS applications and 'mundane' applications, to have to worry about these transformations. The solutions should be transparent, having been taken care of by the data service resident in the GDI.

In designing and building a GDI, however, the developers need to be aware of the potential georeferencing problems. Mistakes made at this stage will inevitably lead to systematic errors in the integration of data sets, which in turn may affect the quality of the decision processes the GDI is designed to support. In many cases, the errors and discrepancies caused by the different datums may be neglected; in others, they may be of crucial importance. In all cases, it is therefore necessary to assess the situation carefully in the early stages of the GDI development in a particular application domain, and take specialist advice.

Bibliography

'Geodesy for the Layman', at *http://www.nima.mil/geospatial/geospatial.html*

A list of datums can be found at: *http://www.nima.mil/geospatial/products/GandG/historic/hdatums.html*

A comprehensive set of map projections and associated transformation software can be found at *http://kartoserver.frw.ruu.nl/html/staff/oddens/mapsat14.htm*

Transformation software (MADTRAN) is available from several sites on the Internet, e.g. from *http://www.ntis.gov/fcpc/cpn5198.htm* or from *http://www.nima.mil/geospatial/products/GandG/madtran/madtran.html*

GEORGIADOU, Y. (1999). 'The impact of GPS technology on National Mapping Organisations', Invited Paper 7.3, Proceedings of Cambridge Conference, 19–23 July 1999, Ordinance Survey, UK, pp. 12.

HOOIJBERG, M. (1997). *Practical Geodesy using Computers*. Heidelberg, Germany: Springer-Verlag.

SEEBER, G. (1993). *Satellite Geodesy*. Hawthorne, New York: Walter de Gruijter.

12

Photogrammetry and remote sensing in support of GDI

Gottfried Konecny

12.1 Introduction

As the largest cost in a GDI is arguably that of building and maintaining the databases, an overview of two major geospatial data acquisition technologies is presented here. The technologies and their platforms are introduced in Sections 12.2 to 12.4, and the data acquisition techniques for vector and raster data are described in Sections 12.5 and 12.6. The acquisition of digital terrain models is discussed in Section 12.7, and thematic data acquisition from different platforms in Section 12.8. Parcel-based data systems are of particular significance to a large variety of civic applications, both in urban and in rural areas. The parcel database and the database of administrative areas derived from it form an integral part of the foundation data in the GDI. The role of photogrammetry and remote sensing in building and maintaining these parcel-based systems is presented in Section 12.9. Data acquisition decisions are usually a balance between cost, required accuracy, and timeliness. Hence, the cost–performance of various photogrammetric and remote sensing products are compared in Section 12.10.

12.2 Remote sensing and photogrammetry

Remote sensing is a technique for obtaining information about distant objects without direct contact. The information is obtained from the object's reflections of force fields by sensors that react to electromagnetic radiation spatially separating the reflections. The directional separation of the reflections permits us to accumulate the signals in the form of images. In the visual spectrum, these have been generated for 150 years by lens objectives on to an image plane and fixed on photographic emulsions. Today, they can be accumulated by digitally separated signals as image pixels. The images can be used directly for image

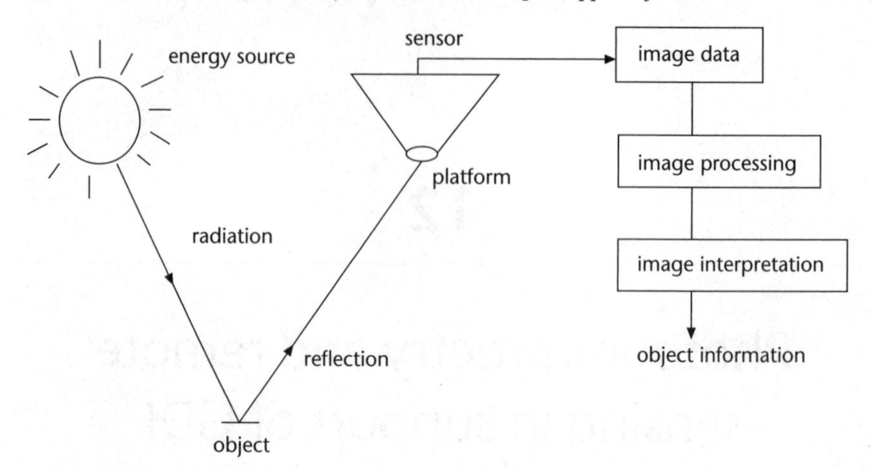

FIGURE 12.1 Principles of remote sensing.

interpretation so that objects can be detected, recognized, and classified (Figure 12.1).

In photogrammetry two different images are taken from different viewing stations to make a 3D geometric model of the image objects. The process of reconstructing these objects geometrically, based on accurate measurements made in the images, is called photogrammetry. This is therefore a special application of remote sensing (Figure 12.2).

12.3 Geospatial data infrastructure in relation to photogrammetry and remote sensing

Geospatial information systems offer the possibility of generating a digital geometric and semantic model of an object. Objects cover the Earth's surface, and since knowledge of the object distribution on the Earth's surface is essential for planning, managing, and monitoring the natural and human environment, spatial information systems are of great scientific, technical, and economic importance.

A modern geospatial information system is based on the alternative technologies available to geoinformatics (or geomatics) which are described in this book. It employs hardware and software for input, manipulation, storage, and output of digital data for user-specified spatial information products. The data in raster, vector, and alphanumeric form constitute the most costly, and thus most important, component of the system (Figure 12.3). The system must also contain an administrative component stressing the purpose and the application (Figure 12.4). The three major components of a system are the hardware and software, data, and administration, consuming approximately 10 per cent, 80 per cent, and 10 per cent of the total effort (see, for example, Dangermond, 1997).

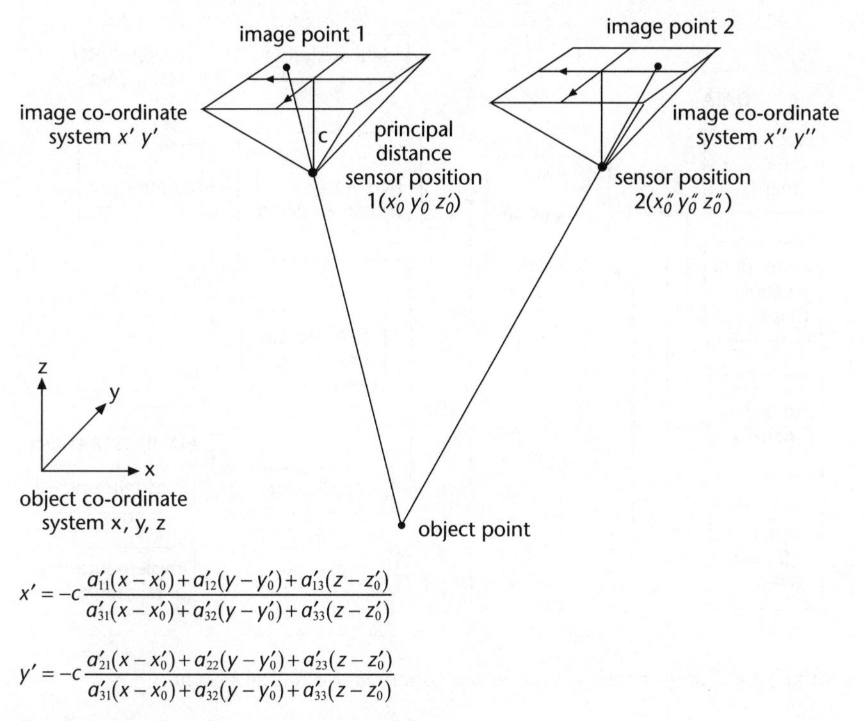

These equations are called collinearity equations.
$a_{11}, a_{12} \ldots a_{33}$ are the coefficients of a three-dimensional orthogonal rotation matrix representing the direction cosines of the space angles between the relative axes.

FIGURE 12.2 Principles of photogrammetry.

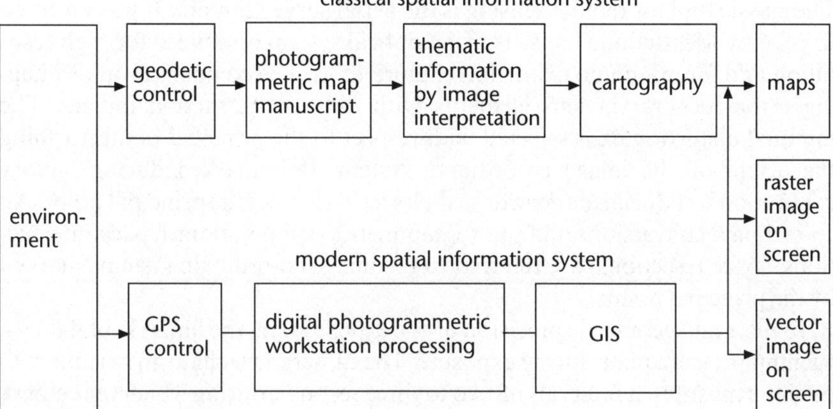

FIGURE 12.3 Classical and modern geospatial information system.

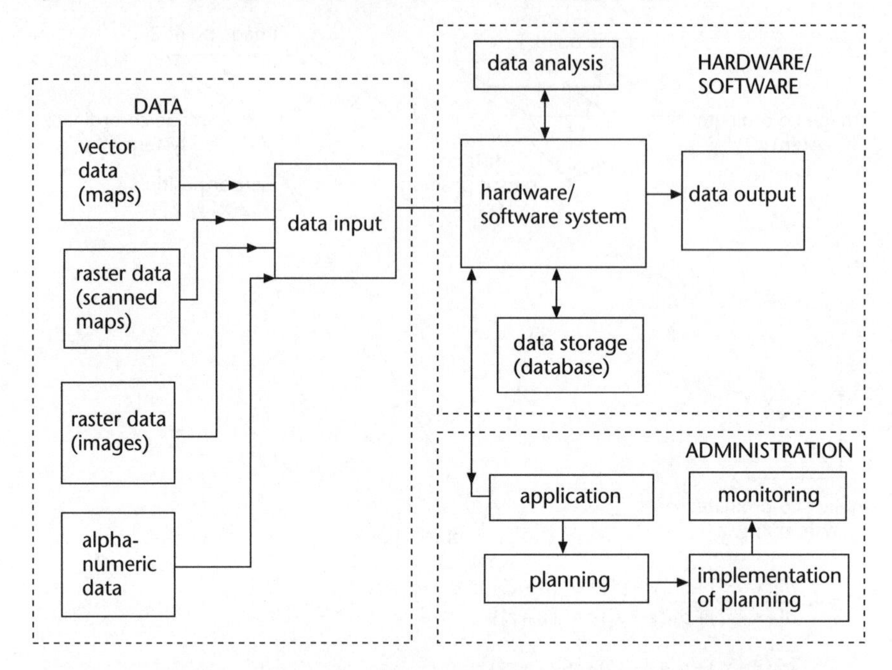

FIGURE 12.4 Components of a modern geospatial information system.

These systems are extensively applied to routinely produce data, which can be made accessible through the GDI architectures presented in Chapter 9. To control the contribution to the overall cost, the data to be surveyed need to be carefully defined using the techniques described in Chapter 10, for example.

12.4 Sensors and platforms

The classic tool for remote sensing is the aerial survey camera. It was invented in 1916 by Messter and consists of an optical system optimized for high resolution and for minimum distortion (Figure 12.5). The compensation of imaging errors occurs via lens elements with different refraction indices. The minimal distortion $d\tau$ is selected with respect to the principal point, forming the origin of the image co-ordinate system. It is marked during factory calibration by fiducial marks whose intersection defines the principal point. All co-ordinate conversions and photogrammetric computations (space intersections, space resections) are referred to the image co-ordinate systems marked by the principal points.

To maintain geometric precision of the exposed film, the film is kept flat by a vacuum pressure plate during exposure. The camera's mechanism permits successive exposures at intervals of two to three seconds, during which the camera must transport the film, apply the vacuum to the pressure plate, and perform the exposure lasting from 1/100th to 1/1000th of a second. The successive expo-

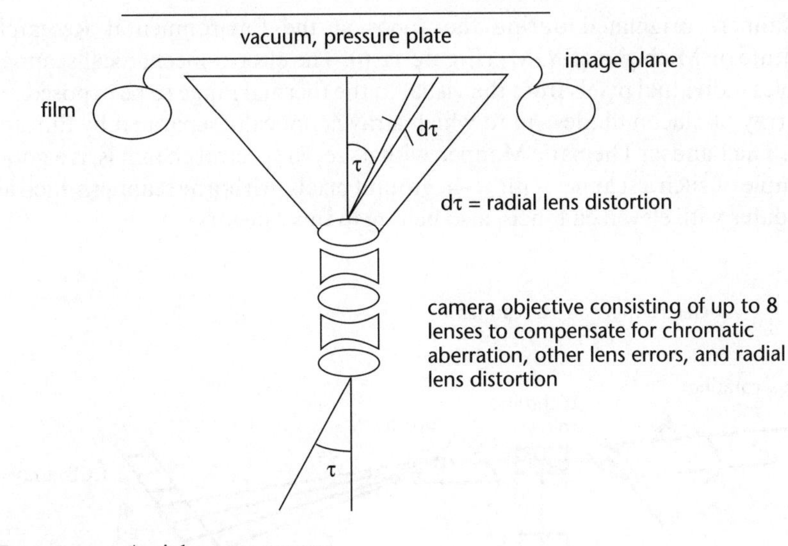

FIGURE 12.5 Aerial survey camera.

sures must overlap by at least 60 per cent to permit stereo coverage (each overlap is called a stereomodel). This allows two models to be connected longitudinally by transfer points and the strips to be assembled for stereo restitution. An adjacent strip usually overlaps laterally by 20–30 per cent to cover a photogrammetric block in a survey area. Recent improvements to aerial survey flights entail determination of the exposure station co-ordinates (x'_o, y'_o, z'_o) via airborne differential GPS. Furthermore, image motion compensation allows the shift of the image plane by an amount commensurate with the aircraft speed, so that long exposures (1/100th second) are possible on high-resolution, low-sensitive film with 60 lp/mm. A prerequisite is that the camera is carried in a gyro-stabilized mounting, so that changes in the aircraft's attitude do not affect the long exposure. GPS navigation allows us to restrict the overlap conditions to a very small limit (e.g. 10–15%). It also allows control of the location of adjacent models from strip to strip (see, for example, Hartfield, 1997).

Aerial photography is restricted by film sensitivity to the visual and near-infrared electromagnetic spectrum. Black and white aerial photos usually contain images in the combined visual (400–700 nm) wavelength spectrum. Infrared images are exposed in the 500–900 nm wavelength spectrum. Colour films are three-layer films sensitive to blue (400–500 nm), green (500–600 nm), and red (600–700 nm). False-colour infrared film is sensitive to green (500–600 nm), red (600–700 nm), and infrared (800–900 nm), which becomes blue, green, and red on development. Infrared film is widely used in vegetation studies and camouflage detection because of its sensitivity to infrared radiation, which reveals the status of healthy vegetation.

Scanners originated during the 1960s at the Environmental Research Institute of Michigan (ERIM) (Figure 12.6). The electro-mechanical scanner enables individual pixels from the visual to the thermal range to be exposed on an array of silicon diodes, on to which a ray is optically separated by diffraction. The Landsat Thematic Mapper, with its seven spectral channels, is a good example of such a scanner with 30-m ground pixels. Airborne scanners, such as Daedalus with eleven channels, also belong to this category.

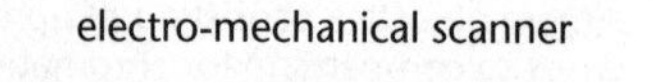

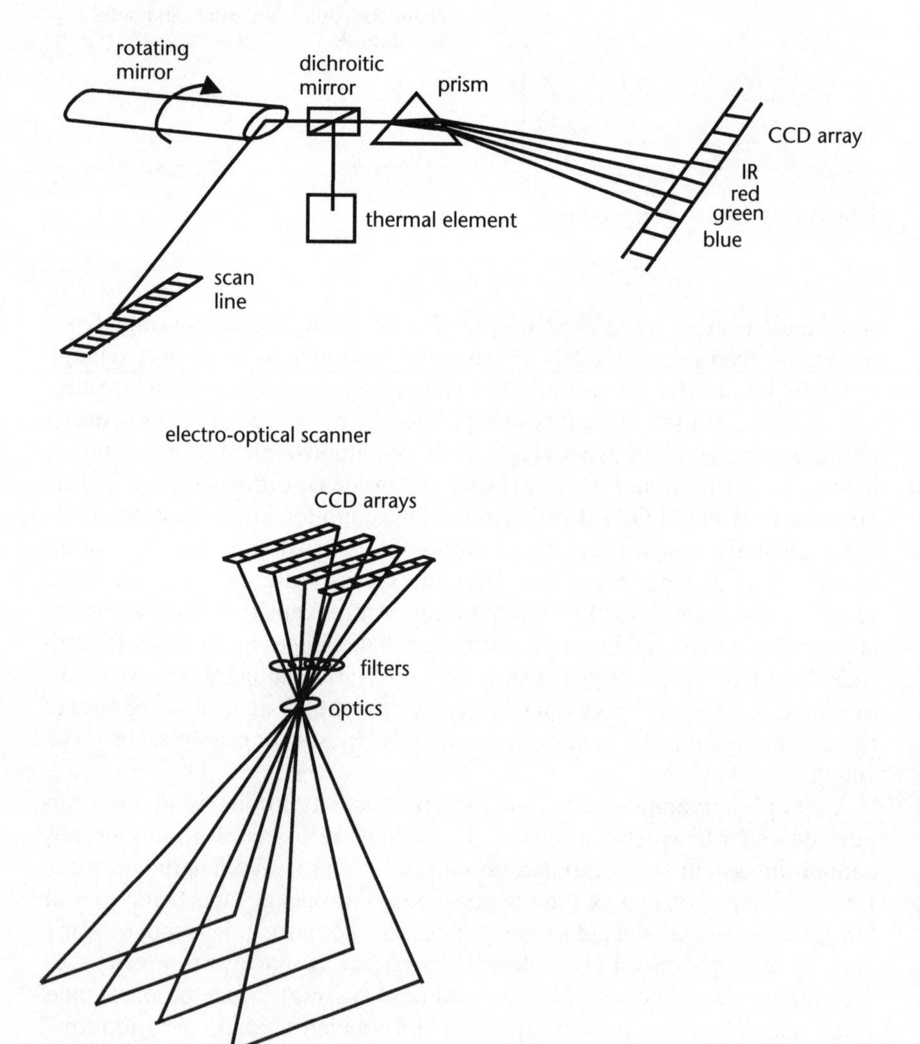

FIGURE 12.6 Electro-mechanical and electro-optical scanners.

Electro-optical scanners used in the *Système probatoire de l'observation de la terre* (SPOT), for example, enable us to record different images in specific spectral channels by optical projection on to a linear array perpendicular to the flight path. If several arrays are filtered, these can be combined to colour or false-colour digital images at 20-m ground pixels. The multi-spectral images can be merged by image processing with 10-m ground pixels using a single, double-resolution, panchromatic channel of the same area. Charged coupled device (CCD) arrays are sensitive to the visible and near-infrared regions of the electromagnetic spectrum, but not to the thermal infrared bands.

More recently, radar sensors have been developed primarily because of their promise of good quality images under conditions in which optical sensors are not effective, for example through cloud cover.

As the solar (passive) radiation is generally too weak in the microwave range, microwave remote sensing is carried out by radar systems generating their own energy. The energy has a wavelength of a few centimetres (*X*-band = 3 cm). It is transmitted from an antenna in short, successive pulses. The antenna is switched alternately from transmission to reception, permitting reception of backscattering pulses from the terrain points. The pulse duration determines the resolution in range direction, while the time difference between transmission of the pulse and reception of the backscattered signal, with a known velocity of electromagnetic wave propagation, yields the double distance between the antenna and terrain point.

Brute force radar systems (SLAR) only use the time information. The greater the length of the antenna, the narrower the antenna beam becomes, and the more it approximates a plane. Since long antennas are impractical because of energy and space constraints in aircraft and satellites, coherent radar systems (synthetic aperture radar (SAR)) are preferred. They constitute the current standard.

Aircraft are still a significant platform for sensors, complementary to satellites. An aeroplane makes possible a systematic coverage of the terrain with regular overlap conditions and restitution of the images into 3D terrain information by means of photogrammetry. Flight plans can be executed accurately with modern navigation systems (GPS). The aeroplane carries the survey camera in a mounting and the camera lens points towards the terrain at a near-vertical direction through a camera window. The flying altitude depends on the type of aeroplane used: simple, motor-driven aeroplanes can reach an altitude of up to 5 km with oxygen support for the crew, while pressure-cabin aircraft with jet engines can fly at a height of 12 km, and supersonic aircraft at 20 km or more. The aeroplane velocity sets the lower limits of flights, since it must be possible to meet the repeat cycle of successive exposures necessary to achieve the longitudinal overlaps. Thus, image scales from about 1 : 3,000 to 1 : 130,000 can be achieved with aerial surveys.

Satellite platforms have existed since the launch of Sputnik by the former USSR in 1957. A few years later, Russian and US satellites were carrying imaging sensors. Nowadays, satellite-imaging systems are operated by many national and international space agencies in the USA, Japan, Russia, India, France, Germany, China, and Canada.

The orbit of a satellite depends on its height above the terrain. Geostationary orbits operate in an equatorial plane at a height of 37,000 km. They rotate once per day in synchrony with the Earth and appear to stay in a fixed position relative to the Earth. Geostationary orbits are ideal for communication satellites and for weather satellites (Meteosat, GOES, GMS, Insat). Earth observation satellites (Landsat, SPOT) operate at lower heights of 400–800 km and these rotate around the Earth in about 90 minutes. A sun-synchronous orbit is generally chosen, so that each equator crossing always takes place at the same time of the day (e.g. at 10.30 a.m.).

Depending on the swath width of the sensor, repetition rates for the same area can be achieved from 12 hours (NOAA) to 17 days (Landsat) to 28 days (SPOT). Cloud cover restrictions for optical sensors often do not permit cloud-free images to be obtained (for NOAA satellites, a global coverage can be obtained in 15 days). In temperate zones, optical remote sensing satellites mostly do not permit images of the same area to be repeated in less than three months. Satellites with radar systems (ERS, Radarsat, JERS) do not have such restrictions. The prerequisite is that the imaged pixels can be transmitted at a sufficiently high rate (at least 100 Mb/s). This is best achieved by locating receiving stations around the world, unless temporary on-board storage is provided.

Orbiting laboratories, such as Space Shuttle or MIR, have orbital heights between 200 km and 400 km above the Earth. Their images must take variations of the sun's angle into account. The potential for achieving a global coverage is summarized in Figure 12.7. A summary of recent and upcoming high-resolution missions is shown in Table 12.1. Figure 12.8 shows a high-resolution image produced by a Russian KVR 1000 camera at 2-m pixels (DD5).

12.5 Vector data acquisition by map digitizing and photogrammetry

Digital vector data can be obtained most easily and cheaply by vector digitization of existing maps. Manual digitizers vectorize successive positions of the cursor, at regular distance intervals, while the cursor is moved over the tablet on to which the map is projected. Digitization of maps can be achieved much faster when the map is scanned in a raster scanner. The lines are converted into binary pixels (black or white) using thresholds. A subsequent automatic or interactive raster-to-vector conversion program can derive vectors for the black pixel sequences of the scanned map. More expensive map scanners can do this by colour separation using filters. Map digitization is only useful if the maps are reasonably up to date.

The United Nations Secretariat conducts official surveys of the world's map inventory and updates it regularly. The latest survey, conducted in 1993, on the

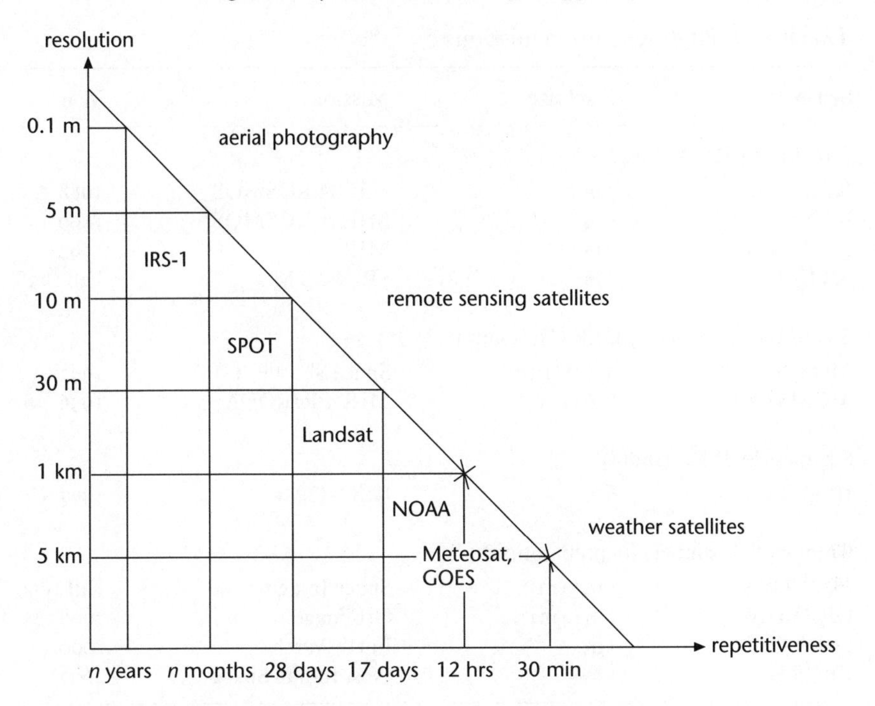

FIGURE 12.7 Resolution and repetitiveness of remote sensing missions.

status of mapping in the world is shown in Table 12.2. Table 12.3 shows the annual update rate of these maps. One-third of the Earth's land areas are mapped at a scale of 1 : 25,000, approximately 65 per cent at 1 : 50,000, and 95 per cent at 1 : 200,000. The average age of the 1 : 25,000 maps is 20 years, the 1 : 50,000 maps are nearly 50 years old, and the 1 : 200,000 maps are 30 years old. There are no figures available for maps at larger scales (e.g. for urban areas).

Analogue maps in the scales of 1 : 50,000 have, in almost all instances, been produced by photogrammetry. They have been the source for the map scales at 1 : 100,000 and smaller. In nearly all cases these maps have been digitized and thus formed the basis for geospatial information systems of the world at small and medium scales. A fully digital, global coverage is only available in the form of the 'Digital Chart of the World' at the scale of 1 : 1,000,000, published by NIMA, USA. A digital map of the world at the scale 1 : 250,000 is now in preparation. Germany is currently the only country to have full 1 : 25,000 digital coverage in the form of the ATKIS data model. Most other countries are still in the process of data conversion.

The production of maps was originally carried out by plane tabling, and other ground survey methods. In Germany, for example, this process took nearly 100 years and was completed around 1930. Plane tabling was cumbersome and costly, and this led in the late 1930s to photogrammetric stereoplotting becoming the standard mapping method throughout the world. The bulk

TABLE 12.1 High-resolution missions

Sensor	Pixel size	Mission	Year
Cameras by RKA (Russia)			
KFA 1000	5 m	MIR & KOSMOS	1988
KFA 3000	5 m	MIR & KOSMOS	1990
KVR 1000	2 m	MIR	1992
KVR 1000	2 m	SPIN-2 TM	Feb 1998
Stereo Line Scanner by DLR (Germany)			
MOMS 02-D	4.5 m (15 m)	Space Shuttle	1993
MOMS 02-P	6 m (18 m)	MIR-PRIRODA	1996–98
Scanners by ISRO (India)			
IRS 1C	6 m	IRS 1-C	1997
Commercial scanners (in preparation)			
IKONOS 1	1 m (4 m)	Space Imaging	Fall 1999
ORBVIEW 3	1 m (4 m)	Orbimage	2000
Quick Bird	1 m	EarthWatch	2000
EROS B	1 m	West Indian Space	2000

of the work was carried out after the Second World War with analogue stereoplotters. With these instruments the spatial 3D orientation of two overlapping aerial photographs can be accomplished manually. The resulting model can be viewed by means of anaglyphs or stereoscopes.

Helava invented analytical plotters in 1957, which were capable of recording model and image co-ordinates with limits of 3–5 μm with reference to the rest of the image, while a point may be visually recognized with a precision of ±5 μm. Analytical plotters, like analogue plotters, are capable of recording vector information. They are one of the means of generating vector information in a stereomodel.

On-screen monoplotting is a technique for obtaining vector data from a geometrically correct photo (orthophoto) on the computer screen. Thus, digital orthophotos may be used for updates of existing digital vector databases, which are superimposed on the digital ortho-image.

Updating of digital databases via workstations is also possible in stereo, using digital stereo workstations. A digital stereo workstation generates corresponding images on the display screen at sequential intervals of 50 msec each. These are displayed as polarized images that can be viewed with polarization filters or 'electronic shutters', achieved by the 'Crystal Eyes' principle in stereo.[1]

[1] A pair of stereo images is viewed using spectacles permitting switching of the 50-msec images on the screen alternatively to the left and right eyes, in synchronization with the screen. If the switching of the two images is faster than 15 Hz the human operator sees the images in stereo.

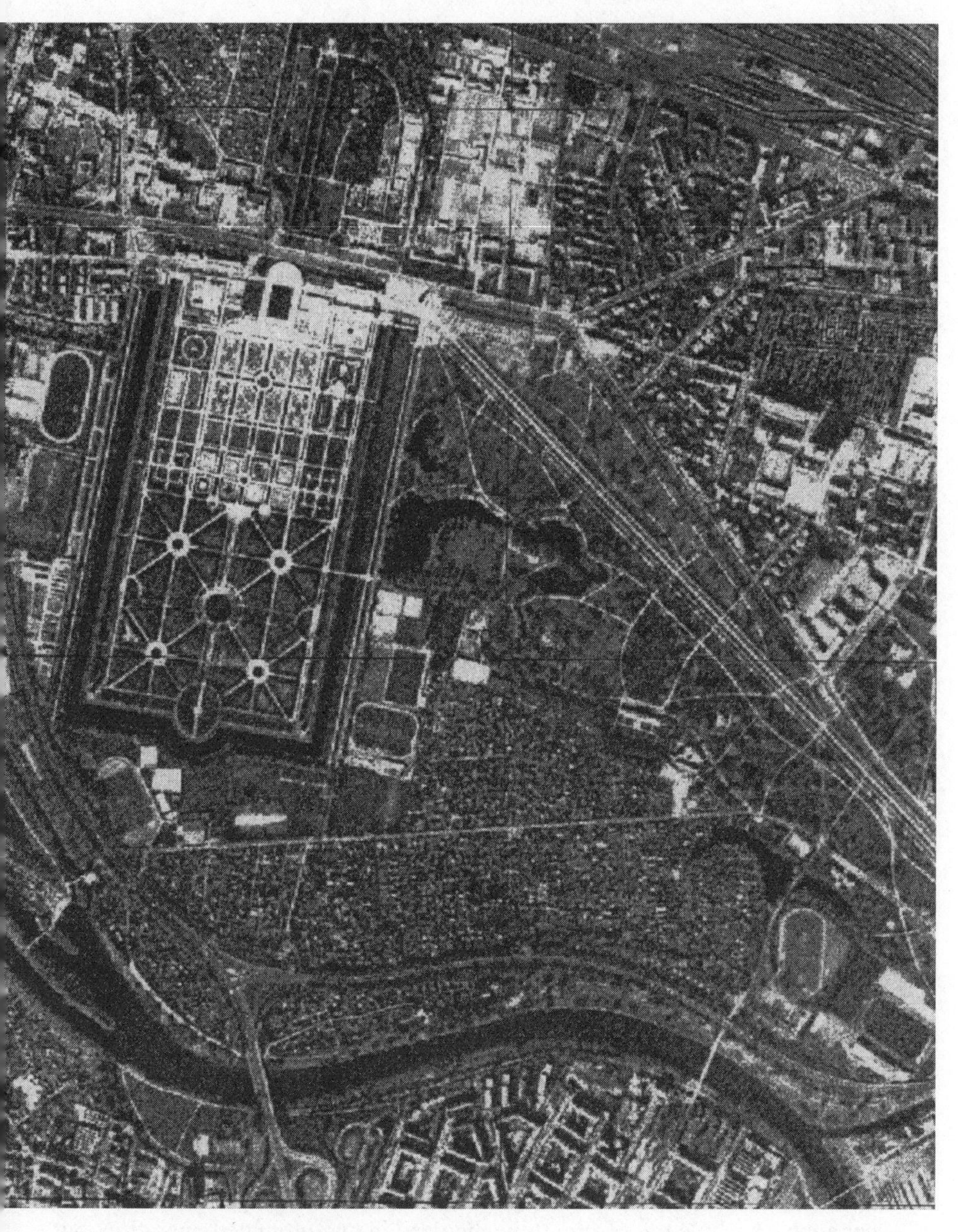

FIGURE 12.8 KVR 1000 (DD5) image of Hanover.

TABLE 12.2 Status of mapping in the world (1993), in percentage of area covered

Region	1:25,000	1:50,000	1:100,000	1:200,000
	%	%	%	%
Africa	2.9	41.1	21.7	89.1
Asia	15.2	84	66.4	100
Australia and Oceania	18.3	24.3	54.4	100
Europe	86.9	96.2	87.5	90.9
Former USSR	100	100	100	100
North America	45.1	77.7	37.3	99.2
South America	7	33	57.9	84.4
World	33.5	65.6	55.7	95.1

TABLE 12.3 Annual update rate of maps

Region	1:25,000	1:50,000	1:100,000	1:200,000
	%	%	%	%
Africa	2	2.5	4.1	1.2
Asia	4	0.8	0	2.2
Australia and Oceania	2.2	1.8	0.1	0.5
Europe	7.5	6.4	7	8.3
North America	4.8	3.1	0	6.3
South America	0	0.8	0	0.4
World	5	2.1	0.7	3.3

A stereo workstation is programmed using software corresponding to the analytical plotter principles (see, for example, Gruen, 1997, and Petrie, 1997).

12.6 Raster data acquisition by photogrammetry

Aerial photographic negatives or diapositives can be raster scanned by special scanning devices such as the SCAI scanner made by Zeiss/Intergraph, the Leica scanner, or the Wehrli scanner Rastermaster RS1. The scan raster can be as small as 7 μm (or 7.5 μm, or 12 μm). The SCAI scanner and the Leica scanner enable a whole roll of film to be digitized in an automated operation, but only the RS1 can digitize from cut film. After the digital conversion of grey level information in the film to picture elements (pixels), the raster image can be subjected to all the facilities of digital image processing such as rectification to produce an orthophoto image. Digital image processing allows us to geometrically distort or rectify the image (see, for example, Lee and Thorpe, 1997).

The collinearity equations of Figure 12.2 are functions valid for aerial photographs. A pixel x_i, y_i is chosen (together with an appropriate height z_i from a digital terrain model) to calculate the corresponding image point $x'_i\ y'_i$ whose grey level d_i is transferred to the output pixel (orthophoto pixel). There are a variety of computational techniques which improve these transformations. Depending on the distortions of the imagery (e.g. Landsat), the collinearity equations may be replaced by a simpler function, such as a second-degree polynomial:

$$x'_i = a_o + a_1 x_i + a_2 y_i + a_3 xy + a_4 x^2$$
$$y'_i = b_o + b_1 x_i + b_2 y_i + b_3 xy + b_3 y^2$$

with the coefficients to be calculated from control points and their identified image points. Each sensor type (SPOT, Radar, etc.) accordingly has its own rectification function. The resultant ortho-image corresponds to the required geometry of a GIS referenced to a specific datum on a reference ellipsoid (such as WGS84 on ITRF) and a chosen map projection (e.g. UTM or 3° Transverse Mercator). As elevations are usually based on the geoid as a vertical reference, a conversion for standard orthometric heights to ellipsoidal heights on the basis of our knowledge of geoidal undulations is required (see Chapter 11).

12.7 Digital elevation models by photogrammetry

A digital elevation model (DEM) usually consists of a rectangular grid based on a reference system and a specific map projection (e.g. UTM or 3° Transverse Mercator). For the grid points the elevation is recorded as an attribute.

When photogrammetric map manuscripts or topographic maps are available, the elevation distribution of the terrain is usually depicted by contour lines. In analogue stereoplotters, the operator follows lines of equal elevation, the contours. Recording devices may register them directly in digital vector form. The digitized contours may be used to interpolate elevations at the grid points. A variety of interpolation models has been used for this purpose, resulting in a variety of DEM interpolation programs. This method introduces distortions in the terrain model because the linear interpolation between contours ignores bumps and breaks in the terrain that fall within the contour interval. More sophisticated models allow for the introduction of breaklines in a discontinuous terrain as required in highway construction or terrain modelling in mountainous areas. No smoothing takes place over such breaklines, thus providing a more truthful rendition of the terrain.

In analytical plotters or in digital workstations, the sequence of regular grid points may also be precalculated and set in a stereomodel for operator measurement of the height by stereo observation. Such an interactive method has the advantage of being able to eliminate ground points which are hidden by houses or trees.

Since operator measurement of contours or of regular grids is a cumbersome and time-consuming operation, attempts have been made to determine heights automatically by digital image correlation (also called image matching). The principle of image matching is based on calculation of the correlation coefficient for two matrices of pixels of a reference image and that of the corresponding stereo pair image. The promise of this technique can be illustrated as follows.

Standard topographic maps usually show only ground elevations, but not building heights. New requirements by the telecommunications industry have raised the need for the creation of 3D city models by rapid means. While semantic modelling techniques to extract features such as houses, roads, and other objects automatically are still at the research stage and not yet available for operational use, some progress has been made in generating 3D city models by a combination of technologies. For example, a vector-based GIS may allow areas of existing buildings to be predefined in 2D (see Figure 12.9), and only those areas are then selected for 'brute force' digital image correlation methods to determine automatically the missing heights of the buildings in comparison to the ground heights contained in the GIS (see Figure 12.10). Oblique views of the schematic building distribution of a city thus become possible in an effective manner via ray tracing algorithms (see, for example, Ackermann, 1997, and Lang and Foerstner, 1996).

The computation of digital elevation models by SAR interferometry is now being tested and further explored. The technique follows the principle of computing an interferogram on the basis of phase difference information of two radar signals emanating from two different locations separated by a known distance, the base.

The main problem with radar interferometry from two different orbits (or satellites) is that the length and the orientation of the base are not well known. Thus, the antenna positions must be estimated from orbital data resulting in ambiguities in the results. At the same time, multiple reflections, radar shadows, and foreshortening disturb the interferometric fringes from which the computations are made.

Current radar systems, such as ERS-1, ERS-2, JERS 1, Radarsat, and a combination of ERS-1 and ERS-2 when flown in a one-day sequence as a tandem mission, lead to interference ambiguity problems, yielding accuracies of ±15 m for a 100–300 m base line in rather flat areas without vegetation, with discrepancies of up to 100 m in mountainous and forested areas. The longer the base line, the more noise that is expected in the interferometric signals, which ultimately limits the achievable accuracy. On the other hand, a short base length of less than 100 m diminishes the geometric accuracy; unfortunately, such a digital elevation model is not very useful.

When two interferograms are used as difference interferograms, however, in the form of 'differential interferometry', the result is useful for detecting surface elevation changes of a few millimetres or centimetres, e.g. in ice movements, plant growth, or volcanic activity. It is hoped that the Shuttle Topographic Radar Mission will bring improvements in the near future. Two separate radar systems are to be mounted on a beam that can be extended up to 60 m. The accurate position of the two radar systems will be monitored by

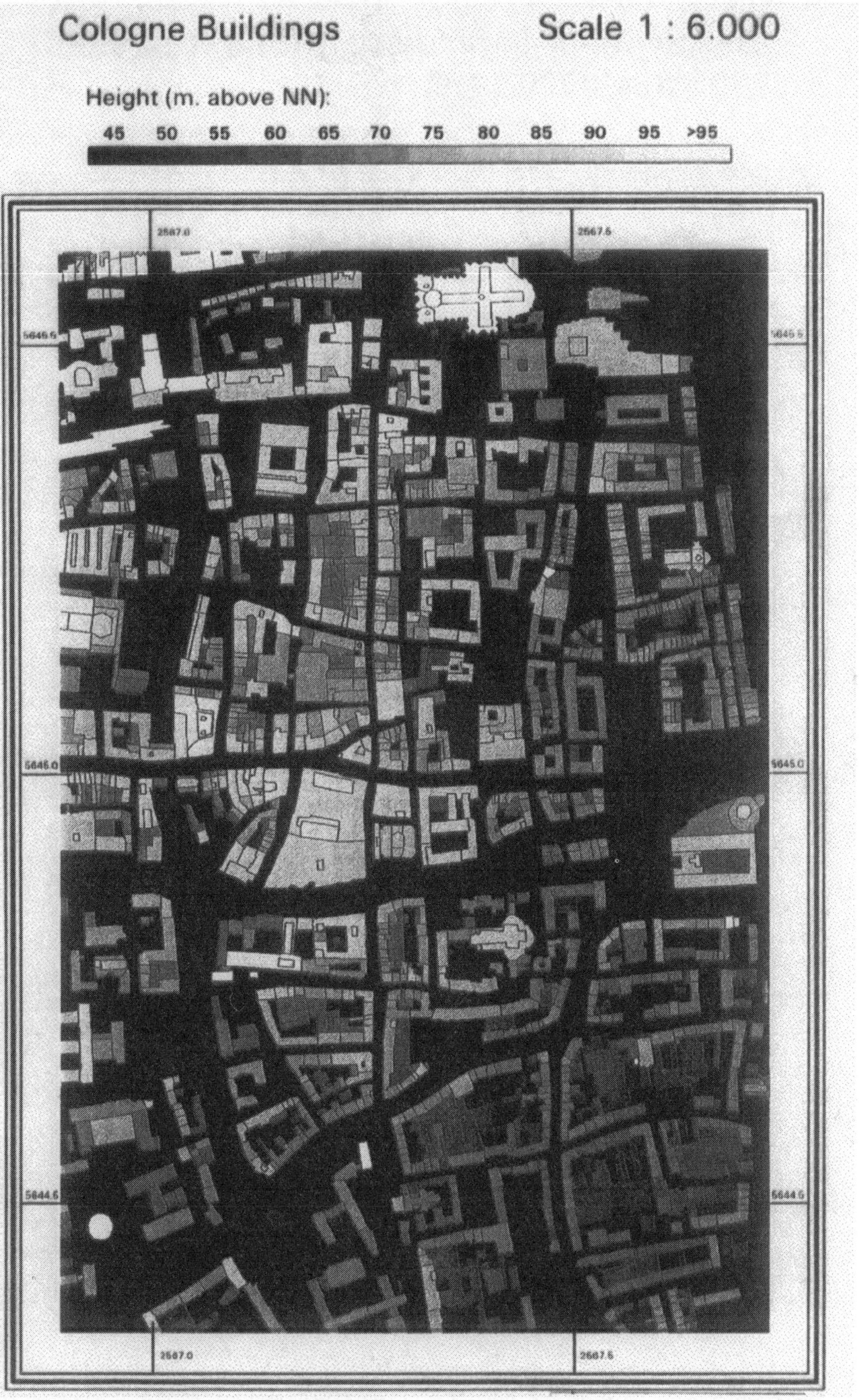

Cologne Buildings

Scale 1 : 6.000

Height (m. above NN):

45 50 55 60 65 70 75 80 85 90 95 >95

2567.0

2567.5

5645.5

5645.0

5644.5

FIGURE 12.9 GIS extracted buildings.

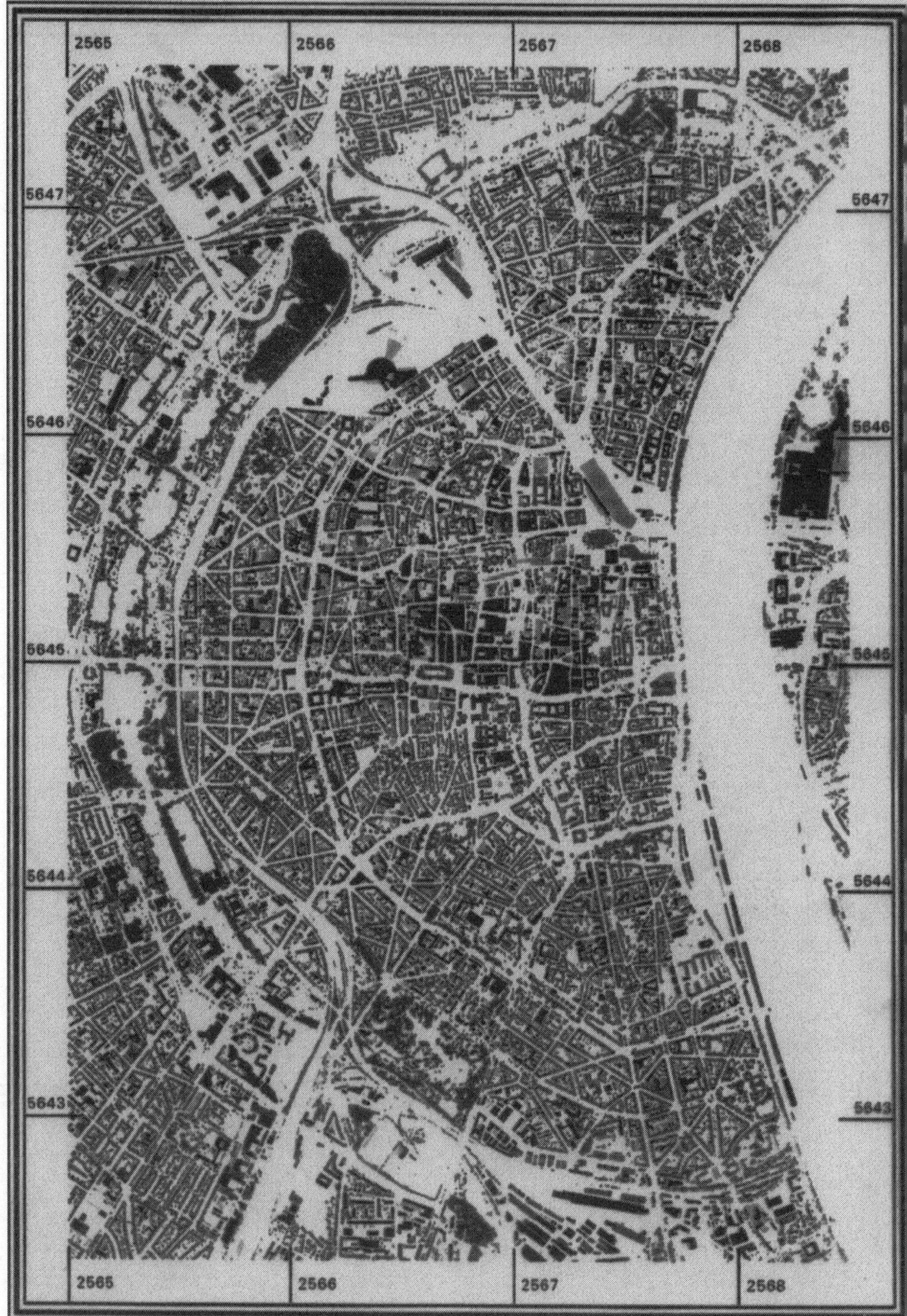

FIGURE 12.10 Image correlation results for buildings.

differential GPS, and the base length will be measured within a few millimetres by laser techniques. This may bring the expected accuracy of digital elevation models down to ±5 m. Within a 10-day shuttle mission, a significant part of the land mass could be covered by interferograms, offering an economic measurement capability for digital elevation models.

12.8 Thematic data acquisition by remote sensing

Thematic data are best collected for specific object classes (roads, forest areas, agricultural crop areas, settlement areas, etc.). Objects are defined by their boundaries with the possibility of forming subclasses with sub-boundaries, all arranged in a topologically structured manner. For objects for which boundaries have been graphically defined and identified by unique numbers, thematic information becomes an attribute in a relational database (e.g. in ESRI's Arc/Info or in Intergraph's MGE or MGA). Such object attributes can be derived by visual image interpretation from multi-spectral (colour/ false colour) and textural signatures. Context information may aid in the interpretation.

Multi-spectral and multi-temporal image classification may be automated by statistical analysis. After geometric co-registration of the images, the pixels of corresponding geometry in both images are recorded in a multi-dimensional space, as shown in Figure 12.11 for a 2D example. Images of water pixels will show low reflections in all spectral channels; they will all be accumulated in the lower left corner of the diagram. Coniferous forest will show low reflections in the green spectrum, but due to the water-filled cell structure of vegetation it will

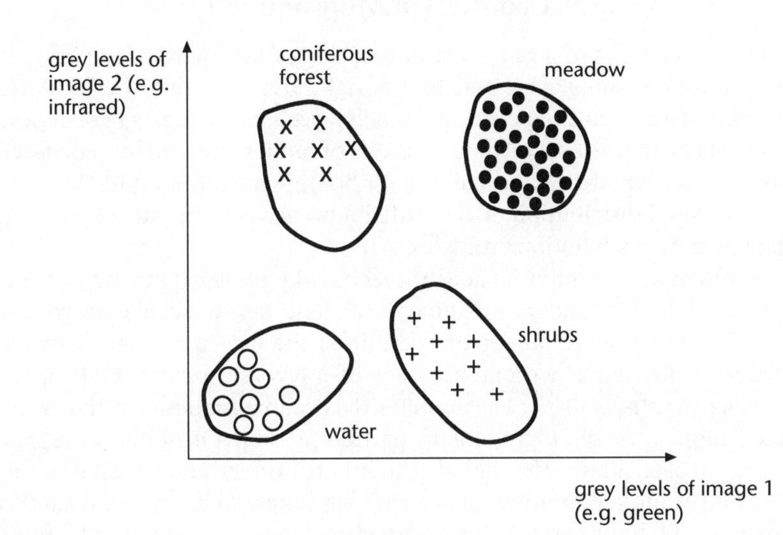

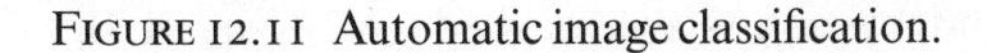
FIGURE 12.11 Automatic image classification.

show a high reflectance in the infrared, shown in the upper left corner of the diagram. Low infrared but high green levels will indicate dry shrubs.

In the sampled space, statistical boundaries between clusters must be found to distinguish the four classes shown in the example. This is done most easily by probability ellipses around the class clusters in the 2D example shown. Any pixel not falling into the cluster ellipses is shown as a fifth category of 'other classes'. Landsat™ enables hyper-ellipsoids to be generated in seven dimensions for the seven spectral channels available. The number of dimensions may be expanded to fourteen when two multi-temporal images are available at different time intervals.

The classification algorithm becomes very computation intensive if all fourteen channels are used. It is therefore better to reduce the number of channels by making a principal component analysis, in which only the most significant components of the hyper-ellipsoids are used for classification, thus reducing the information to low-correlation principal components. More independent object classes may thus be differentiated.

Under certain circumstances, knowledge-based systems can be applied and semantic networks may be utilized in automatic verification of object classes. They may be used to monitor changes in geometry and classification of objects, in particular. If an object has been defined by geometry and attributes in an existing GIS, then new images of the object may be geometrically superimposed on to the database. The spectral content and the texture properties of the imaged object may be verified using a hypothesis that the object still agrees with the original classification in content and geometry. This allows a method to be set up for automatic change detection (see Figure 12.12).

12.9 Cadastral photogrammetry

The cadastre consists of a geometric description of land parcels and attribute information concerning ownership or user rights. Depending on the local situation, other data, such as land use, soil quality, occupancy, mortgages, or other encumbrances may also be included. Parcels of land referenced to a consistent co-ordinate system are a natural entity for holding attribute values necessary when the spatial distribution of these attributes needs to be part of decision-making or civic consultation processes.

Early attempts to establish cadastral records by photogrammetry resulted in signalization of boundary monuments, if these were available. Large-scale (1:1,000 or 1:2,500) aerial surveys permitted the determination of the co-ordinates of the signals with an accuracy of a few centimetres. Only in rare cases, where fences, walls, or hedges mark the occupation limits of the parcels is direct mapping via photogrammetric plotters possible. Orthophoto mapping technology (particularly the digital orthophoto) offers new possibilities for reaching an economic solution for determining cadastral boundaries, but then adjudication of boundaries using photography or orthophotography in the field becomes necessary. The boundary points are marked on copies of the

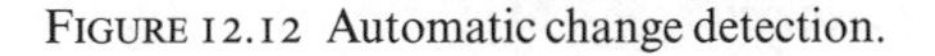

FIGURE 12.12 Automatic change detection.

photo or orthophoto. The neighbours of each parcel agree on this document by signing it on the location of the identified boundary point.

In some developing countries, where there may be large parcels, high-resolution satellite images offer possibilities for a quick and economic solution for establishing a 'graphical cadastre' for public or even for private purposes.

12.10 Cost considerations

Remote sensing and survey costs depend on map scale or resolution. Table 12.4 gives a summary of the costs per square kilometre for various German technical co-operation projects. The costs include acquisition of imagery, image processing and interpretation, as well as the required ground data supplementation and the preparation of a final product. Surveys at scales of 1 : 50,000 or smaller may utilize current satellite imagery at an image data cost of less than US$1/km^2. Larger scale survey costs depend on the photo scale and cost between US$4/km^2 (1 : 80,000) and US$40/km^2 (1 : 4,000) on average. In general, project costs are one to two orders of magnitude higher than imagery costs. Vector mapping costs, requiring supplementary ground data acquisition, are the most expensive item. By comparison, orthophoto mapping leads to more reasonable costs. The requirements of global GIS at the scale of 1 : 200,000 can generally be met for a cost of less than US$10/km^2 using satellite

TABLE 12.4 Summary of costs for technical co-operation projects

Field	Type	Scale	Imagery	Cost (US$/km²)
Agriculture	Phenol. change	1:1,000,000	NOAA	80
Bio-material	Biomass change	1:1,000,000	NOAA	80
Forestry	Forest mapping	1:250,000	MSS	6
Geology	Reconnaissance	1:100,000	TM	20
Forestry	Forest development	1:100,000	TM	20
Irrigation	Watershed mapping	1:100,000	TM	10
Regional planning	Planning study	1:100,000	TM	25
Land use	Land use mapping	1:100,000	TM	13
Bio-material	Biomass inventory	1:100,000	TM	20
Erosion	Vegetation cover	1:100,000	TM	20
Desertification	Change detection	1:100,000	TM	35
Food security	Cultivation inventory	1:100,000	TM	25
Environment	Environment inventory	1:100,000	TM	50
Regional planning	Feasibility study	1:50,000	Spot-XS	40
Environment	Risk zone mapping	1:50,000	KFA 1000	150
Urban development	Urban change	1:50,000	KFA 1000, Spot-P	45
Topography	Base map	1:50,000	aerial photo	120
Geology	Photogeology	1:25,000	aerial photo	50
Transport	Road design	1:20,000	aerial photo	180
Topography	Orthophoto	1:12,000	aerial photo	24
Water supply	Base map	1:10,000	aerial photo	800
Forestry	Forest inventory	1:10,000	aerial photo	350
Land use	Land use mapping	1:10,000	aerial photo	520
Bio-material	Energy study	1:10,000	aerial photo	250
Transport	Photogr. map	1:10,000	aerial photo	700
Cadastre	Orthophoto map	1:10,000	aerial photo	400
Topography	Base map	1:5,000	aerial photo	2,000
Topography	Orthophoto	1:5,000	aerial photo	78
Cadastre	Photogr. or survey map	1:2,000	aerial photo	10,000
Cadastre	Orthophoto	1:2,000	aerial photo	1,000
Topography	Orthophoto	1:1,000	aerial photo	800
Urban cadastre	Base map	1:1,000	aerial photo	20,000
Urban cadastre	Multi-purpose cadastre, utilities, topography	1:500	aerial photo	40,000

TABLE 12.5 Prices of digital map data (Lower Saxony, Germany)

Map data	US$/km^2
Raster-scanned maps	
1:500,000	0.01
1:100,000	0.15
1:50,000	0.60
1:25,000	2.50
1:5,000	30
Vector data	
1:10,000	35
1:1,000	1,000
Digital elevation models	
1:50,000	2
1:5,000	65
Digital orthophotos	
1:10,000	30
1:5,000	80
1:1,000	800

images. The requirements of regional GIS at a scale of 1:50,000 will cost about US$100/km^2 on average, whereas local GIS at scales between 1:1,000 to 1:10,000 will cost between US$1,000–10,000/km^2, unless orthophoto mapping is used at between US$30–800/km^2.

Surveys and mapping administrations are beginning to market their base products. Table 12.5 lists the current prices of the state survey authority of Lower Saxony in Germany.

12.11 Conclusion

One of the reasons for including a description of several classical techniques of photogrammetry in this book is because much topographic source data used in the GDI environment depends on topographic maps constructed using these techniques. Modern remote sensing and photogrammetry offer a wide variety of new methods for geospatial data acquisition. Improvement in the precision and cost-performance continue to be made: in airborne methods by the forward-motion compensation in cameras, and in satellite platforms by the application of increasingly higher resolutions. By far the most significant development in terms of cost and quality performance is the integration of differential GPS techniques (see Chapter 11, pp. 183–5) with 'sensing' processes. As can be seen from the cost-performance figures, the optimization of technical choices is important and has ongoing consequences for the budgets, and consequently for the ultimate cost of the GDI.

The optimization process requires specialized expertise which developers of GDI must have at their disposal to ensure that the right choices are made for the purposes identified. More recently, military high-resolution remote sensing technology has been released for commercial exploitation by the private sector in the USA. As a result, several corporations have emerged with plans to provide a complete suite of commercial data acquisition services, including value-added product services partly based on these new facilities. The pricing structures of these services are still unclear but it is hoped that healthy competition will bring them down to interesting levels (see, for example, Li, 1998). Aside from the impact these commercial services may have on national surveying and mapping agencies, it will be interesting to watch this development closely for other jurisdictions.

Bibliography

ACKERMANN, F. (1997). 'Digital Terrain Models—New Techniques, Demands, Concepts', IAPRS Commission 3/4 Workshop, 32: 3–4W2, Stuttgart.

DANGERMOND, J. (1997). 'Synergy of Photogrammetry, Remote Sensing, and GIS', Photogrammetric Week '97, Stuttgart: Heidelberg, Wichmann, pp. xi–xvi.

GRUEN, A. (1997). 'Digital Photogrammetric Stations Revisited: A short list of unmatched expectations', *Geomatics Information Magazine* (GIM) January, pp. 20–7.

HARTFIELD, P. (1997). 'Higher Performance with Automated Aerial Triangulation', Photogrammetric Week '97, Stuttgart.

LANG, F. and FOERSTNER, W. (1996). '3D-city modelling with a digital one-eye-stereo system', IAPRS, 31/4: Vienna.

——(1996). 'Surface reconstruction of man-made objects using polymorphic mid-level features and generic scene knowledge', IAPRS, 31/B3 Commission 3: Vienna.

LEE, G. and THORPE, J. (1997). 'USA's National Digital Orthophoto Program', Photogrammetric Week '97, Stuttgart.

LI, R. (1998). 'Potential of High-resolution Satellite Imagery for National Mapping Products', *Journal of Photogrammetric Engineering and Remote Sensing*, 64/12: pp. 1165–9.

PETRIE, G. (1997). 'Digital Photogrammetric Workstations: A perspective on suppliers and users'. *Geomatics Information Magazine* (GIM) July, pp. 18–23.

13

Access to GDI and the function of visualization tools

Menno-Jan Kraak

13.1 Introduction

'Geospatial data' immediately makes us think of maps. Only maps provide an instantaneous overview of the relations among geographic objects. They help users to gain a better understanding of geospatial patterns. Maps can show the distribution of a country's population, of environmentally sensitive areas, or any other phenomena that can be physically located. In the framework of GDI, maps have different, but important roles to play. They can also act in their traditional function of presenting geospatial data; land use maps, geological maps, and traffic maps are examples, and as such they represent the information held by an organization. They can be part of a geospatial search engine to find and access geospatial data. Here the map functions as a user interface for locating data sets on a particular region. They can also act as an index to other data available on a particular topic.

Any map, static or dynamic, on screen or on paper, complex or simple, is created during a process called cartographic visualization. This process is considered to be the translation or conversion of geospatial data from a database into graphics (Figure 13.1). Cartographic methods and techniques are applied during this process. These can be considered as a kind of grammar that allows for the optimal design, production, and use of maps, depending on their application. The process is guided by the questions, 'How do I say what, to whom, and is it effective?' The questions contain four key words: 'How' refers to the cartographic methods and techniques; 'what' refers to the geospatial data; 'whom' refers to the map audience and the purpose of the map; and 'effective' reflects the usefulness of the map (see Kraak (1998) for more details).

Cartography, the discipline studying all aspects of maps, has changed enormously. A huge number of maps are created these days, but only a few are created as a final product. Admittedly, these are the maps most people see, and

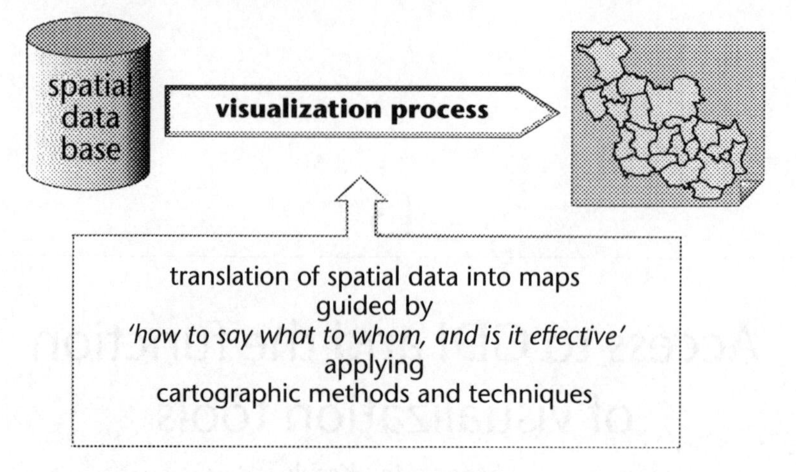

FIGURE 13.1 The cartographic visualization process.

they often serve a wide audience. But most maps are created for a single purpose only: to display alternative routes between A and B, to show the contents of a data set, or to view the intermediate results of a geospatial analysis. They can be created during any phase of the geospatial data handling process, which involves the acquisition, storage, manipulation, and visualization of geospatial data in the context of a particular application.

Users' attitudes to maps have also changed. Cartographers, or even geoinformatics specialists, no longer create most of our maps. The new 'cartographers' have different expectations of a mapping environment because of their experiences in their own discipline environment. This wider use has exposed mapping to external influences. Developments such as scientific visualization and the World Wide Web have introduced two important keywords to cartography and the GDI world: interaction and dynamics (Hearnshaw and Unwin, 1994; Longley *et al.*, 1999; MacEachren and Taylor, 1994; McCormick, DeFanti, and Brown, 1987). New users want to be able to click on the map, and expect an immediate reaction. These developments are also forcing cartographers, as well as others in the GDI world, to adopt a more demand-driven approach, instead of the traditional, supply-driven one. Users with access to geospatial databases around the world can now compile their own maps. Morrison (1997) referred to this trend as the democratization of cartography.

This trend means different visualization strategies are being used (Figure 13.2). On the one hand, the visualization process is part of the traditional realm of cartography. It deals with the well-structured approach to presenting a well-known geospatial data set. Cartographers predominantly create the resulting maps. On the other hand, the visualization process stimulates the new cartography. Maps are created to explore geospatial data sets, to support a geoscientist in solving a problem. They stimulate what has been called the visual thinking process. This type of cartography is called exploratory cartography.

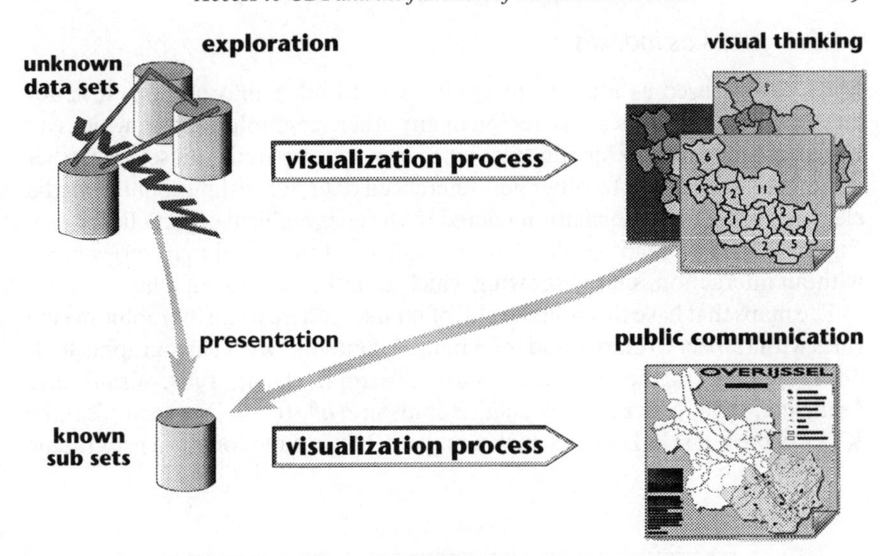

FIGURE 13.2 Cartographic visualization strategies, thinking, and communication.

This chapter will elaborate on map function in the GDI environment. Its structure reflects the procedure that we might follow in an attempt to solve a geo-problem. It will first address the problems of finding and getting the data needed, and the tools required to further assess the data. This phase of exploratory cartography is followed by the actual 'manipulation' of the data. A final section will deal with the presentation of data. The chapter ends with a description of a potential, integrated, web-based example.

13.2 Geospatial data before use

Here we refer to maps used to find and retrieve geospatial data, which is needed to solve a particular geo-problem. This approach should be interpreted from a user's perspective. It involves data access. With respect to GDI, the World Wide Web (WWW) plays a prominent role. This can be both on a local level (Intranet) or a global level (Internet). From a provider's perspective, these maps are tools for offering their geospatial data or derived products. Within an organization, assuming the data exist, the search for data will only be successful if the GDI is properly organized. Searching is often based on the data's metadata description (see also Chapter 9, pp. 146–8). To find data externally can be more complicated, although the principles are the same. Maps have three major functions in this process of finding and retrieving geospatial data:

- maps can function as an index to the data available;
- maps can be used to pre-view available data;
- maps can be part of a search engine.

13.2.1 Maps as indexes

Maps can be used as indexes, to guide users to other information. It is, for instance, possible to click on a region or any other geographic object, which can lead to a list of links referring to other maps, or geospatial data sets. Another option could be a link to other geo-referenced data, for instance, multi-media elements or textual information related to the geographical object clicked (see Figure 13.3). The maps could also function as a traditional map series index without interaction, simply showing which map sheets are available.

The maps that have this role should, of course, adhere to cartographic design rules with respect to clarity and information contents. Most cartographic textbooks provide further information on this topic (Dent, 1985; Kraak and Ormeling, 1996; MacEachren, 1994; Robinson *et al.*, 1995). One example is the KINDS project (URL 1), in which a map of Britain functions as a geographic

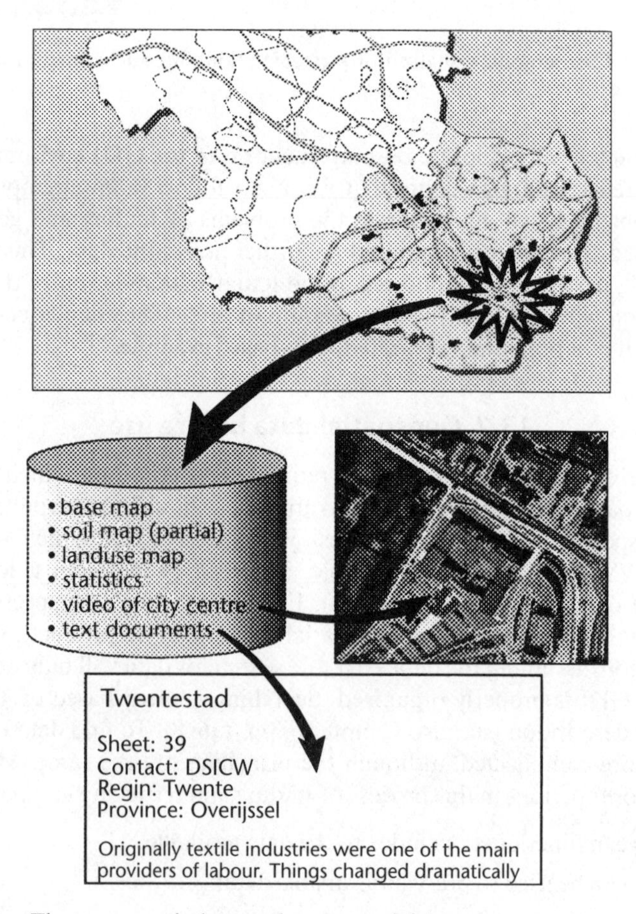

FIGURE 13.3 The map as an index to other geospatial data.

interface. Clicking on any region will lead to a list of the available data sets for that particular area. Another example is the Exite Travel website (URL 2), in which a world map provides links to more detailed regional maps, right up to selected individual cities.

13.2.2 Maps for preview

The quest for geospatial data can also be served by maps that visualize the data that can be obtained. Several organizations have a wide range of geospatial data, and can offer the user a preview to help determine the data's suitability. Some of these preview maps not only show the geospatial data, its extent and attributes, but also some aspects of the data quality (lineage, accuracy, etc.). This may further help the user in deciding on the data's fitness for use. In some cases, sample data are offered as teasers. The user can retrieve a small data set and experiment with the data before obtaining the 'real' data set (Figure 13.4).

In this context, an interesting example is the service offered by the Map Library of Penn State University (URL 3), where one can obtain data from the Digital Chart of the World (DCW) in ESRI's ArcInfo format. Maps and lists guide users to a country. Next users can select the data layers required. This

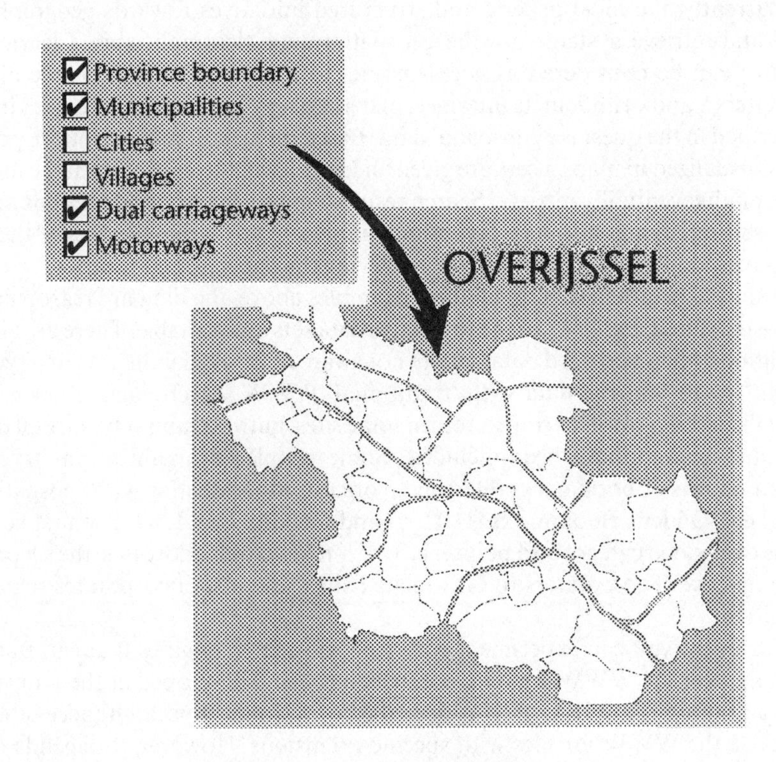

FIGURE 13.4 Maps as a preview to the data to be obtained.

selection is returned in a static map showing the user's choice. If acceptable, the data can then be downloaded. The design of most preview maps is often rather crude. The mapping is done on a 'quick and dirty' basis. The US Bureau of the Census offers a well-designed 'preview' example with its Tiger map files (URL 4). It allows you to select your own map contents.

13.2.3 Maps to search

Maps can also function as part of a geospatial data search engine. In general, the map represents the location component of the data. The attribute and/or temporal component are often complemented by textual search. Imagine one is looking for a land owner data set of a particular region between 1950 and 1960. The user can indicate the region on a map, the theme 'land ownership' can be given in a text window, and the period '1950–1960' selected on a time-line. The engine will provide the location and information on the data, if available. The contents and design of these engines is limited to basic topography for orientation purposes only. An example of this approach is the Alexandria Digital Library (URL 5). It is a prototype 'distributed digital library' for geo-referenced materials and access is via a map-oriented graphical user interface.

Currently, the most generic and structured initiatives towards geographic find-and-retrieve systems are the geospatial data clearinghouses. Clearinghouses can be considered as access points to a country's geodata (see also Chapters 8 and 9) and, in its interface, maps can play any of the three roles just described in the quest for geospatial data. Based on metadata description, possibly visualized in maps, users are given an idea of what to expect from the data they might eventually retrieve. Search results can be a list of organizations and data sets satisfying the query. Hyperlinks allow users to access the sites of these organizations, where conditions for retrieving the data can be found.

It should be noted that, in all three examples above, the data are reasonably well-structured, e.g. links to the respective data sets are available. There are also situations when you need data but are not aware of its availability. At first, you might try to find the data with 'traditional' WWW search engines, such as AltaVista (URL 6). They might return some sites, but you cannot be sure all the possible answers have been included. Some people or organizations try to maintain lists of bookmarks referring to organizations dealing with geospatial data, e.g. Oddens Bookmarks (URL 7) and Geolink (URL 8). The first contains over 5,000 categorized pointers to map-related web addresses, the second contains over 800 pointers to GIS-related sites, and also incorporates search facilities.

Currently, you can experiment with true geo-search engines that can track data all over the WWW, e.g. Geobot, which is being developed at the University of Edinburgh (URL 9). It is described as a 'robot' to search (accessible) parts of the WWW for files with specific extensions. However, it depends on how providers make their data accessible as to whether they can be reached.

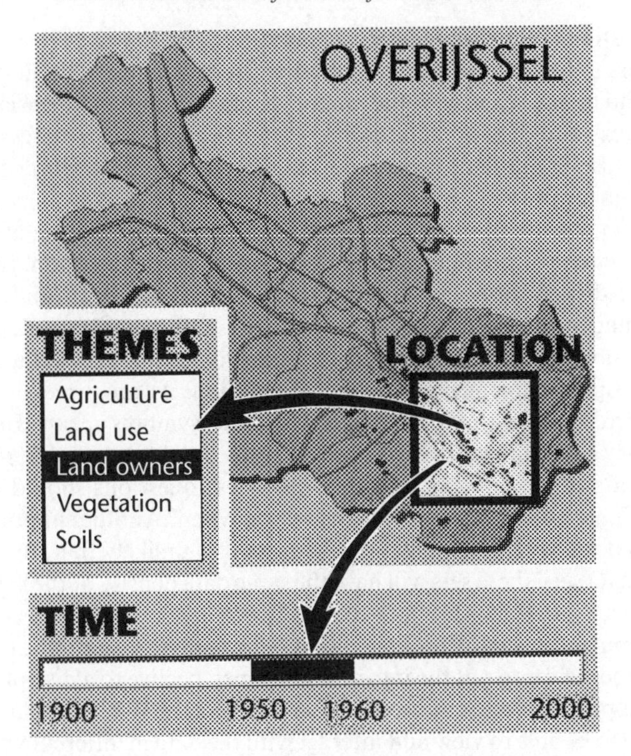

FIGURE 13.5 The map as a location component of a spatial data search engine.

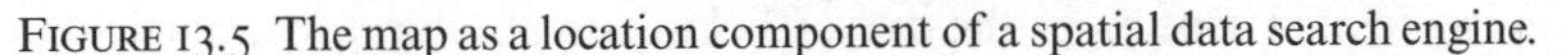

GIS vendors also offer some WWW search mechanisms, e.g. ESRI's ArcExplorer (URL 10).

A technique called data mining has been developed for organizations with vast amounts of data (Berry and Gordon, 1997). It enables huge amounts of data to be searched for patterns and relationships. It is easy to imagine that a similar approach adapted to the special characteristics of geospatial data could be developed for GDI. Plewe (1997) has described many of the technical aspects of finding and retrieving data from the WWW.

13.3 Geospatial data during use

Although operations executed to find and retrieve geospatial data are definitely part of exploration, this section will treat the typical cartographic functionality of 'manipulating' the data to determine its significance. This manipulation can be done locally, or somewhere over the Internet.

What do we need? To answer this question, it makes sense to see what tools geoscientists require to solve their problems. However, we can only make assumptions from a geospatial data point of view, since it is not possible to judge or estimate the discipline-dependent function as well. The tools should allow users to look at geospatial and other geo-referenced data in any combination, and at any scale, with the aim of seeing or finding geospatial patterns. In their 'play and display' actions, geoscientists are trying to assess the data's fitness for use in different geospatial analysis operations, or just to select the correct data for presentation purposes.

Figure 13.6 gives an impression of the tools needed. Starting at the upper left-hand corner of the figure and going in a clockwise direction: the elementary basic display options to pan, zoom, scale, transform, and rotate are required under all visualization circumstances, and should be independent of the dimensionality of the displayed geospatial data. It is obvious users should also have options for navigation and orientation. At any time, users should know where the view is located and what the symbols mean. In a typical exploratory environment there is always the need to have access to the geospatial database to get answers to questions. These questions should not necessarily be limited to a simple What? Where? or When? Another important set of functions deals with generalization, because it is unlikely that, in combining different data sets, these sets will have the same data density or the same level of abstraction.

To stimulate visual thinking, some alternative visualization methods enabling the viewer to put the information in perspective to a discipline's traditional mapping approach is essential. Considering the WWW and multimedia, users have to be able to view and interact with the data in different windows, all representing related aspects of the data. Clicking on an object in a particular view will show its geospatial relations to other objects or representations in all the other views, and could prove helpful in gaining a better or different insight into the geospatial data at hand. In particular, temporal data can be visualized by means of animation. Trends become easily visible. However, animation can also be used for non-temporal data, for example, a flight over a terrain model shown in combination with other data layers. It seems obvious that these functions should be available with close links to 'traditional' GIS functionality.

All maps created in this exploratory environment should adhere to some map etiquette. Since all the tools are used in a real-time interactive situation, a map appearance as you might expect for a presentation purpose is not likely. However, if we have to gather information from maps, we have to adhere to some 'rules' to be able to transfer that information. At the end of the process, it will be difficult to determine whether the map's intended message is influenced most by the map's design, or whether it is a result of the geoscientist's visual thinking process.

13.4 Geospatial data after use

This section offers an overview of the methods and procedures available for creating a map to function in a public presentation environment. Although the section's title suggests it is a final stage, in fact it reflects the production of maps.

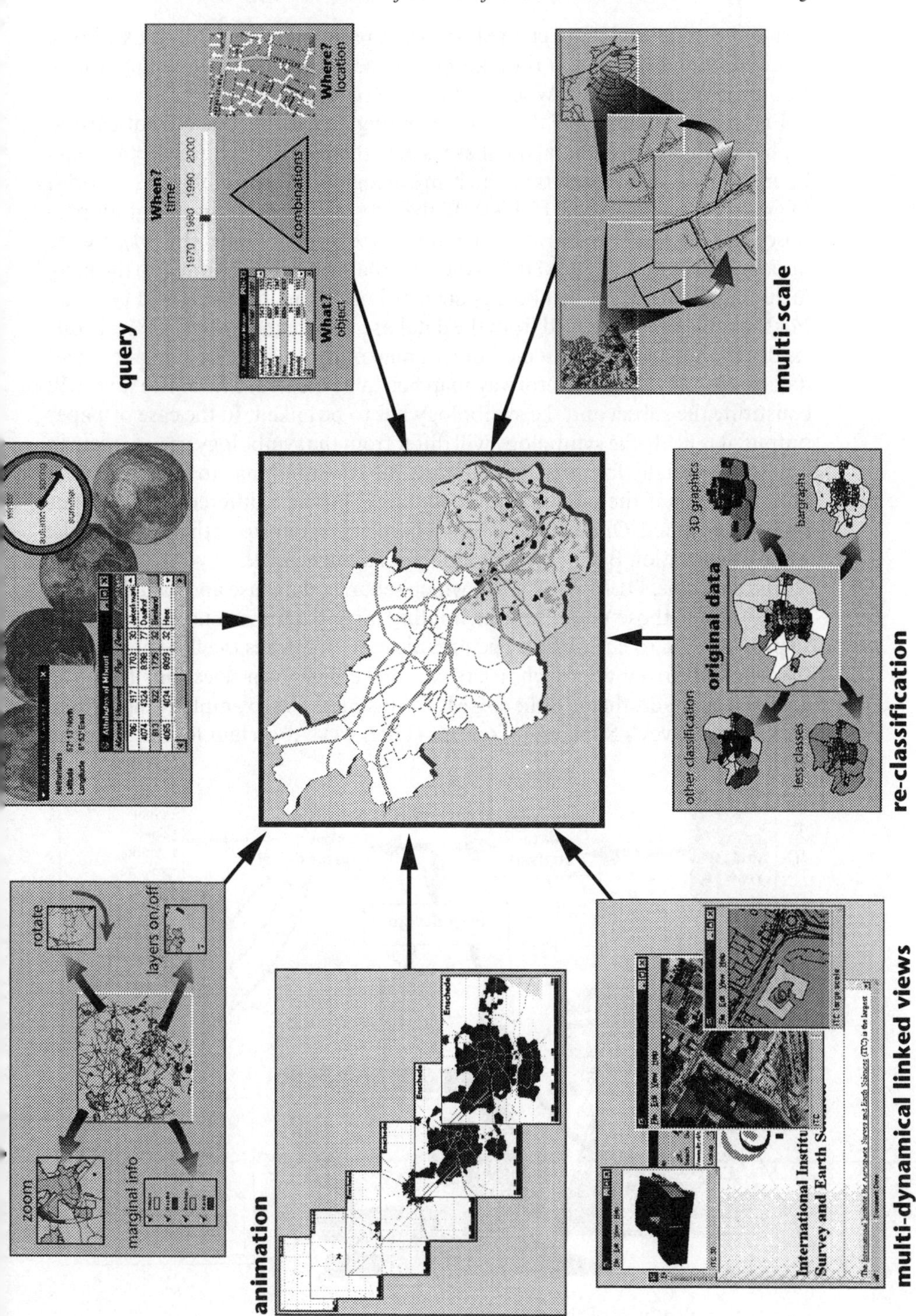
zoom
rotate
layers on/off
marginal info
animation
query
When?
time
1970
1980
1990
2000
combinations
What?
object
Where?
location
multi-scale
original data
other classification
3D graphics
less classes
bargraphs
re-classification
multi-dynamical linked views

While the maps are being created, the data represented are still in use. For a more elaborate discussion on map design, see Kraak and Ormeling (1996), Robinson *et al.* (1995), or Wood and Keller (1996).

Figure 13.7 summarizes the entire mapping process for this type of cartography, in which a geospatial database is also the starting point. Selections can be made from this database, which in cartographic terminology is called a Digital Landscape Model (DLM). If the subset is used for visualization purposes, it is often referred to as the Digital Cartographic Model (DCM), see the inset in Figure 13.7. A DCM holds the data that will finally be put into the map, including cartographic symbology such as line colour, font size, and textures. For the same subset derived from the database, the contents of a DCM will differ depending on the output medium, or map purpose. Imagine a detailed road database, from which a motorway map has to be created. The motorways will constitute the subset and the symbology has to be linked. In the case of paper output, it is likely the symbology will differ from the symbology used for screen output. Especially for on-screen display, the resolution has to be considered. However, even if the same output medium is chosen, a different DCM might have to be created. Organization A might specify colours to fit their house style, while organization B may prefer a different colour scheme.

The concepts, DLM and DCM, which separate database and visualization, work well for those organizations holding large databases and whose aim is to provide data to customers. The organization organizes the flow of data. In the demand-driven approach, users can make their own selections from the database and add them to the DCM, for instance. An example is the British Ordnance Survey's Superplan concept (URL 11). At certain locations in the

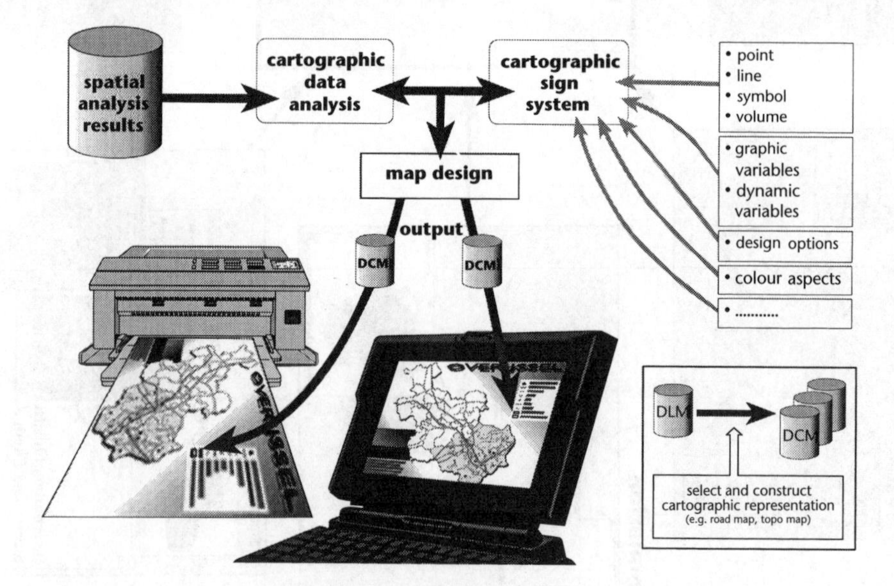

FIGURE 13.7 Synthesis of the modern mapping process.

country, users can select data from the Ordnance Survey databases and make a plot, while selecting their own symbology. In an exploratory environment, the concepts of DLM and DCM are difficult to maintain. Users want to have direct access to the database during the visualization process. The images will stimulate their visual thinking, and will immediately result in user responses to queries and map images to add additional data, to change classification schemes, etc.

In creating a communicative map, we have to execute a cartographic data analysis. This will reveal the nature of the data. Will the map contain qualitative or quantitative data? Are we dealing with ordered data? When data characteristics are known, it is possible to link them to the cartographic sign system. If, for instance, absolute numbers have to be visualized, the symbols used should reflect this; it makes no sense to use symbols that differ in colour. Yellow will not be perceived as being more significant than violet. It is more appropriate to choose symbols that vary in size, since these will offer the user a notion of quantities. For each type of data, symbols can be chosen to allow the user to perceive the map content. Of course, a map can still become fairly complex if it contains too many themes. A traditional topographic map is an example of a complex map, as are most inventory types of maps, such as those for geology, soils, and vegetation. However, if the map audience is less skilled, or there is a limited time frame for viewing the map, such as during TV news broadcasts, the map contents should be adapted and kept simple. Most maps presented on the WWW by providers to attract customers need to be well designed.

13.5 Maps at work: online integration

The effects of the democratization of cartography will definitely change our views on maps and map-making. Maps will become individual products. Users are not always interested in very accurate maps. As long as the maps fit their purpose, the users will be satisfied. Although people have easier access to data and will be creating more maps themselves, there will still be an important role for geoinformatics experts: if not in creating communicative maps, then in facilitating how to do so, or in providing tools for exploratory cartography. Experts will also be able to inform people about new possibilities such as the WWW.

What would an ideal environment look like? There is definitely a role for the National Geospatial Data Clearinghouse. A clearinghouse aims to become a geoplaza, a sort of shopping mall for all the available geospatial data. However, most clearinghouses do not provide service at this level and are just (valuable) shop fronts. Their web pages should allow users to select a location and a theme, and to be given an overview of the data on offer with links to the providers.

However, to add extra value to the clearinghouses for both users and for the participating organizations, it would be a challenge to combine the clearinghouse concept with the concept of a national atlas. A national atlas offers a coherent view of the physical and social aspects of a country. The atlas as such is not only a mechanism to provide information, but could also become part of the search engine—another entrance to the geospatial data—and function as a teaser to the wealth of data offered by the participating organizations. As can been seen in Figure 13.8, the participating organizations will cover all the themes normally found in a national atlas. These include the organizations responsible for the foundation data, the base topography on different scales. Organizations such as geological and soil surveys can offer their data as framework data. Although all their data will have a metadata description, it is unlikely that all of it will be made available to a clearinghouse. However, they do depend on foundation data for creating their own databases. Those organizations offering application data use both foundation and framework data, and add their own data to it, for instance, to create and execute environmental impact assessment studies. Their final products are often single use only, and although also described by metadata, they offer only a small proportion of their information to the clearinghouse. In other words the shareability is very limited (see also Groot, 1997, p. 291).

For a national atlas, the idea is that each organization would make up-to-date data available on a certain level of aggregation. The national survey could offer data for a country's base map, while the other organizations offer data on the themes they represent. Since the idea is that this information is offered

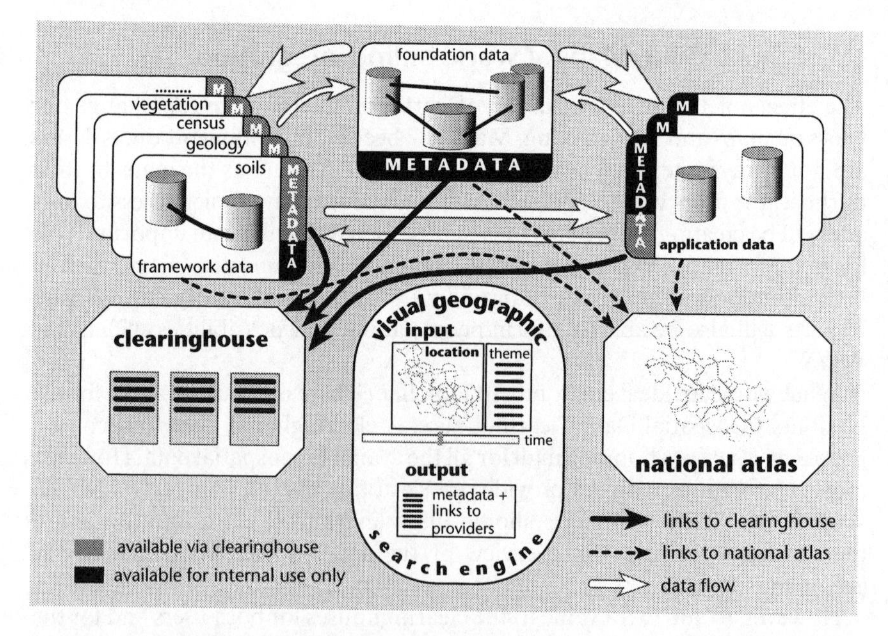

FIGURE 13.8 Relationship between the providers of foundation, framework, and application data, the clearinghouse, and the national atlas.

freely, the level of detail of the information will be limited. More detailed information can be obtained from the respective organizations and should be accessible via links in the atlas and/or pointers at the geoplaza. It is up to the organization whether the data are freely available or only under commercial conditions. These sites could thus provide access to additional data and knowledge. Different formulas for exploitation can be thought of for making it available.

Ideally, the online atlas would present the data in communicative maps designed by professional cartographers, while experts from the organizations offering the data could write a short description explaining the maps' geospatial patterns. Users could freely download the maps and/or the data for their own purposes. An example of this is the National Atlas of Canada (*http://cgdi.gc.ca/frames.html*). Both the map and description would provide an opportunity to take users to additional information on the mapping methods applied or to the organization's web site via a hyperlink.

However, from a practical point of view, it would probably be better if the maps could be generated directly by the participating organizations, according to a national atlas standard. This would guarantee up-to-date data and maps. It would be interesting to see if the design rules could be part of, or derived from, the metadata needed for the clearinghouse, to avoid double effort in creating and maintaining such maps by the participating organization. However, the simple design of a static map is not what most visitors want to find in an online national atlas. They have a different attitude to maps than those looking at a traditional paper atlas. These new users are probably eager to click on the maps, and they expect action. Not only should the data be free, it should also be interactive. As such, a national atlas would offer an up-to-date overview of the country's characteristics and would increase the value of the GDI. This idea could be implemented on both an international, as well as a more regional, scale.

Relatively new players in the mapping field, for example, Microsoft's Encarta Virtual Globe (URL 12) have also discovered atlases in relation to the WWW. Its sophistication shows how organizations with the resources and knowledge of information technology can generate a geo-information product that can now be found in many households. Its vast amount of data is cleverly integrated on a single CD-ROM, and this low-priced product is a source for exploration even for the experienced user. Data sets such as the Digital Chart of the World have been incorporated. The atlas, which has a web browser-type interface, claims to give access to more than 1.2 million geographic objects. Name placement algorithms and generalization procedures are applied effortlessly. Users can create, within limits, their own maps. This electronic atlas combines maps with encyclopaedic geographic text and multi-media elements and statistics all hyperlinked together. Web links exist for many geographic objects. Clicking on those will connect users with the Encarta web site, where a web page for a country or city contains maintained links to tourist information, weather, etc.

13.6 Conclusion

This chapter has presented the role of cartographic visualization within the framework of access to the GDI. This visualization concerns interactive and dynamic online maps. The maps must fulfil various roles: they present geospatial data and they function as part of the GDI interface for indexing, previewing, and searching for geospatial data. In this process, the maps are tools that will stimulate the users' visual thinking during their exploration of the data and their solving of geo-problems.

Bibliography

BERRY, M. J. A. and GORDON, L. (1997). *Data Mining Techniques: For Marketing, Sales, and Customer Support*. New York: J. Wiley and Sons.

DENT, B. D. (1985). *Principles of Thematic Map Design*. Reading, Massachusetts: Addison Wesley.

GROOT, R. (1997). 'Spatial Data Infrastructure (SDI) for Sustainable Land Management', *ITC Journal*, 3/4: p. 291.

HEARNSHAW, H. M. and UNWIN, D. J. (eds.) (1994). *Visualization in Geographical Information Systems*. London: J. Wiley and Sons.

KRAAK, M.-J. (1998). 'Exploratory Cartography: Maps as Tools for Discovery', *ITC Journal*, 1: pp. 46–54.

——and ORMELING, F. J. (1996). *Cartography, the Visualization of Spatial Data*. London: Addison Wesley Longman.

LONGLEY, P., GOODCHILD, M., MAGUIRE, D. M., and RHIND, D. (eds.) (1999). *Geographical Information Systems: Principles, Techniques, Management, and Applications*. Cambridge, UK: Geoinformation International.

MACEACHREN, A. M. (1994). *Some Truth with Maps: a Primer on Design and Symbolization*. Washington, DC: Association of American Geographers.

——and TAYLOR, D. R. F. (eds.) (1994). *Visualization in Modern Cartography*. London: Pergamon Press.

MCCORMICK, B., DEFANTI, T. A., and BROWN, M. D. (1987). 'Visualization in Scientific Computing', ACM SIGGRAPH *Computer Graphics*, 21/6: special issue.

MORRISON, J. L. (1997). 'Topographic mapping for the twenty first century', in RHIND, D. (ed.), *Framework of the World*. Cambridge UK: Geoinformation International, pp. 14–27.

PLEWE, B. (1997). *GIS Online: Information Retrieval, Mapping and the Internet*. Santa Fe: OnWord Press.

ROBINSON, A. H., MORRISON, J. L., MUEHRCKE, P. C., KIMERLING, A. J., and GUPTILL, S. C. (1995). *Elements of Cartography*. New York: J. Wiley and Sons.

WOOD, C. H. and KELLER, C. P. (eds.) (1996). *Cartographic Design: Theoretical and Practical Perspectives*. New York: J. Wiley and Sons.

URLs

URL 1	KINDS	*http://cs6400.mcc.ac.uk/kinds/imageMap.html*
URL 2	EXITE TRAVEL	*http://www.city.net/*
URL 3	PENN STATE DCW	*http://www.maproom.psu.edu/dcw/*
URL 4	TIGER MAPSERVER	*http://tiger.census.gov/cgi-bin/mapsurfer*

URL 5 ALEXANDRIA *http://www.alexandria.ucsb.edu/*
URL 6 ALTAVISTA *http://www.altavista.digital.com/*
URL 7 ODDENS *http://kartoserver.frw.ruu.nl/html/staff/oddens/oddens.htm*
URL 8 GEOLINK *http://www.gislinx.com/*
URL 9 GEOBOT *http://www.geo.ed.ac.uk/~anp/public/gisrbt_sch.htm*
URL 10 ESRI'S ARCEXPLORER
http://www.esri.com/base/products/arcexplorer/arcexplorer.html
URL 11 BRITISH ORDNANCE SURVEY'S SUPERPLAN
http://www.ordsvy.gov.uk/products/computer/superpln/index.html
URL 12 ENCARTA'S VIRTUAL GLOBE
http://www.encarta.com/ewa/default.shtm

14

Human resources issues in the emerging GDI environment

David Coleman, Richard Groot, and John McLaughlin

14.1 Introduction

In an early paper defining GDI, McLaughlin (1991) identified the three stages of geospatial database development which have taken place since the 1960s:

Stage 1 (c.1960–1980) witnessed the first uses of computers in surveying and mapping, the first efforts at automating land records, the first GIS software being developed in government laboratories and universities, and the first attempts to build urban and regional information systems.

Stage 2 (c.1975–2000) saw computer-based geospatial information systems come into their own for administrative, facilities management, and planning purposes, and witnessed the rapid emergence of commercial GIS and digital mapping software. With a heavy emphasis on program-driven geospatial database loading, most of the progress in data sharing and integration was made *within*—rather than *between*—organizations.

Stage 3 (1990 and beyond) is now focusing on linking databases into distributed geospatial information networks, on developing application software and decision support tools to exploit more effectively the information available, and on building a much broader information services industry. New institutional arrangements, alluded to in Stage I and largely dismissed in Stage II, are finally beginning to emerge.

With organizations across the developed world now settled into this third stage, markets for geospatial technologies, data sets, and their applications have grown rapidly in the past decade. The demand for skilled persons familiar with their operation and management has also increased. However:

1. senior managers in many organizations are confused about the types of skills they should aim to obtain (in new staff or by re-training current staff) in order to make optimum use of these new tools;
2. employees interested in both job security and future employment prospects are concerned about what skills they should obtain through further education or on-the-job training; and
3. professional associations are concerned about the impacts of technology in reducing, or at least changing, the roles, responsibilities, and authority of their members in the workplace and in the community.

The positioning and mapping services now offered by these technologies were traditionally provided by professionals in such related fields as surveying engineering, geodesy, cartography, GIS, remote sensing and photogrammetry, and others. More recently, the umbrella terms 'geomatics' and 'geoinformatics' have been used in an attempt to amalgamate the more traditional disciplines under one blanket definition. However, converging interests and skills from a number of disciplines make the roles and responsibilities of such professionals blend with those educated in computer science, geography, and other disciplines. Moreover, while such people may be well-trained in operations and technical management, they do not often have the business management, policy understanding, and organizational skills required for project leadership.

This chapter builds on the explanations in 'Who wants a GDI?' (Chapter 2) and the nature of the technologies described in Chapters 8–13. Specifically, it examines the changing requirements of organizations searching for new staff and/or upgraded skills at both the technical and leadership levels. After discussing the nature of the challenges facing these organizations, we offer some possible approaches and strategies for consideration.

14.2 The challenges

14.2.1 Economic pressures

Economic pressures have forced employers in both government and industry to re-examine their human resources requirements. While investment in technology rose dramatically over the past decade, the severe recession of the early 1990s decimated the ranks of middle managers and process technicians in large corporations and government organizations alike. By 1995 40 per cent of the organizations in the 1979 Fortune 500 list of companies no longer existed (Tapscott, 1996).

In employment terms, recovery from the latest recession was the weakest ever. Despite record profits being posted by large organizations, layoffs from manufacturing, clerical and middle management staff have continued. Saddled with the perception that their original missions have been accomplished, cuts and redirection of government control surveys, base mapping programmes, and the mapping of natural resources have meant a reduction in technical management staff and fewer government-funded contracts. Since improvements in technology and workflow processes have improved employee productivity, small businesses in the geomatics sector have focused on technical upgrading rather than new job creation in response to new opportunities (Smith Gunther Associates, 1996). Finally, executives in the larger user community—the resource industries, county and municipal governments, utilities and services firms which now employ GIS and GPS technology in their daily work—have also been forced to come to grips with personnel requirements for such positions.

14.2.2 Technological advances

In little over ten years, positioning, mapping, and geospatial analysis technologies have improved and proliferated to the point where they are no longer regarded as specialist tools. In Lance McKee's terms (see *2.1.4*) they have become mundane, mere appliances to a generation of end-users born after the introduction of micro-electronics, personal computers, communication satellites, and the emergence of the information society. When used in combination with the increasing array of geospatial data sets and ubiquitous telecommunication networks now in place (and moving into the wireless domain), these appliances promise to provide future users with a variety of services related to positioning, navigation, and real-time visualization of selected thematic data.

Nowhere is this more apparent than in statistics reporting the rapid growth in organizations now making use of GIS and GPS technologies. Dataquest, Inc., a US-based market research firm, estimated the 1993 market for GIS-related hardware and software at US$1.8 billion, up almost 900 per cent from 1987. Subsequently, the GPS Industry Council estimated the size of the market for GPS-related products at US$1.3 billion in 1995, and expected it to climb to over US$8 billion by the year 2000. By 1998, the GEOID Network estimated the world-wide market for geomatics products and services to be approaching a value of US$10 billion in annual sales with an annual growth rate of 10 per cent (GEOID, 1998). Even if we consider the probable overlaps in calculating these estimates, they do not include the tremendous growth of the services sectors in GIS, GPS, remote sensing, ocean mapping, and traditional surveying, much less the burgeoning consumer data products and information access industries now emerging.

As the requisite technology moves increasingly into the domain of mainstream DBMS, and as the investment in geospatial database assets continues to grow, day-to-day control of the data management function in the 'enterprise information management' models of large organizations is moving from the end-users to the information systems groups. Partly in response to this change, a 'diffraction' of the technology is taking place which is effectively splitting familiar GIS software product lines into three components:

1. *shrink-wrapped*, *single-user*, PC-based desktop mapping systems (e.g. ArcView and GeoMedia);

2. *high-end, enterprise-wide* GIS incorporated within sophisticated, multi-user database management systems (ESRI's SDE, Oracle's Spatial Data Cartridge and others); and finally
3. *mini-applications* prepared for a wide variety of end-users using modular GIS tool-kits (e.g. ESRI's MapObjects™, USL's CARIS++™ and others) which are embedded into a variety of larger applications within the enterprise.

The changing emphasis in user organizations (i.e. from data collection to database maintenance), taken in combination with this diffraction in GIS software supply, has major implications for the demands which will be made on human resources development:

- Fewer of the large number of positioning and mapping specialists and professionals will be required for planning and managing the initial data collection effort.
- There will be a growing requirement for specialists familiar with GIS as it operates within the context of a larger mainstream DBMS installation;
- The demand for GIS 'power users' familiar with a specific software package will be supplanted by increasing demand for geospatially literate application programmers capable of developing (and maintaining) a collection of smaller, customized applications within the organization.

These points imply the need for a changing emphasis from geomatics professionals towards those with an information systems background. However, while computer science recruits bring stronger and more up-to-date programming skills to the organization, they often have neither the geospatial literacy nor the proper understanding of the characteristics and limitations of the technology and data sets involved. Moreover, since the more generic skills associated with database administration are currently in high demand, organizations often find it difficult to retain these specialists once they obtain the requisite knowledge and experience.

14.2.3 Management and institutional issues

While the technical challenges associated with building GDIs at the enterprise, national or even global level are certainly impressive, they pale in comparison with the political, administrative, and human relations issues involved in reaching consensus on data sharing within and between organizations, countries, or regions. Representative discussions of these problems may be found in Palmer (1984), Mapping Sciences Committee (1994), Onsrud and Rushton (1995), and Coleman and McLaughlin (1998), among others.

With this in mind, there are three significant sets of specialized organizational and process management skills required to facilitate the development and implementation of GDI as defined in earlier chapters of this book. These include:

- experience in establishing and refining the organizational and regulatory structures, policies, and processes required;

- familiarity with quality management, business process redesign, and workflow management practices, and insights necessary to meet the new demands for database maintenance and transaction-based updating;
- experience in the design and management of service facilities and ensuring the integrity of related transaction management, contract management, and error management processes.

Basic training in such matters is usually found in social and behavioural science courses, with further specialization usually required in such disciplines as law, business administration, industrial engineering, telematics, and others.

As a result of traditional organizational practice—or sometimes just in the absence of any viable alternatives—organizations have opted to place a professional manager (MBA or equivalent) in the position of leadership and rely on technical support staff to provide the technological insights required. This strategy has produced mixed results, with relative success or failure depending largely on the personalities and circumstances involved in each situation.

As demands evolve, reliance on the combination of a general business background supported by a strong technical operations staff may no longer be enough. Unless an organization's internal leadership has a clear understanding of its requirements and direction, there will be an increasing reliance on technology vendors to indicate how and where to move next.

Who is the geomatics/geoinformatics professional capable of managing these processes? What is the appropriate mix of economic, professional, and academic building blocks required to prepare professionals for leadership in both instigating and accommodating change in the development and management of GDI? How must an organization's view of its human resources requirements, in terms of both new hirings and training of existing personnel, change to accommodate evolving technologies, expanding market demands, and heightened expectations? Before answering these questions, we discuss some of the alternative courses now being offered by educational institutions.

14.3 Alternative approaches to meeting the challenges

14.3.1 Undergraduate degree courses

As discussed in Coleman (1998), courses in GIS, 'Geomatics' and 'Geoinformatics' at both the community college and university levels in North America, Europe, and Australasia have all attempted to provide some balance between the push of rapidly advancing technology and the pull of new social and environmental demands on the community at large and the professions in particular. However, no single group of graduates from computer science, geomatics, engineering, surveying engineering or surveying technology courses has shown itself to be immediately productive on graduation *and* also adaptable within a changing market-place.

Geomatics employers today find their new graduates from one or more of four different sources:

- technical and community college courses in GIS applications and programming;
- computer science courses at the university and technical college level;
- university courses in geography, planning, or related disciplines; and
- university courses in surveying, and civil and geomatics engineering.

According to Petersohn (1995), graduates in the first group usually possess strong and immediate job readiness in terms of developing and implementing applications using 'off-the-shelf' GIS software. By comparison, graduates from computer science courses often demonstrate a broader scope than members of the previous group, but their lack of understanding of specific applications means they may require more time to become productive on specific projects.

Petersohn also points out that, in his experience, graduates from cartography, geography, and planning-oriented courses with geomatics components typically demonstrate a stronger appreciation of GIS from an 'end-user' rather than developer perspective. They have a wider outlook than many of the more technically oriented graduates, but still require additional investment in specific skills training in order to be productive.

Graduates of geomatics engineering (and related) courses are often found to be less attractive in terms of immediate job readiness than either the technical school or computer science graduates. However, their stronger multidisciplinary background and their ability to adapt to new problems and situations in the workplace offset this. The greatest return is found in this group after further investment in specific skill training.

Admittedly, these observations are anecdotal and may not represent the findings and experiences of the entire industry. Moreover, the recognition of these and similar characteristics has already resulted in incremental improvements to all three types of courses in many leading institutions over the past five years.

14.3.2 Trends in post-graduate courses at universities and other degree-granting international institutions

Post-graduate courses in surveying engineering, geomatics, geographic information science, geoinformatics, geography, computer science, and related fields have provided positioning, mapping, and GIS specialists to industry and government organizations for over forty years. At universities these courses tended to have a strong emphasis on research at both the Master's and Ph.D. levels. At international institutes such as the ITC, the focus is on mid-career post-graduate education and training, specifically in the geoinformatics and GIS-related disciplines. The students are mature and have a wide variety of undergraduate degrees plus organizational and/or research experience. Although the courses have strong links to the needs of the students' home organizations, the response to the rapidly advancing technology has been to orient the courses towards the mastering of concepts and process design and management. Increasingly, universities and international educational institutions now offer candidates a choice of either a 'research' degree (coursework plus research thesis) or a 'professional' degree (more extensive coursework plus a well-defined project).

In recent years, as the number of courses in the market-place has increased, there has been significant pressure on many universities to reduce the length of these courses, especially those based on coursework, from two to three years down to twelve months. While content varies from place to place, such courses still generally attempt to provide a balance of conceptual theory, team projects, and hands-on experience with selected equipment and software. Examples of such courses may be found at the University of Edinburgh in the UK and the ITC in The Netherlands.

14.3.3 College diploma courses

In Canada, community colleges are beginning to offer an alternative to post-graduate degrees with one-year post-graduate diploma courses of their own. Rather than cater to the colleges' traditional markets, these diploma courses are designed especially for university graduates from natural resources, planning, civil engineering, and business administration departments. Courses and projects focus heavily on building up a high level of hands-on application and development experience with specific software packages. The standard of student is typically high since, depending on the institution, the entry requirements would be similar to any post-graduate course. Rather than moving into research or management positions, however, graduates from such courses often initially assume key programming position roles with GIS and remote sensing software firms. Nova Scotia's College of Geographic Sciences (COGS) and the British Columbia Institute of Technology are both excellent examples of institutions providing such courses.

Academic critics of such diploma courses suggest these are simply 'white-collar' counterparts of the heavy-equipment training courses in the construction and transport industries. In the sense that these courses build tool-based 'know-how' skills and rely on the student's previous education for bridging and critical thinking, they are correct. However, given the record of employment of the graduates, these courses definitely satisfy a need and are attracting widespread support in the market place. Even if existing computer science courses could fill the gap alone, such graduates would not possess the system-specific skills required to be immediately productive.

In all these courses, there exists a continuing tension in terms of striking the right balance between understanding of technology versus people versus processes; procedural versus conceptual understanding, and confidence in being able to 'handle the present' versus an ability to 'prepare for the future'. In the 'zero-sum' environment characteristic of fixed-length courses in universities and community colleges everywhere, a given amount of subject material must be dropped in order to introduce or extend new concepts or technologies.

14.4 Discussion

Hundreds of university departments world-wide now offer some level of GIS or geomatics-related education focusing on either the technology itself or its application to a particular discipline. Given the different strengths and limitations of each educational institution—as well as the unique backgrounds of the students involved—it is reasonable to assume that students will emerge from these courses with an outlook and combination of skills which fall somewhere within the triangular framework shown in Figure 14.1.

- At one apex, individuals with a strong *Geographic Information Science and Applications Development* focus are interested in the technologies, processes, operations, and data set integration considerations necessary to acquire and build large geospatial databases, assess their overall reliability, create special-purpose geospatial models and analyses; and produce meaningful, high-quality output to a wide variety of end-users.
- Individuals at the *Computer Science, Telematics, and Systems Development* apex typically have a strong interest in the development and smooth operation of large databases and corporate information systems within an organization.
- Individuals at the *Management and Policy Implementation* apex examine the managerial aspects of implementing the new technology, including such issues as standards and network implementation, organizational and policy requirements, legal issues, economic considerations, and information technology course evaluation.

Different combinations of experience and interest in each of these areas will 'plot out' to a different location within the triangle. Naturally, very few individuals will possess talents and interests focused in just one of these areas. Moreover, a number of educational institutions are now taking the lead in

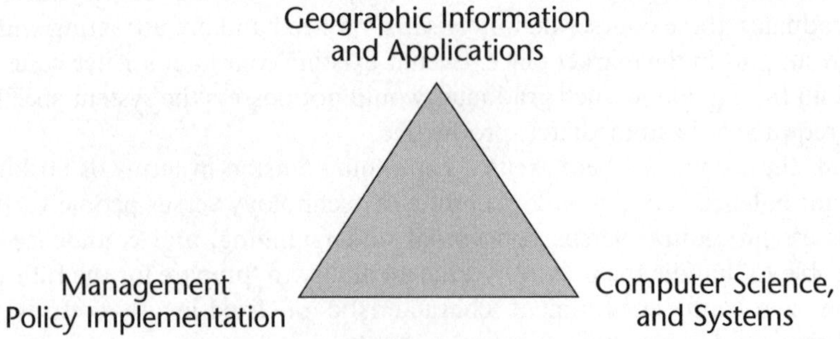

Geographic Information
and Applications

Management
Policy Implementation

Computer Science,
and Systems

FIGURE 14.1 Education and skills framework for individuals working in GDI.

developing courses (and graduates) that will fall at different points within this framework. For example:

- Depending on their previous backgrounds, graduates of land information management courses at institutions like the University of New Brunswick, the ITC, and the University of Melbourne may fall in the upper or lower left-hand side of the triangle. Graduates of more technically oriented geomatics or geoinformatics courses at the same universities may move closer to the upper middle of the triangle.
- Graduates from a diploma course at COGS or BCIT would fall in the right-hand side of the triangle, with their ultimate position depending on whether their earlier education and experience was more in computer science or an application area.
- MBA graduates from courses with a technology management focus would fall somewhere slightly above the base of the triangle (and probably a bit left of centre).

This all points to organizations needing to re-think their recruitment practices. Once hired, both an employee's capabilities and the needs of the organization will change over time as experience is gained and the needs of the organization evolve. Therefore, it would be unwise for an organization to develop a job description for a new position that, when plotted into the above framework, was too close to one corner of the triangle. However, a quick review of many advertisements confirms this practice is still alive and well—particularly in the recruiting of senior technical managers in many geomatics organizations and GIS installations.

These 'mismatches' in terms of hiring have resulted in both: (*a*) software specialists sometimes being moved into general management ranks too quickly; and (*b*) non-technical professional managers being required to make technical decisions with respect to purchases and future directions with little or no knowledge of the operational implications involved in the decision. In both cases, people 'in one part of the triangle' who were hired on the basis of one set of skills and experiences are being forced elsewhere within the triangle by virtue of the responsibilities and decisions asked of them. Unless the proper advance preparation takes place, this shift can result in poor performance and employee–employer dissatisfaction.

Understandably, there will be situations where an organization needs a person with well-defined skills, plotting at one apex of the triangle. However, in most cases, employers must:

1. develop stronger interviewing and hiring criteria that identify the combination of talents, education, and experience required to meet the short-term requirements of the organization;
2. 'hire from the middle'—that is, look more for a balance of skills and realize that good people will develop as the position requires;
3. determine the potential 'speed and direction' of movement of an

employee's actual responsibilities within this framework over the medium term; and

4. determine the most appropriate combination of mentorship and training (both internal and external) that would encourage this development.

14.5 Conclusion

The evaluation made in this chapter of the human resources issues in GDI has serious implications for surveying engineering departments all over the world. Even if they have changed their names to include the words geomatics or geoinformatics they have not always created the necessary renewal in their education courses to reflect the integrated nature of these new professions. In some cases, the content of their courses has hardly changed at all but merely continues under the new name.

One of the consequences of a commitment to education and training in geoinformatics is to recognize that surveying, photogrammetry, geodesy, and cartography are supporting disciplines as are computer science, commercial, legal, and management disciplines. Increasingly, these will also include subjects from other fields such as operations management, business process redesign, system development methods, workflow management etc.

Surveying engineering departments still give in-depth education and training, especially in the first four disciplines, and students become a surveyor, photogrammetrist, geodesist or cartographer. But under the influence of the 'democratization' of the practice of these disciplines, the demand for those specialists is declining. Many departments struggle with low intake although this is also due to demographic shifts. In North America we see that undergraduates find employment in the surveying field but that at the graduate level they tend to get involved more successfully in the broader area of geomatics. Careers for specialists in say photogrammetry are developing better in the industrial and non-topographic fields than in the survey-related ones. But in these new applications those graduates have significant competition from physics, electronics, and mathematics graduates who are comfortable with image processing techniques.

The essence of the geoinformatics professional is that he or she can purposefully integrate many subjects such as those which have been presented in the chapters of this book. In fact, it is a design profession marrying technical, legal, financial, and social science subjects in a problem-focused way. This raises two key questions for educators:

1. How much should geoinformatics professionals have to know about the individual subjects to be effective in their field?
2. Is geoinformatics a post-graduate education of say one or two years to which individuals with undergraduate degrees in related technical and geospatial data application disciplines can be admitted?

The first question is difficult but must be answered by employers with the educators. The second question is answered in North America generally by a geomatics course that is focused on that subject from the start at the undergraduate level. In Europe the trend appears to be more towards geoinformatics being a specialised field which is taught at the post-graduate level. It should be noted that in North America the technical colleges tend to have post-graduate courses which train people with undergraduate degrees from many different fields in specific systems competences. Hence, in part influenced by the market, in part by tradition and the legal degree granting position in different educational jurisdictions, the answers are not unequivocal.

Whatever the solution, it should meet the perceived demands of the employment market and, in our perception, deliver professionals and technologists that fit within the triangle presented in Figure 14.1. Those professionals should have the flexibility to move their career closer to one of the apexes depending on their undergraduate work or interests developed along their career track. At the same time it should be recognized that too heavy an emphasis on a seamless link up with the current employment market may be at the expense of critical understanding and independent development in university graduates. These capacities are necessary to adapt to rapidly developing concepts and technologies and are important for meeting the demands for life-long learning and remaining competitive. This also suggests that the spin-off from new geoinformatics and geomatics courses and from associated courses at the colleges of technology must be short, mid-career, upgrading courses to meet the demands in the continuing education market.

Bibliography

Coleman, D. J. (1998). 'Applied and Academic Geomatics into the Twenty-first Century', *Geomatica*, 52/1: pp. 11–24.

—— and McLaughlin, J. D. (1998). 'Defining Global Geospatial Data Infrastructure (GGDI): Components, Stakeholders And Interfaces', *Geomatica*, 52/2: pp. 129–43.

Geoid (1998). *Geomatics for Informed Decision Making (GEOID)*, Proposal for a Canadian Network of Centres of Excellence in Geomatics. URL: *http://www.crg.ulaval.ca/an/network/frProposal.htm*.

Mapping Sciences Committee (1994). *Promoting the National Spatial Data Infrastructure Through Partnerships*, US National Research Council. Washington, DC: National Academy Press.

McLaughlin, J. D. (1991). 'Towards National Spatial Data Infrastructure', *Proceedings of the 1991 Canadian Conference on GIS*, Ottawa. Canadian Institute of Geomatics, Ottawa, Canada, March, pp. 1–5.

Onsrud, H. J. and Rushton, G. (1995). *Sharing Geographic Information*, Centre for Urban Policy Research, The State University of New Jersey, New Brunswick, New Jersey.

Palmer, D. W. (1984). *A Land Information Network for New Brunswick*, M.Sc.E. thesis, Department of Surveying Engineering, University of New Brunswick, Fredericton, New Brunswick, Canada.

Petersohn, C. (1995). *Geomatics Education for the Private Services Sector*, Proceedings of the 1995 Atlantic Institute 'Think Tank' Conference, Laval University, Quebec, Canada, 1–2 June 1995, pp. 46–57.

Smith Gunther Associates Limited (1996). *Canadian Geomatics: Benchmarking Directions For Growth, Industry Sector Outlook*, Contract Report prepared for Industry Canada, Ottawa, March 1996.
Tapscott, D. (1996). *The Digital Economy*. New York: McGraw-Hill.

15

Four cases

David Finley, Karen Siderelis, Don Grant, and Arnold Bregt

15.1 Service New Brunswick: the modernization of a land information service—David Finley

15.1.1 Background information

Information is now recognized as one of an organization's most valuable resources (Branscomb, 1986; McLaughlin and Coleman, 1990; Melody, 1981). In the case of an organization concerned with land, good and timely land information allows for greater efficiency in managing its natural resources. This is particularly important to a small rural province such as New Brunswick, Canada, whose economy has traditionally relied heavily on its natural resources. New Brunswick, in fact, was a pioneer in both recognizing its land information shortcomings and subsequently working towards the development of its land information infrastructure, with a view to providing easy public access to its databases.

During early European settlement in New Brunswick, land was often regarded as a virtually free good, and neither the quality nor maintenance of the cadastral arrangements were considered important (McLaughlin, 1991). The deed registry system, adopted from British methods of tracking rights and interests in land, was established shortly after the Province of New Brunswick was formed in 1784 (Nichols, 1993). Although occasional attempts to modify the registry system were made, the system remained largely unchanged until after the Second World War (NRC, 1980). Post-war growth brought to light the inadequacies of the existing registration practices, and thus the need for reform (Doig and Patton, 1994). In particular, the magnitude of the social and economic costs incurred by continued public reliance on the rudimentary registry system were becoming manifest. These costs became most apparent in land transfer and land valuation processes, and in the provision of information for resource management purposes (McLaughlin, 1991; Ogilvie, 1997). This led to a growing awareness in New Brunswick of the importance of having an integrated data bank of information related to land (Dale and McLaughlin, 1988; Roberts, 1976).

15.1.2 Developing the information infrastructure and building the database

New Brunswick has a thirty-year history of examining and adopting innovative approaches to sharing and distributing the land information maintained by its various provincial government departments (Coleman, 1988; Finley *et al.*, 1998*a*; Palmer, 1984; Strutz, 1994). As early as the 1960s, land information experts acknowledged the importance of an integrated data bank of information relating to land. In addition, it was recognized that in order for this information to support improved decision-making, it would have to be easily accessible and that the public should be involved in the process (McLaughlin, 1991).

During the late 1950s and 1960s, there were a number of initiatives already underway aimed at addressing the growing need for improved collection and management of land information. Early efforts also explored the idea of building computer-based mapping systems and automated land databases (McLaughlin, 1991). These early land information system (LIS) designs were based on large, centralized databases using the technology of the day—mainframe computers (Roberts, 1976). But at that time plans exceeded both budgets and capabilities of the government agencies. Instead, government departments began developing isolated, project-based electronic databases, and organizational and institutional issues impeded the realization of the centralized database design.

A shift in the LIS design became necessary, and by the 1980s, technology had advanced enough to support a shift from an LIS model to the concept of a land information network (LIN) (Palmer, 1984). Databases developed and maintained by departments and housed at departmental locations, rather than in a centralized data bank, were linked to other databases by the common parcel identifier. The success of the network model meant that participating government departments had access to shared information. However, the public still had no convenient access mechanism. While technology advances of the 1990s have finally made it possible to complete the original, albeit modified, LIS model, perhaps the fundamental reason for New Brunswick's current success was the government's strategic decision to create an environment where sharing information is feasible.

The province instituted a Land Information Policy in 1989 that, for the first time, provided clear guidelines for the collection, storage, retrieval, dissemination, and use of land information (GIC, 1989; Simpson, 1990). To carry out the objectives of this policy, the functions of land registration, property assessment, and mapping in New Brunswick were amalgamated under a single entity—Service New Brunswick (SNB) (Coleman, 1989; Nichols, 1993). While SNB inherited many of its information resources from earlier initiatives, it devoted much effort to the development and conversion of the information into digital data sets which now form the foundation for the provincial land information infrastructure (Arseneau and Ogilvie, 1994). SNB's focus is now on the maintenance and improvement of these data sets, integration of the base information with thematic information from other government departments, and dissemination of this integrated information.

In this regard, SNB has leveraged Internet technology to provide a convenient mechanism for public access to government data sets (Arseneau *et al.*, 1997). Prior to this time, it was not a simple exercise to gather comprehensive information relating to land. Individuals had to check with numerous government departments and agencies to ensure they had all the relevant information pertaining to a particular land parcel.

In August 1996, SNB implemented one of the first commercially available, online land registry systems in the world, providing access to integrated data sets (Dawe, 1996; Plewe, 1997). This land information browser allowed clients to access non-confidential, parcel-based information and maps residing at a password-protected SNB Internet site (Finley *et al.*, 1998*a*). In January 1997, the service was enhanced and extended to include three environmental data sets supplied by the New Brunswick Department of the Environment.

It is SNB's intention that as additional relevant data sets are developed, they, too, will be linked to land parcels and integrated into this enhanced browser. SNB also anticipates that the functionality of the enhanced browser will be further extended, and made readily available to the public—in effect becoming an electronic *Land Gazette*, allowing government organizations to electronically serve public notice of time-limited interests in land, such as mortgage sales and quieting of titles (Finley *et al.*, 1998*b*) (see Figure 15.1.1).

15.1.3 Standards

While the importance of information standards have long been recognized in New Brunswick, until the creation of SNB, the organizational framework was not in place to support a co-ordinated provincial approach to standards. Part of SNB's responsibilities included the development and implementation of New Brunswick land and water information standards. SNB organized an inter-departmental 'Standards Advisory Committee' with broad representation to ensure input from experienced line departments and other interested organizations. Limited-term working groups were created and outside expertise was retained to provide technical expertise where required (Ogilvie, 1991). The committee produced a 'living' standards manual, which is still in effect today, and which is updated to accommodate new issues as they arise. SNB has also taken a proactive approach when developing or improving digital data sets. Part of this approach includes the development of quality control procedures to ensure all provincial data sets meet the same standards.

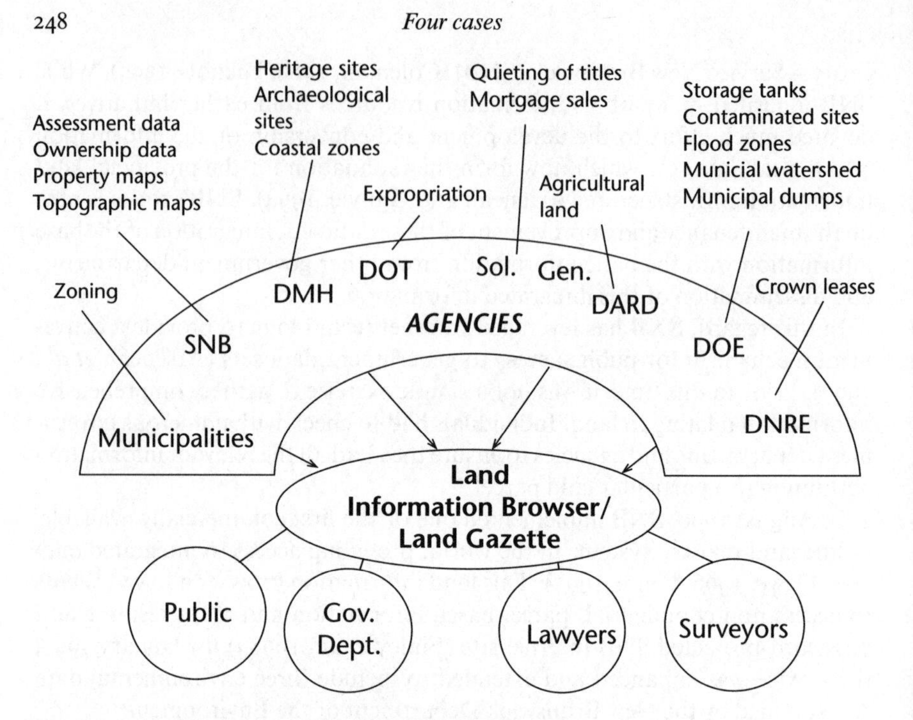

FIGURE 15.1.1 New Brunswick's electronic *Land Gazette*.
Note: The list of represented government departments or agencies is not exhaustive but includes:
SNB—Service New Brunswick
DMH—Department of Municipalities and Housing
DOT—Department of Transportation
Sol. Gen.—Department of the Solicitor General
DARD—Department of Agriculture and Rural Development
DOE—Department of Environment
DNRE—Department of Natural Resources and Energy

15.1.4 Management structure

The creation of SNB allowed for the amalgamation, for the first time, of all basic land information activities under one umbrella. These activities previously fell under the responsibility of the County Registry Offices of the Department of Justice; the Assessment Branch of the Department of Municipal Affairs and Environment; and the New Brunswick portion of the Land Registration and Information Service, an agency of the Council of Maritime Premiers (Simpson, 1990) (see Figure 15.1.2). SNB is headed by a President and governed by a Board of Directors, comprising members from industry, education, and government (GIC, 1992; 1993).

The agency operates as a Crown Corporation providing both basic geographic information to government departments and agencies, and geographic

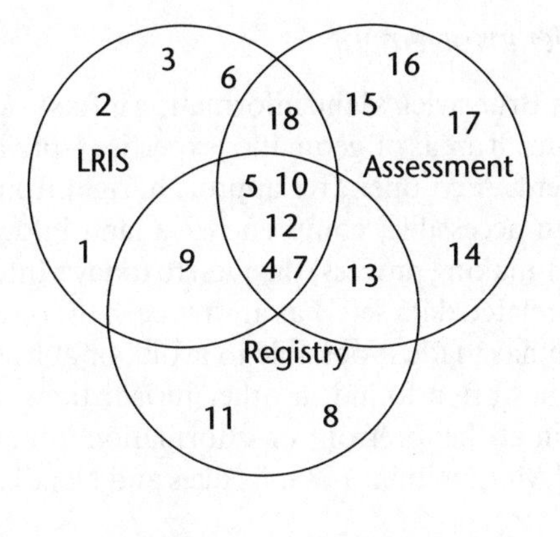

Type of Land Information

1 Aerial photography	7 Deeds	13 Property values
2 Survey control	8 Chattels	14 Property improvements
3 Basemaps	9 Encumbrances	15 Building information
4 Property maps	10 Subdivision plans	16 Market analysis
5 Parcel index files	11 Corp. registry	17 Appraisals
6 Map distribution	12 Land titles	18 Maps/PATS

FIGURE 15.1.2 Types of land information/agency responsibility (after Simpson, 1990).

information to the public; it also fosters private sector development. Operating as a Crown Corporation means that (GIC, 1989):

- SNB has a business orientation, complete with revenue incentive and flexibility not present in a government department model.
- The requirements to maintain public records continue to be satisfied.
- Corporation structure and control are tailored to SNB's own unique requirements.
- Separate budgeting and accounting procedures are in place.
- Government maintains full control over policy and standards.
- Staff are employed within the public service framework.

15.1.5 Financial structure

SNB was established to operate as a business. Its first annual report stated that the Corporation was to become financially self-sufficient within five years. As a result, SNB operates with a business plan in which return on investment is a primary determinant in setting priorities. The general strategy has been, and continues to be, to commercialize those products and services which yield sufficient returns, in order to enable SNB to carry out development projects leading to a more responsive and efficient operation overall. The Corporation maximizes leverage from development resources by participating as a partner on projects, and wherever practical, by tendering projects to the private sector and redirecting development resources to priority projects (GIC, 1992).

15.1.6 Stakeholder involvement

Stakeholders in New Brunswick's land information infrastructure development effort have come from all areas of geomatics expertise—private industry, academia, and government. A co-operative approach, right from the initial vision (establishment of an accessible, comprehensive land information system to improve the decision-making process) through to today's Internet-based access mechanism to land-related data sets, has always been the order of the day.

This co-operation has made it possible to avoid, or at least reduce, some of the institutional issues often found in other jurisdictions, and allowed New Brunswick to remain at the forefront of information infrastructure development (Coleman and McLaughlin, 1988; Loukes and Nandlall, 1990).

15.1.7 Measures of success

While New Brunswick is a small, rural province and its experiences in developing a land information infrastructure may not be readily replicated elsewhere, its history does provide some important lessons. For example, much of its success to date may be attributed to:

- *The importance of a long-term strategic vision and high level political support.* The province had a vigorous agenda for more than two decades related to building and linking its geospatial databases, and providing effective public access to them. This agenda has been strongly supported politically, especially by a former Premier who gave the highest priority to enhancing the information infrastructure of the province.
- *The importance of a lead agency.* For more than a decade, New Brunswick has had a lead agency (SNB) for (1) designing and implementing the GDI concept; (2) co-ordinating the development of standards and protocols; (3) building and sustaining core data sets; (4) providing online public access.
- *A focus on key priorities.* The province has concentrated on building key data sets of particular importance to the economic and social development of the province. This has included parcel-based data sets in support of reforming the province's land administration systems and selected geospatial data sets required for effective resource management (especially in support of integrated forest management practices).
- *The importance of a business focus.* The development and maintenance of the geospatial data infrastructure in New Brunswick is driven by a multi-year business plan. The lead agency, SNB, is required to be self-sufficient. This has contributed to the emphasis on well-documented business cases for data and networking priorities, and on the funding strategies for ongoing upgrading and maintenance of the infrastructure.

15.1.8 Summary

In New Brunswick, land information infrastructure development efforts have built upon earlier initiatives and past successes, which laid a firm foundation.

Service New Brunswick (SNB) is now completing the infrastructure by providing convenient public access to integrated data sets of land-related, government information. As this happens, an environment of participatory democracy will be fostered as everyone will have equal access to more comprehensive land information, and thus the ability to make better informed decisions based upon that up-to-date information.

15.2 Geospatial data infrastructure in North Carolina—Karen Siderelis

15.2.1 Introduction

It would be somewhat misleading to describe the development of the geospatial data infrastructure in North Carolina as a sequence of planned, integrated events—in fact, the infrastructure as it exists today is the result of many disparate activities that have come together through fortuity and opportunism. It now forms a robust and highly collaborative data infrastructure. That being said, the State of North Carolina's long-standing focus and commitment to technology certainly underpins the success of its geospatial data infrastructure and has provided a unifying framework for implementation.

The evolution of the North Carolina geospatial data infrastructure generally follows the historic trends described in this chapter with slight exception. The use of digital geospatial data and GIS technology did not take a foothold in North Carolina until the 1970s. Consequently, that decade possessed a combination of the characteristics of both the 1960s and the 1970s described in this chapter.

For the sake of simplicity, the history of building the infrastructure can be tied symbolically to four entities: the Centre for Geographic Information and Analysis (CGIA), the Land Records Management Program, the Geographic Information Co-ordinating Council (GICC), and the Corporate Geographic Database. The CGIA was established in 1977 as a result of the Land Policy Act, which provided the initial impetus to create an information system to support land use planning and resource inventory. CGIA operates as a service centre, and its early years involved a major emphasis on the practical application of GIS technology to problems and issues facing the state. From the beginning, applications were undertaken largely to support decision-making about critical issues rather than simply to automate day-to-day business operations. While the agency recognized a compelling need for a statewide digital topographic base map, there was insufficient financial or political support to address this need, and the focus remained on one-off, innovative applications of technology until well into the 1980s.

Simultaneously with the CGIA's efforts at the state level, the Land Records Management Program (also created in 1977) provided technical and financial assistance to local government in developing cadastral information systems to advance the modernization of land records. While the results of this programme have been a boon to recent efforts towards a statewide geographic information infrastructure, early efforts were mostly approached county-by-county. Many individual counties created valuable, comprehensive cadastral databases according to common standards and specifications, but there was little effort to integrate these databases into a statewide data resource.

Consistent with trends described in this chapter, the 1980s brought about increased awareness of the benefits of statewide, intergovernmental collaboration and formalization of the notion of the 'North Carolina Corporate Geographic Database'. In the early 1990s the state embarked on the goal of a true statewide geospatial data infrastructure. In 1991 the Governor, by executive order, created the Geographic Information Co-ordinating Council, designated CGIA as a lead agency for co-ordination of geographic information, and placed the Centre within the organization of the Governor's Office of State Planning. Henceforth the GICC and associated co-ordination structure have guided the implementation of a statewide geographic information infrastructure.

15.2.2 Management structure

Figure 15.2.1 displays the management structure of the Geographic Information Co-ordinating Council. As shown, the GICC is a standing committee of the Information Resources Commission, the state's regulatory body created by the legislature and responsible for all information technology in the state. The Council is a policy body consisting of elected and appointed heads of state agencies, representatives of users, and representatives of the various public and non-public sectors with an interest in geographic information and technology. The GICC is staffed by CGIA and supports three user committees and two special advisory committees. There also is a Management and Operations Committee, whose membership includes the Council chair and vice-chair, and each committee chair.

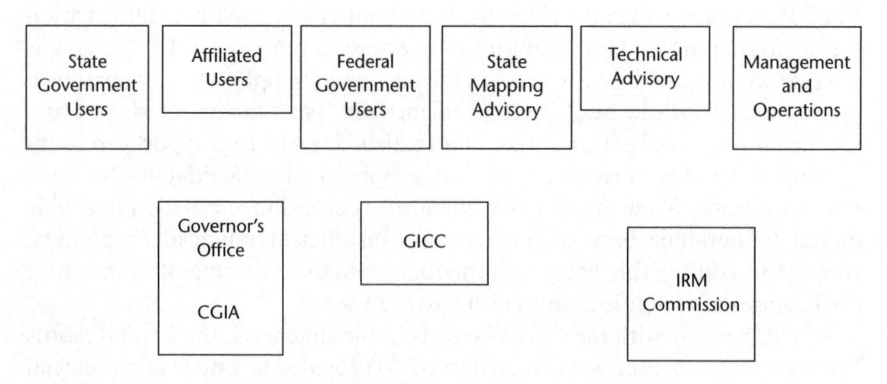

FIGURE 15.2.1 Management structure of North Carolina's Geospatial Data Infrastructure.

15.2.3 Stakeholders

The membership of the GICC is designed to be representative of the major stakeholders of the geographic information infrastructure and to provide a forum for stakeholders to jointly oversee statewide initiatives. The members of the GICC include:

- Secretary of the Department of Environment and Natural Resources
- Secretary of Transportation
- Secretary of Administration
- Secretary of Commerce
- Secretary of State
- Secretary of Health and Human Services
- Secretary of Crime Control and Public Safety
- Commissioner of Agriculture
- Superintendent of Public Instruction
- Head of an at-large state agency, appointed by the Governor
- State Budget Officer
- State Planning Officer
- President of the Community College System
- President of the University of North Carolina System
- Representative of the State Government User Committee
- Representative of the Affiliated User Group
- Representative of county government, appointed by the Governor
- Representative of municipal government, appointed by the Governor
- Representative of federal government (in NC), appointed by the Governor
- Representative from Lead Regional Organizations, appointed by the Governor
- Representative of non-government sector, appointed by the Governor.

Ex officio members include representatives from the Information Resources Management Commission, the League of Municipalities, and the Association of County Commissioners. The current chair of the GICC, appointed by the Governor, is the Senior Advisor for Science and Technology.

15.2.4 Purpose of the geospatial data infrastructure

The purpose of North Carolina's geographic information infrastructure is best expressed in a vision statement that forms the basis of the 'Strategic Plan for Geographic Information Co-ordination in North Carolina'. The vision statement, and indeed the entire plan, was developed through consensus of a broad community of stakeholders in the state and reads as follows:

The state of North Carolina aims to have a statewide framework for geographic information operational by the year 2000. That framework will enable North Carolinians to take the availability of geographic information for granted, in the same way that they take good roads and clean water for granted.

The foundation is a comprehensive statewide database whose content, accuracy, and scales have been determined through consensus and in recognition of the critical uses to which it will be applied. While any user may have a unique view of the database and it ostensibly may be physically distributed and maintained by a variety of data custodians, it will appear to users as a consolidated, integrated database.

The database will be accessible over a network to all sectors of the State including government agencies, utilities, private firms, schools, universities, and individual citizens. The network will operate at high speed and support a diversity of information types including voice, data, and video. Access to the database through the network will be priced fairly and economically.

Standards and procedures will ensure that the database contains no unnecessary redundancies or inconsistencies, and that data are adequately and uniformly documented. Security measures will be implemented to protect confidential/restricted data and to limit access to any user's esoteric, local data.

Users will be supported by service centre(s) that provide GIS production and consulting services, technical support, education, training and outreach, and clearinghouse functions.

Innovative partnerships and co-operative agreements between agencies will be in place to ensure that the geographic information infrastructure endures and continues to meet user needs.

Although the 'Strategic Plan for Geographic Information Co-ordination in North Carolina' was finalized in 1994, the vision statement has stood the test of time and continues to exemplify the purpose and goals of North Carolina's geographic information infrastructure.

15.2.5 Costs and financing

Owing to the diversity of stakeholders and contributors to the North Carolina programme, it is virtually impossible to measure the complete costs of development and implementation. Likewise, there has been no single source of funds to support the infrastructure nor has there been a unified financing strategy. However, the following anecdotal evidence provides insight into both the costs and methods of financing the information infrastructure.

CGIA, from its inception, has operated on a cost recovery basis. Its sole source of funds is receipts collected as user fees. The costs of the co-ordination functions provided by CGIA are treated as overheads and incorporated in the CGIA rates. Annual revenues of the CGIA programme are approximately US$2 million.

The Land Records Management Program has been funded through state appropriation. However, the amount of appropriation allocated from grants to local government has varied widely from year to year. In the early years, grant moneys often approached US$1 million per year but, unfortunately, no moneys have been provided for grants in recent years. It is estimated that approximately US$5 million has been provided through the programme as seed grants to local government, encouraging about US$30 million in local spending.

15.2.6 Success factors

From many perspectives, the North Carolina geospatial data infrastructure is a tremendous success story. This success can be attributed to a number of factors: the commitment and creativity of individual people at all levels in the state who exploited opportunities to create an information infrastructure; an emphasis on using the infrastructure for decision support and to address critical issues facing the state; enthusiastic focus on multi-lateral collaboration; and the spin-off benefits of being located in a technologically progressive state where political support for technology initiatives is high.

15.3 The Public Sector Mapping Agencies of Australia —Don Grant

15.3.1 Meeting a national need

In 1993 the public mapping agencies of the Commonwealth, the States and the Territories of Australia agreed to co-operate as a consortium, to be known as the Public Sector Mapping Agencies (PSMA), in response to the Australian Bureau of Statistics' 1996 Census Mapping tender. For historical reasons, there was no complete digital map coverage of Australia available from any agency at the scales required by the Australian Bureau of Statistics (ABS).

The PSMA won the data supply portion of the ABS tender in 1993 and developed a digital map data set to meet the needs of the ABS. The data set is a multi-resolution data set sourced from the Commonwealth, States and the Territories, incorporating high-resolution topographic and cadastral information into one detailed, accurate, and authoritative national spatial database, as shown in Figure 15.3.1. It was considered as a realistic start to the data component of the National Geospatial Data Infrastructure.

One of the major challenges faced by the PSMA was the inconsistency between the various data sets. The States and Territories did not hold their data in the same system, format, or specification, and not all jurisdictions held equivalent map coverage. Data description tables, as shown in Figure 15.3.2, were adopted to characterize the PSMA data by jurisdiction, to highlight differences between data, to indicate compliance with ABS requirements, and to provide ABS and Candata, a cartographic enhancement company, with a working specification upon which to design the mapping system. The

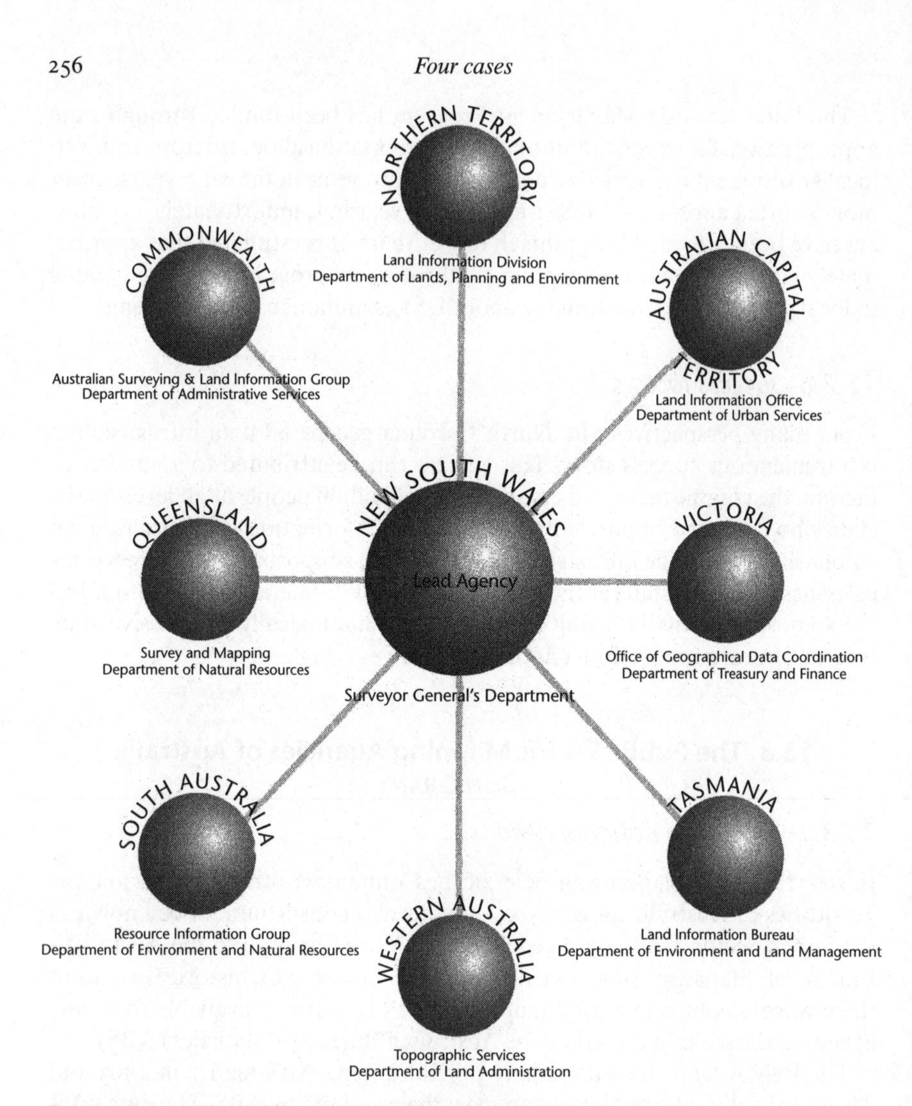

FIGURE 15.3.1 The public sector mapping agencies.

differences created a complex situation and highlighted the need to identify a uniform national approach to the collection and storage of digital data. This would enable all parties to maximize benefits and utilize all the resources available. As the lead agency, the Surveyor-General's Department of New South Wales examined and documented all the disparate specifications, and defined one common specification. Guidelines have been set for the data set.

Three categories were delineated across Australia, as shown in Figure 15.3.3. The categories were described as:

1. *Urban*: covering capital cities and major populations. Data was sourced principally from digital cadastral databases and other large-scale coverage provided by the States and Territories.

NSW

FEATURE	C (%)	PA (m)	LC (%)	LA	L(s)	ID	OTHER SOURCE	REMARKS
(a) Highway	95	5	95	99	DCDB	Y		
(b) Main Road	95	5	95	99	DCDB	Y		
(c) Sealed	95	5	95	99	DCDB	N		

QLD

FEATURE	C (%)	PA (m)	LC (%)	LA	L(s)	ID	OTHER SOURCE	REMARKS
(a) Highway	95	5	95	99	DCDB	Y		
(b) Main Road	95	5	95	99	DCDB	Y		
(c) Sealed	95	5	95	99	DCDB	Y		

ACT

FEATURE	C (%)	PA (M)	LC (%)	LA	L(s)	ID	OTHER SOURCE	REMARKS
(a) Highway	100	0.1	99	99	A1	N	NO	
(b) Main Road	100	0.1	99	99	A1	N	NO	
(c) Sealed	100	0.1	99	99	A1	N	NO	

VIC

FEATURE	C (%)	PA (m)	LC (%)	LA	L(s)	ID	OTHER SOURCE	REMARKS
(a) Highway	95	5	95	100	DCDB	N	Topo dig.	C/L
(b) Main Road	95	5	95	100	DCDB	N	Topo map	&
(c) Sealed	95	5	95	100	DCDB	N	ESMAP	casing

TAS

FEATURE	C (%)	PA (m)	LC (%)	LA	L(s)	ID	OTHER SOURCE	REMARKS
(a) Highway	99	5	99	95	Atlas	Y		
(b) Main Road	99	5	99	95	Atlas	Y		Road or street
(c) Sealed								

SA

FEATURE	C (%)	PA (m)	LC (%)	LA	L(s)	ID	OTHER SOURCE	REMARKS
(a) Highway	95	5	95	99	Topo	N	DCDB	From 1:2,500
(b) Main Road	95	5	95	99	Topo	N	DCDB	& 1:10,000
(c) Sealed	95	5	95	99	Topo	N	DCDB	C/L & casing

WA

FEATURE	C (%)	PA (m)	LC (%)	LA	L(s)	ID	OTHER SOURCE	REMARKS
(a) Highway	95	3	95	95	SCDB	N	DMR	Classified by
(b) Main Road	95	3	95	95	&	N	DMR	surface
(c) Sealed	95	3	95	95	TDB	Y	DMR	

NT

FEATURE	C (%)	PA (m)	LC (%)	LA	L(s)	ID	OTHER SOURCE	REMARKS
(a) Highway	95	1	95	99	Photo	Y	From 1:2,500	
(b) Main Road	95	1	95	99	Photo	Y	&	
(c) Sealed	95	1	95	99	Photo	Y	1:10,000	

FIGURE 15.3.2 Extracts from the PSMA data description tables showing variations in data characteristics for urban roads across all jurisdictions. C = Completeness; PA = Positional Accuracy; LC = Logical Consistency; LA = Label Accuracy; L(s) = Main Source; ID = Identified Separately.

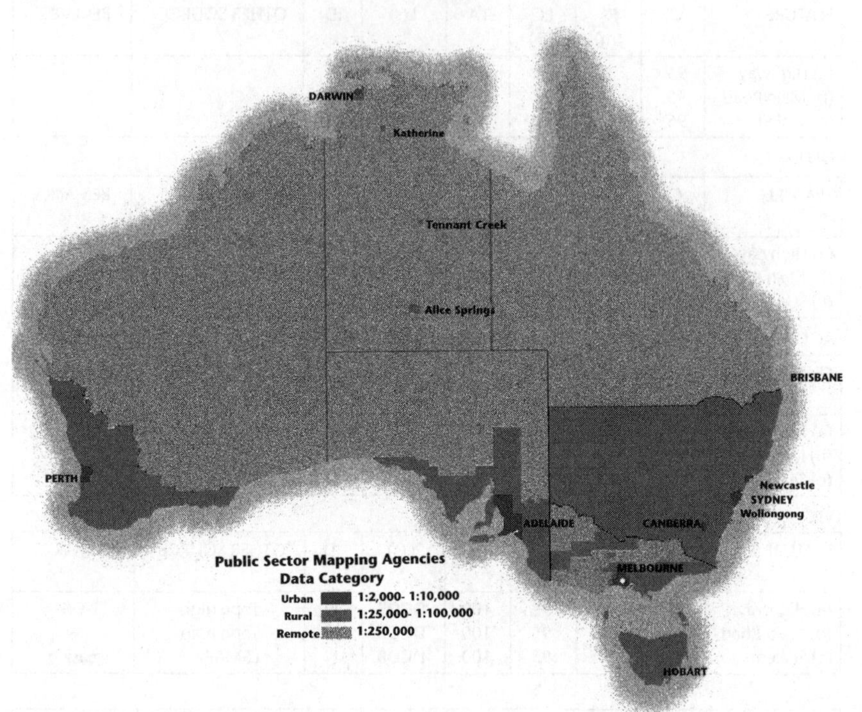

FIGURE 15.3.3 PSMA data categories across Australia.

2. *Rural*: States and Territories provided digitized topographic map data at scales between 1:25,000 and 1:100,000, augmented by data from the Digital Cadastral Database (DCDB) within townships having a defined street pattern.

3. *Remote*: the principal source of data in this zone was GEODATA TOPO 250K, augmented by large-scale topographic data and DCDB data within townships having a defined street pattern.

In each data category, data quality attribute tables were compiled to indicate completeness, positional accuracy, logical consistency, label accuracy, and source. A list of attributes is recorded in Figure 15.3.4.

15.3.2 Task management

The other major challenges for the PSMA were to co-ordinate data delivery according to the strict schedule set by the ABS, cope with changes to the schedule, sort out technical difficulties, and maintain liaison between all the organizations involved. The orchestration of nine agencies providing spatial data of varying quality over the Australian landscape was a major logistic undertaking.

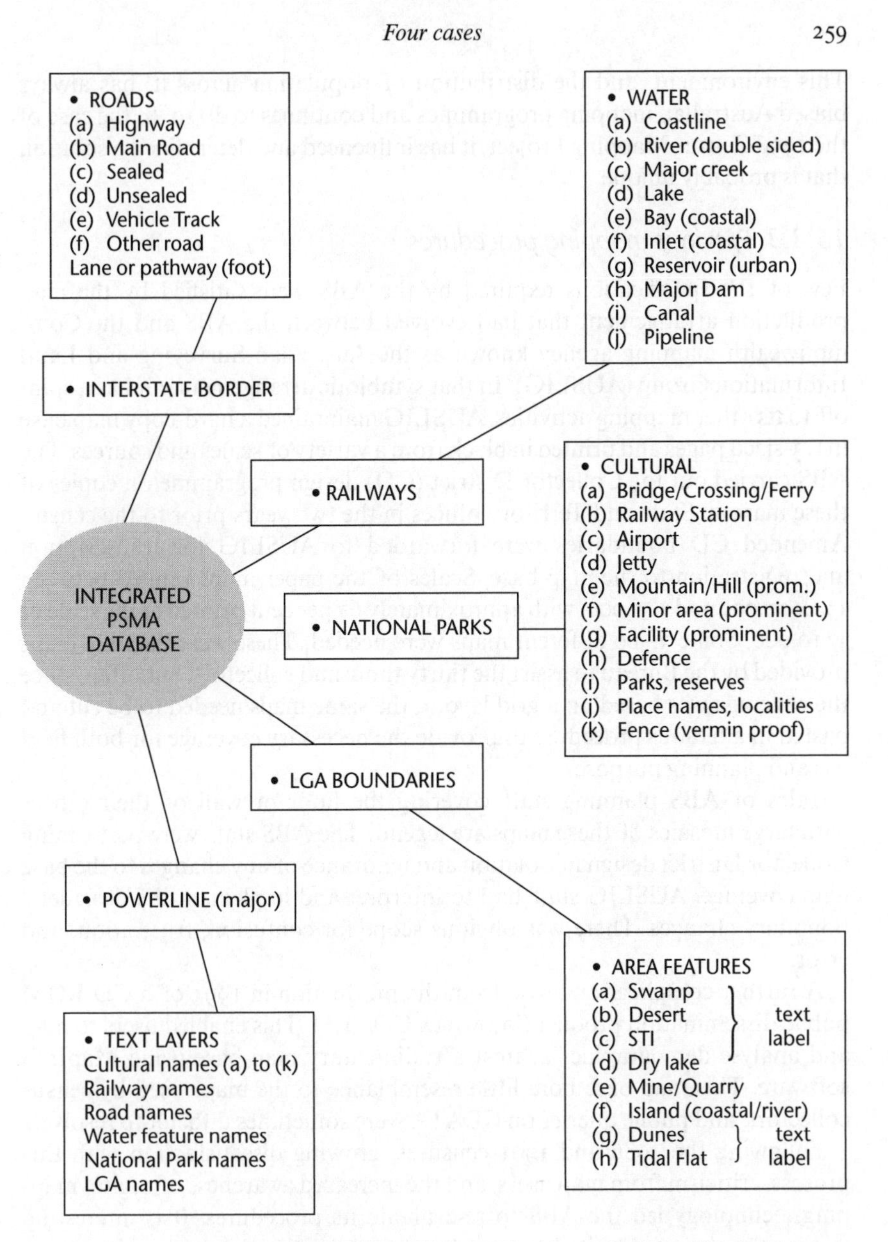

FIGURE 15.3.4 PSMA feature list for the Census Mapping Project Database.

Australia cannot claim to be the first to see its census collection agency drawn towards the vortex of GIS activity. David Rhind (1991) has provided an excellent overview of British and American efforts in this regard. Australia can claim, however, that it has one of the most difficult and challenging census collection environments, presenting formidable difficulties for census mapping.

This environment, and the distribution of population across it, has always biased Australian mapping programmes and continues to do so. In the case of the 1996 Census Mapping Project, it has influenced and determined a solution that is probably unique.

15.3.3 Previous mapping procedures

Few of the specifications required by the ABS were satisfied by the map production arrangement that had evolved between the ABS and the Commonwealth mapping agency known as the Australian Surveying and Land Information Group (AUSLIG). In that symbiotic arrangement, and as a spin-off to its other mapping activities, AUSLIG maintained a hard-copy map base in C3-sized pages and printed in black from a variety of scales and sources. The ABS carried out the Collector District (CD) design programme on copies of these maps in State and Territory offices in the two years prior to the census. Amended CD boundaries were forwarded to AUSLIG for transcription and registration to the map base. Scales of the paper maps ranged between 1 : 1,000,000 and 1 : 1,000, with approximately 67 per cent printed at the scale of 1 : 10,000. Some 7,500 different maps were needed. These were the only maps provided by the Bureau to assist the thirty thousand collectors, but often, since the coverage was based on a grid layout, the same maps needed to be cut and pasted in a tedious procedure to provide the necessary coverage for both field use and planning purposes.

Tales of ABS planning staff covering the floor or wall of their offices with large mosaics of these maps are legend. The ABS staff were performing Collector District design in isolation and ignorance of any changes to the base map coverage. AUSLIG staff had to interpret and implement the designers' boundary changes. There was obvious scope for confusion, frustration, and error.

A further complication arose from the production in 1991 of a CD-ROM public dissemination product (known as CDATA). This enabled users to map and analyse demographics against a rudimentary map base using MapInfo software. This map base bore little resemblance to the maps used by census collectors, and public queries on CDATA were sometimes difficult to resolve.

Following the 1986 and 1991 censuses, growing dissatisfaction with this process, criticism from map users, and the increased awareness of digital mapping technology led the ABS to re-evaluate its procedures. It is interesting to note the close similarity between these procedural and mapping shortcomings and those listed by Rhind (1991) that led the British Census Bureau to re-examine its census mapping procedures.

15.3.4 Future applications

The quality and value of the new census data set (a sample is displayed in Figure 15.3.5) has since enabled the PSMA to sell the data commercially and license its use to numerous value added resellers (VAR). In September 1998

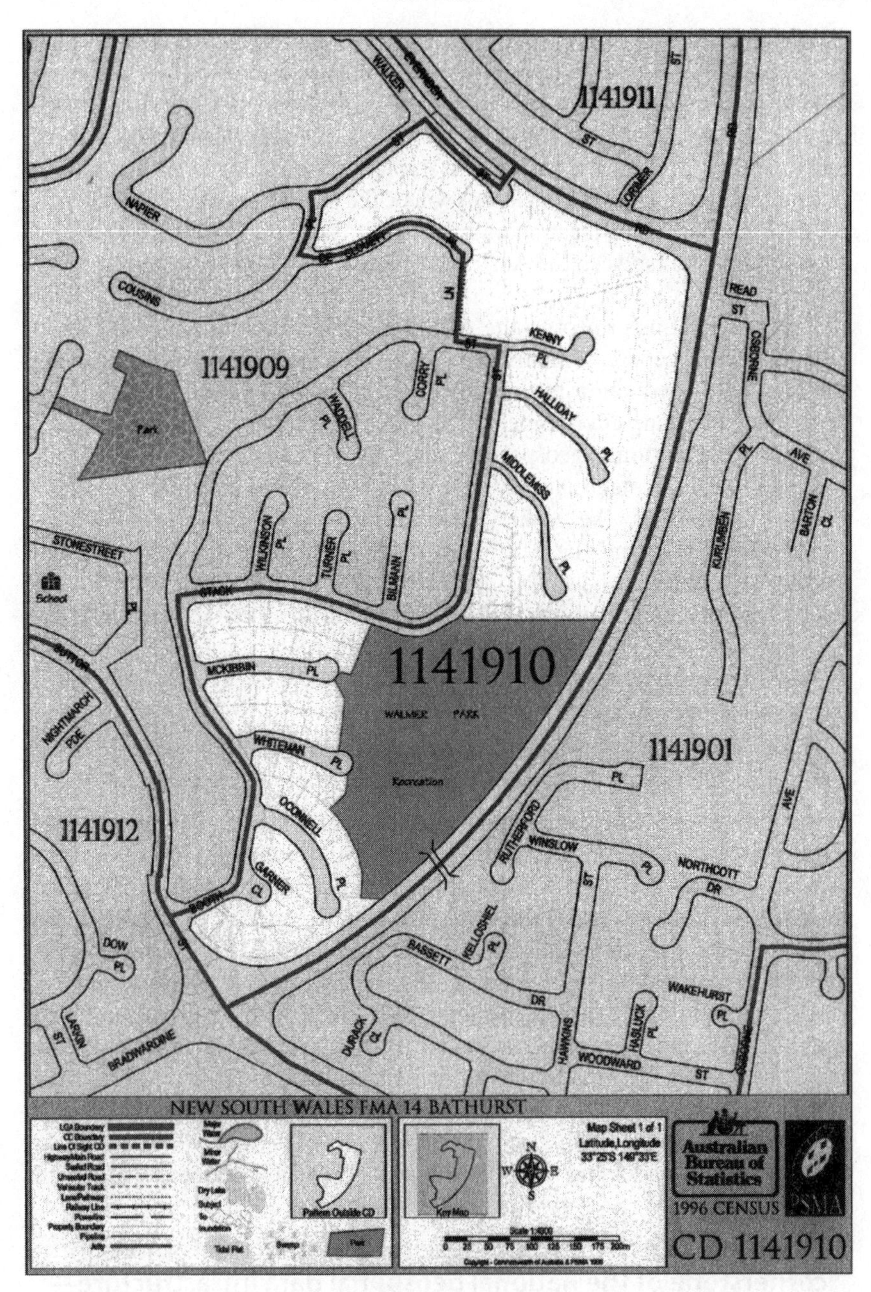

FIGURE 15.3.5 Sample of the Public Sector Mapping Agencies data set.

alone, the PSMA was approached by over fifteen valued added resellers, including banks and agricultural suppliers, with a view to using the database as the foundation for overlaying other information to support business applications. The more unusual applications were for psychometric mapping, agricultural produce tracking, and intelligent transportation systems. In the future, it is expected that business applications will be the primary marketing area of the PSMA. Joint ventures creating additional data sets are also being developed with Telstra, and between the Australian Electoral Commission and Australia Post.

The value of this project to the ABS has become apparent as ABS census designers have become familiar with PSMA data, because of the time savings achieved and a reduction in the cost per head of population made possible by the new mapping environment. The value of the project for creating Edition 1 of the national geospatial framework is also evident through the successful business negotiations with telecommunications carriers, national transport operators, the Sydney Organizing Committee for the Olympic Games (SOCOG), and the Australian Electoral Commission (AEO). Commercial organizations have also recognized that this product, which has the socio-economic data from the ABS aligned to it, is the de facto authoritative national data set. The database not only represents the first edition of a national topographic database but could also be the basis of a national cadastral database and a national digital road network.

15.3.5 Conclusion

In itself, the creation of the national data sets for the Census Mapping Project (CMP) was a major achievement. However, the architects of the scheme believed it would become the foundation for a national geospatial data infrastructure. It is accepted that the data sets, which comprise the CMP, do not constitute all the fundamental information layers needed for sound and comprehensive national decision-making. But already additional layers are being built because of the existence of the CMP. These include urban and rural addressing for emergency despatch and property taxation mapping. A final conclusion, and a warning to the policy makers and standards creators, is that all the proselytizing in the world will not produce a GDI. Whilst essential for the envelope into which a geospatial data infrastructure will fit, it will not happen without leadership, sound management, and a funding source.

15.4 The Dutch clearinghouse for geospatial information: cornerstone of the national geospatial data infrastructure— ARNOLD BREGT

15.4.1 Introduction

The development and implementation of the Dutch national clearinghouse for geospatial data is presented as an example of one step in the evolution towards a National Geospatial Data Infrastructure (NGDI).

In The Netherlands, a more structured thinking on the organization and development of the national geospatial data infrastructure started to appear around 1992, when the RAVI (the Netherlands Council for Geographic Information) released a policy document, related to property (parcel based) information (*Structuurschets voor de vastgoedinformatie*; RAVI, 1992). On the basis of this document, several projects were initiated to develop and improve the Dutch NGDI (standards, policy issues, foundation and framework data, pricing, etc.). One of these was the development of a nationwide metadata service for geo-information, called the national clearinghouse for geospatial information (NCGI). The clearinghouse was developed in three distinct phases with government organizations the dominant stakeholders in each phase. The three phases are described here, with particular attention given to the management structure, financing, stakeholders, and technical aspects. For more detailed information on organizational issues see Absil-van de Kieft and Kok (1997). Finally, there is my personal view of these developments and I pay special attention to our experience of the keys to success or failure.

15.4.2 Pioneer period (1995–1996)

The first phase (the pioneer period) can be characterized as 'learning by doing'. A bottom–up strategy was chosen to realize an optimal involvement of the Dutch geo-information sector through the RAVI and to start projects around the concept of a 'clearinghouse'. No financial support was given to these organizations. Ultimately, five projects were started:

- a metadata standards project;
- a project on user interfaces for clearinghouses;
- two projects on the organization of metadata within a specific organization;
- a project to develop a first prototype for the Dutch clearinghouse.

Besides public awareness and experience among the participants, this phase also yielded a prototype clearinghouse (Idefix). A central metadata service was created on the Internet, which offers the functionality to select and view metadata of about 500 data sets. The European metadata standard (CEN/TC 287, 1996) was used to describe the data sets. The overall management structure in this phase was rather loose. There were occasional, informal meetings to discuss the progress and achievements in the different pilot projects. The initiating organizations in this period were:

- Ministry of Housing, Spatial Planning and the Environment
- Ministry of Public Transport and Water Management
- Central Bureau for Statistics
- Cadastre
- National Institute for Public Health and Environment
- DLO Winand Staring Centre

- Provincial authority of Gelderland
- Local authorities of Amsterdam and Utrecht.

The amount of money involved in this phase is difficult to estimate, because the majority of the funds were spent within the pilot projects of the different organizations. A rough estimation is EURO 100,000 for central co-ordination and EURO 500,000 for the pilot projects.

15.4.3 Project period (1996–1997)

In the second phase (1996–7), a professional clearinghouse was built and, at the same time, the number of participants was significantly increased (*http://www.ncgi.nl*). In this phase, a much more formal project structure (Figure 15.4.1) was established. At the centre of the exercise was the project group, which controlled three sub-projects for:

1. System development;
2. Public relations and communication;
3. Organizational issues.

A new design for the clearinghouse was drawn up within the development of the sub-projects. The functionality of the clearinghouse remained more or less the same as in phase 1, but the look-and-feel of the clearinghouse was changed considerably (Figure 15.4.2). The technical structure of the NCGI was also re-engineered to provide a more robust structure.

The goal of the sub-project organizational issues was to attract more participants by advising organizations on the uses and possibilities of metadata and the metadata standard. A model was developed for evaluating the maturity of an organization for participation in the clearinghouse. This model ranks

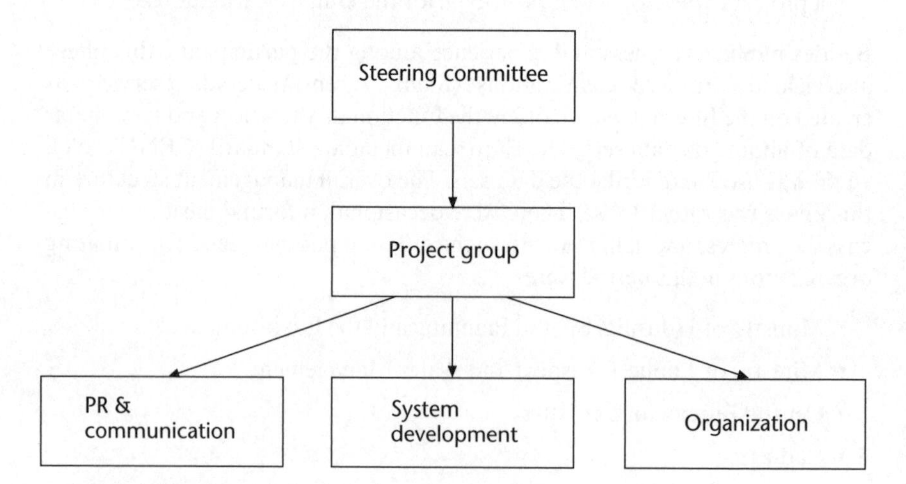

FIGURE 15.4.1 Project structure in phase 2.

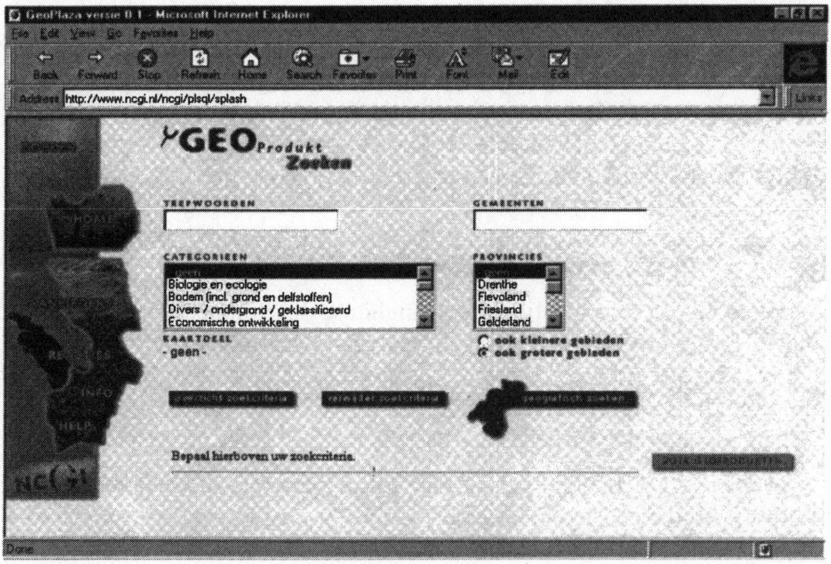

FIGURE 15.4.2 Search screen of the National Clearinghouse for Geo-Information.

organizations on a scale of five, ranging from 1 (NGDI not relevant) to 5 (very mature organization). The ranking of an organization was determined by the following criteria:

- the position of the organization in the field of geospatial information;
- the technical infrastructure within the organization;
- geospatial data infrastructure within the organization.

The activities of the project group were controlled by a steering committee, comprising members of the organizations who initiated the project. The central budget for this phase was about EURO 500,000, jointly provided by the organizations in the steering committee and the government's National Electronic Highway Programme. In addition to this central budget, about the same amount of money was spent by the organizations on supporting activities, such as the preparation of metadata for the NCGI.

15.4.4 Institutionalizing period (1998–)

In the third and current phase (1998–), a separate organization has been set up to maintain and further develop the national geospatial data infrastructure—the NGDI institute. This is an independent organization, lead by a director and a small staff. One of the main reasons for establishing the institute is that a metadata service has an infrastructural character, the continuation of which could not be guaranteed in a project organization. During its short existence, the NCGI institute has initiated various projects to increase the functionality of the site and to attract more participants. The annual budget for maintenance and improvement is EURO 250,000 per year, which is guaranteed by the founding organizations until the year 2000. In the future, it is hoped that the NGDI institute will generate its own income by offering value added services or by taking percentages from financial clearing between suppliers and buyers of geo-information.

15.4.5 Conclusion: A personal view

What has four years of national clearinghouse development in The Netherlands achieved? What were the main successes and mistakes? I will try to answer these questions by reviewing the project and by looking at the effect of NGDI on the geospatial data community in The Netherlands.

There are a number of positive aspects in relation to the project:

Communication. In all phases of the project there was an intensive communication with the geospatial data sector. This has led to wide acceptance and support for the clearinghouse activities.

Bottom–up strategy. The initial development was largely a bottom–up activity. This has resulted in broad support for the clearinghouse.

Maturity model. The maturity model proved to be a practical and simple model to evaluate organizations on their possible participation in the NGDI. Based on the results of the maturity evaluation, practical advice can be given to improve the internal information infrastructure.

The following less positive aspects can also be identified:

System development. For various reasons, the technical development of the clearinghouse site has shifted to a new company at the beginning of each new period. This has had a negative effect on the functionality and robustness of the site.

Centralized concept. The choice for the present centralized concept of the NGDI has the advantage of a relatively simple technical infrastructure. It nevertheless gives rise to a number of problems. The structure of the metadata to be delivered is too rigid, and the update frequency of the metadata is poor. The central organization needed to run the clearinghouse is quite cumbersome and probably not financially sustainable in the long run.

Metadata of organizations. A national clearinghouse can only operate on the basis of existing local metadata services within organizations. Only a few organizations in The Netherlands previously had an operational metadata information system. We underestimated the problems and the time necessary for organizations with no metadata history to deliver such data to the clearinghouse. We found that it takes them at least six to twelve months, which is a significant resources commitment for any organization.

Four years of clearinghouse development has not yet led to an intensive geospatial data market on the Internet in The Netherlands. From this point of view, the clearinghouse has not yet fulfilled its original goals. It may be that the indirect effects of the clearinghouse are at this time in fact more important. The clearinghouse has resulted in the use of a single standard for metadata in The Netherlands. This development has also generated a strong awareness in The Netherlands that metadata plays a crucial role in the design and implementation of a GDI both within and between organizations, within and between application domains. In addition, the project gave a necessary impulse to the discussion on the availability and pricing policy of geospatial data within governments.

Bibliography

ABSIL-VAN DE KIEFT, I. N. and KOK, B. C. (1997). 'The development of a geo-metadata service for the Netherlands', in *Proceedings Third Joint European Conference & Exhibition on Geographical Information*, pp. 1022–32.

ARSENEAU, B. and OGILVIE, M. (1994). 'Maintenance strategy for the New Brunswick digital topographic database', *Proceedings The Canadian Conference on GIS*, GIS '94. Ottawa, Canada, 6–10 June, 2, pp. 1308–12.

——, COLEMAN, D., and MACLEOD, G. (1997). 'The development of a land gazette', *Papers from the Annual Conference of the Urban and Regional Information Systems Association (URISA)*, Toronto, Canada, 19–23 July.

BRANSCOMB, A. (1986). 'Law and Culture in the Information Society', *The Information Society Journal*, 4/4: pp. 279–311.

CEN/TC 287 (1996). *Geographic information-data description-metadata*, Technical Report, draft prEN 12657, Brussels: CEN.

COLEMAN, D. J. (1988). 'Implementing a land information network in New Brunswick'. Unpublished M.Sc.E. thesis, Department of Surveying Engineering, University of New Brunswick, Fredericton, Canada.

——(1989). 'The New Brunswick Geographic Information Corporation: A New Approach to an old Problem', *CISM Journal*, ACSGC, 43/2: p. 199.

——and MCLAUGHLIN, J. D. (1988). 'The Landnet Project in New Brunswick: moving from research to reality', *Proceedings of the 1988 Annual Meeting of the Urban and Regional Information Systems Association (URISA)*, Washington, DC, 1, pp. 80–90.

DALE, P. F. and MCLAUGHLIN, J. D. (1988). *Land Information Management: An Introduction with Special Reference to Cadastral Problems in Third World Countries*. Oxford: Oxford University Press.

DAWE, P. (1996). 'An investigation of the internet for spatial data distribution'. Unpublished M.Eng. Report, Department of Geodesy and Geomatics Engineering, University of New Brunswick, Fredericton, Canada.

DOIG, J. F. and PATTON, B. (1994). *The LRIS Story: A Legacy for the Maritimes*. Halifax, NS: The Council of Maritime Premiers.

FINLEY, D., ARSENEAU, B., MCLAUGHLIN, J. D., and COLEMAN, D. J. (1998*a*). 'The Provincial Land Information Infrastructure for New Brunswick: From Early Visions to Design to Reality', *Geomatica*, 52/2: pp. 165–75.

——, MCLAUGHLIN, J. D., ARSENEAU, B., and COLEMAN, D. J. (1998b). 'Serving the Public: The New Brunswick Land Gazette', *URISA 1998 Conference Proceedings*, Charlotte, NC, 18–22 July, pp. 268–77.

(GIC) Geographic Information Corporation. (1989). *Geographic Information Corporation: A Discussion Paper*, Government of New Brunswick, Canada.

(GIC) Geographic Information Corporation. (1992). *Annual Report 1990–1991*, Government of New Brunswick, Canada.

(GIC) Geographic Information Corporation. (1993). *Annual Report 1992–1993*, Government of New Brunswick, Canada.

LOUKES, D. K. and NANDLALL, N. (1990). 'Developing a Land Information Network for New Brunswick: implementation recommendation', *Conference Proceedings GIS for the 1990s*. CISM, Ottawa, Canada, 5–8 March, pp. 17–40.

MCLAUGHLIN, J. (1991). 'The evolution of the multipurpose cadastral concept: a North American perspective', in BESEMER, J. (ed.), *Kadaster in Perspectief*. Apeldoorn: DKOR.

——and COLEMAN, D. (1990). 'Land Information Management into the 1990s', *World Cartography* XX, pp. 136–48.

MELODY, W. H. (1981). 'The economics of information as resource and product', in WEDEMEYER, D. (ed.), *Proceedings Pacific Telecommunications Conference*, PTC '81. Honolulu, Hawaii, 12–14 January, c7-5-c7-9.

NICHOLS, S. E. (1993). *Land Registration: Managing Information for Land Administration*, Ph.D. thesis, Department of Surveying Engineering, Technical Report No. 168, University of New Brunswick, Fredericton, Canada, 340 pp.

(NRC) National Research Council. (1980). *Need for a Multipurpose Cadastre*. Washington: National Academy Press.

OGILVIE, M. (1991). 'The standards effort in New Brunswick', *Proceedings of the Canadian Conference on GIS*, GIS '91. Ottawa, Canada, 18–22 March, pp. 748–53.

——(1997). 'Automated systems as a tool in providing security of tenure', *Conference Proceedings in Kusasa CONSAS '97: Surveying Tomorrow's Opportunities*. Durban, South Africa, August, pp. 25–8.

PALMER, D. (1984). *A Land Information Network for New Brunswick*, M.Sc.E. thesis, Department of Surveying Engineering, University of New Brunswick, Fredericton, Canada.

PLEWE, B. (1997). *GIS Online: Information Retrieval, Mapping, and the Internet*. Santa Fe, New Mexico: OnWord Press.

RAVI (1992). *Structuurschets voor de vastgoedinformatie*. Amersfoort: RAVI (in Dutch).

RHIND, D. W. (1991). 'Counting the people: the role of GIS', in MAGUIRE, D. J., GOODCHILD, M. F., and RHIND, D. W. (eds.), *Geographical Information Systems: Principles and Applications*, vol. 2. Harlow: Longman/New York: John Wiley and Sons, pp. 127–37.

ROBERTS, W. (1976). 'An operational multipurpose cadastre'. Paper presented at the Forsyth County Lands Records Information Systems Workshop, March. LRIS Collected Papers, Department of Geodesy and Geomatics Engineering, University of New Brunswick, Fredericton, Canada.

SIMPSON, R. L. (1990). 'The Maritime Land Information Corporation', *Proceedings of AM/FM International Conference XIII*. Baltimore, 13–26 April, pp. 353–8.

STRUTZ, S. (1994). *Development of a Data Distribution Model for an On-Line Real Property Land Information Network*, M.Eng. Report, Department of Geodesy and Geomatics Engineering, University of New Brunswick, Fredericton, Canada.

16

Advancing the GDI concept

John McLaughlin and Richard Groot

16.1 Introduction

In this chapter we review the efforts to date to develop the geospatial data infrastructure concept and examine some of the key building blocks. This book has presented the technical and institutional elements of GDI. It is evident that it is still largely an academic vision to have fully functional access to geospatial data and ensure its responsible use at affordable cost, at either the national or application-domain level. This requires the purposeful integration of these elements. Most jurisdictions still have a primary goal of completing their core database development programmes. In Venezuela, for example, one of the largest mapping projects in the world was recently undertaken by Servicio Autonomo de Geografia y Cartografia Nacional to complete 1 : 50,000 digital coverage south of the Orinoco River.

At the same time, some jurisdictions have advanced to the stage of promoting specific GIS applications, while a much smaller number have begun to address enterprise-wide geospatial data requirements. Moving to the GDI stage still appears quite remote. Nevertheless we believe that the concept is sufficiently advanced to move beyond the academic theorizing, and begin examining the hard, practical issues required for the successful implementation of GDI.

16.2 Progress to date

The concept of geospatial data infrastructure as advanced in this book is a natural extension of telecommunications-based information infrastructure initiatives and the national and regional base mapping paradigms developed in the 1960s and 1970s. Evolving from earlier data sharing and programme co-ordination efforts, the term GDI encompasses the sources, systems, network linkages, standards, and institutional issues involved in delivering geospatially related information from many different sources to the widest possible group of potential users (Coleman and McLaughlin, 1997).

While sharing much of the language of the broader 'information highway' community, the GDI concept has very different roots, focusing not so much on the communications challenges per se but on optimizing both access to and utilization of geospatial databases. Indeed, the data-sharing paradigms which underpin the GDI concept have a history arguably dating back to the last century (e.g. the IMW initiative in 1891; see Chapter 1, p. 6) but re-emerging at least three decades ago with the promise of more flexible implementation due to the potential of Information and Communication Technology:

- In the 1960s, proponents of *integrated mapping* practices advocated the registration, overlay, interpretation, and analysis of different 'layers' or themes of geospatially related data sets to the practical solution of important problems in land use planning, resource inventory, and land administration.
- Through the 1970s, the multi-purpose cadastre concept launched major topographic and cadastral 'base-mapping' mega-programmes to support land administration at the local, state, and federal levels across North America, Australasia, and elsewhere. By putting in place a reliable and comprehensive basic map series, these mega-programmes were intended to reduce perceived duplications of effort in basic mapping by end-users and to encourage greater focus on creation and maintenance of special-purpose 'thematic' layers.
- In the early 1980s, the notion of 'information as a corporate resource' and the *information resources management* movement encouraged individual organizations to implement collective approaches to the collection, management, and sharing of designated hard copy and computer-based data holdings of 'corporate-wide' interest.
- By the beginning of the 1990s the geospatial data infrastructure concept was being proposed in support of accelerating geospatial data exchange standards efforts, selected national mapping programmes and the establishment of nation-wide information networks in Canada (McLaughlin, 1991), the United Kingdom (Rhind, 1992), the USA (Mapping Sciences Committee, 1993), and the European Community (EUROGI, 1996).

The manifestation of these data-sharing precepts evolved from early dreams of centralized 'land information databanks' through the 1960s and 1970s to the vision of more complex distributed land information networks in the 1980s. This vision conveyed the idea of linking together organizations responsible for the management of land-related information in a jurisdiction into a network to form a 'virtual' geographic information system which could be queried in a manner similar to a single database. Hearle (1962) suggested such a concept at the state and local government level in the early 1960s, and researchers in the land administration, geography, and geomatics communities have since examined the institutional and technological issues involved in considerable depth (e.g. Onsrud and Rushton, 1995; Palmer, 1984).

Anne Branscomb (1982) introduced the term 'information infrastructure' to refer collectively to the various media, carriers, and even physical infrastructure used for information delivery. By the end of the 1980s, the term was being used in a much broader context, increasingly viewed as an *enabling agent* (i.e. enabling users to 'plug in' to independent databases). Neil Anderson (1990) extended the definition by suggesting that information infrastructure should possess the following three important characteristics:

1. contents (data), conduit (telecommunications network), and process flow standardization;
2. extensive networking of major sources and users;
3. customization of the network for easy third-party access.

Through the 1980s and early 1990s, technological limitations and long, expensive database-loading programs served to limit the number of jurisdiction-wide, multi-participant land information projects that could legitimately qualify as a GDI initiative. Even so, critical mass has now been reached in a number of more recent enterprise- or jurisdiction-wide efforts.

At least five important reasons account for this acceleration (Coleman and McLaughlin, 1997):

1. *Increasing prominence of geospatial data handling within organizations*: There is now a proven track record of intelligent application of GIS and desktop mapping-based processes to an increasing number of strategic and operational problems in business and government.

2. *Robust, easy-to-use and relatively inexpensive tools*: Desktop mapping and geospatial data viewing software packages (e.g. ArcView, MapInfo, GeoMedia and Maptitude) are now widely available at relatively low cost, and Internet-based spatial data viewing and map-making tools (e.g. CARIS Internet Server, Autodesk Map Guide, GeoMedia Web Browser, and ESRI Internet Browser) are increasingly appearing in the market place.

3. *Ubiquitous data*: The necessary digital geospatial data sets covering *entire* areas of interest have now become more accessible, widely available, and better packaged than ever before. Increasingly, government mapping agencies are under pressure to provide robust, consistent, well-documented, and competitively priced digital data products to end-users and/or value-added repackagers. Especially in the USA, strong competition among value-added resellers has fostered a 'buyer's market' for geospatial data.

4. *Ubiquitous communications*: A critical mass of organizations and individuals have now worked through the technical and cultural transformations inherent in providing employees with desktop access to corporate networks

and, more recently, the Internet. In addition, recent advances in communications and database technology are finally resolving what were serious technical limitations in the management and transfer of large, distributed data sets.

5. *Greater availability of experienced people*: Perhaps most importantly, a large and increasing number of specialists and informed end-users are now familiar with the capabilities and limitations of both the tools and data sets *and* with how they relate to the requirements of their particular business.

In addition, the extensive and inexpensive positioning, tracking, and navigation capabilities of GPS are already pushing the next wave of market developments and are affecting the manner in which end-users view and employ existing geospatial data sets. Perhaps even more than the current mapping and GIS products in the market place, positioning 'appliances' will drive the requirements, demands, and practices of geospatial information users over the next decade.

Clearly, Anderson's second two criteria—dealing with interconnection and easy third-party access—are now being met on a daily basis. But other huge issues remain.

The issue of standardization, for example, is still largely being addressed through a series of ad hoc policies, one-on-one agreements, and proprietary product development or data integration strategies. There are significant efforts underway to overcome the standards' bottleneck—note, for example, the work of Open GIS Incorporated—but these are at a relatively early stage of development. There has also been very little real-world experience to date concerning the challenges of actually building and sustaining GDI. This should begin to change quickly over the next few years as jurisdictions like New Brunswick begin to provide 'best practice' information.

16.3 Lessons learned

While New Brunswick is a small, rural province and its experience will not be readily replicated elsewhere, it does provide some important lessons (see Chapter 15.1). For example, much of its success to date can be attributed to the following factors:

- Importance of a long-term strategic vision and high-level political support. The province had a vigorous agenda for more than two decades related to building and linking its geospatial databases and providing effective public access to them. This agenda has been strongly supported politically, especially by a recent Premier who gave the highest priority to enhancing the province's information infrastructure.
- Importance of a lead agency. Since the late 1980s New Brunswick has had a lead agency responsible for (1) designing and implementing the GDI concept; (2) co-ordinating the development of standards and protocols; (3) building and sustaining core data sets; and (4) providing online public access.

- Focus on key priorities. The province has concentrated on building key data sets of particular importance to the economic and social development of the province. This has included parcel-based data sets in support of reforming the province's land administration systems and selected geospatial data sets required for effective resource management (especially in support of integrated forest management practices).
- Importance of a business focus. The development and maintenance of the geospatial data infrastructure in New Brunswick is driven by a multi-year business plan. The lead agency, Service New Brunswick, is required to be self-sufficient. This has contributed to the emphasis on well-documented business cases for data and networking priorities, and on the funding strategies for ongoing upgrading and maintenance of the infrastructure.

North Carolina also has a long history in building the geospatial data infrastructure concept. As Karen Siderelis notes in Chapter 15.2, its success can be attributed at least in part to the commitment of people at all levels in the state who exploited opportunities to create an information infrastructure; an emphasis on using the infrastructure for addressing critical issues facing the state; enthusiastic focus on multi-lateral collaboration; and the spin-off benefits of being located in a technologically progressive state where political support for technology initiatives is high.

The Australian census mapping project provides an important example of the challenges involved in co-ordinating the efforts of a number of public agencies in building the NGDI components (see Chapter 15.3). While agreement on an NGDI architecture and other technical considerations was a necessary prerequisite, the most significant issues related to leadership (in this case provided by the State of New South Wales), sound management practices, and a source of sustained funding. The project has also demonstrated the benefits of a co-ordinated NGDI approach, as evidenced by the numerous value-added resellers seeking to developing applications on top of the core data infrastructure.

Finally, the Dutch clearinghouse experience (see Chapter 15.4) reinforces both the need for effective communication and a bottom–up strategy for advancing a GDI initiative. It also identifies three distinct steps in the development and provides some cautionary notes.

16.4 Concluding remarks

The essence of the GDI concept is that there is no master architect. There cannot be, nor will there be, a single organization responsible for designing and implementing some kind of GDI blueprint, especially at the national level. Instead, we can imagine an almost organic web of partnerships and relationships evolving purposefully within a given jurisdiction. It will sometimes be pushed by technology, sometimes pulled by market requirements. But at some point there will be a sufficient inter-connectedness of databases, a level of access to the data and use of the data, as well as the maturing interest of stakeholders, to participate and invest in the partnerships required for a nascent NGDI to be recognized. Having said this, the evolution of any GDI concept will most likely emerge from a combination of 'top–down' and 'bottom–up' strategies, the specific mix of which will vary significantly from one jurisdiction to another.

The top–down approach will entail defining strategic goals, assessing priorities, developing implementation plans, and obtaining core funding. It will require some kind of institutional framework (lead agencies, steering committees, working groups for standards, etc., and monitoring arrangements). Examples of outputs which may be expected from the top–down approach are defining the fundamental geospatial data sets, building the clearinghouses, establishing metadata standards and access protocols, and resolving the information policy issues. Core funding for geospatial database development and networking, as well as for subsequent maintenance and upgrading, will also be key to defining the initial directions and priorities. This funding will invariably result from a combination of direct public investment and revenues derived from the sale of information products and services (as described in detail by David Rhind in Chapter 4). While the debate continues over whether one should charge for public information, the reality is that some combination of direct investment and charging for products and services will be required.

The bottom–up approach will recognize the multiple local initiatives to build application-specific and enterprise-wide geospatial databases and will encourage their evolution towards a universal framework for accessing, combining, and using the data through a process of proselytizing, providing incentives (especially shared financing), and imposing regulations (e.g. through standards regimes). Money talks and access to shared funding will be of special significance in advancing the GDI agenda; even more important will be the concrete evidence of the power and potential of an integrated infrastructure.

Bibliography

ANDERSON, N. M. (1990). 'ICOIN infrastructure', *Proceedings of a Forum on the Inland Waters, Coastal and Ocean Information Network (ICOIN)*, The Champlain Institute, Fredericton, Canada, pp. 4–17.

BRANSCOMB, A. (1982) 'Beyond Deregulation: Designing the Information Infrastructure', *The Information Society Journal*, 1/3: pp. 167–90.

COLEMAN, D. J. and MCLAUGHLIN, J. (1997). 'Defining Global Geospatial Data Infrastructure Components, Stakeholders and Interfaces', *Geomatica*, 52/2: pp. 129–43.

EUROGI (1996). 'Towards a European Policy Framework for Geographic Information: a working document'. Working Group report prepared on behalf of the European Organization for Geographic Information.

HEARLE, F. R.(1962). 'A Data Processing System for State and Local Goverment', *Public Administration Review*, 22/3: September, pp. 146–52.

Mapping Sciences Committee (1993). *Towards a Coordinated Spatial Data Infrastruc-*

ture for the Nation, National Research Council. Washington, DC: National Academy Press.

McLaughlin, J. D. (1991). 'Towards National Spatial Data Infrastructure', *Proceedings of the 1991 Canadian Conference on GIS, Ottawa, Canada.* Canadian Institute of Geomatics, Ottawa, Canada, March, pp. 1–5.

Onsrud, H. J. and Rushton, G. (1995). *Sharing Geographic Information.* New Brunswick, New Jersey: Center for Urban Policy Research, Rutgers University Press.

Palmer, D. W. (1984). 'A land information network for new Brunswick'. Unpublished M.Sc.E. thesis, University of New Brunswick, Fredericton, Canada.

Rhind, D. (1992) 'The information infrastructure of GIS', *Proceedings of the Fifth International Symposium on Spatial Data Handling*, 1. University of South Carolina, Columbia, South Carolina, pp. 1–19.

Index

academic sector, expenditure 42
access:
  right to, government owned data 36
  and use of GDI 217–30
Active Control System 185
addressing, Internet layer 125–6
Adler, R. 71
aerial photography 198–202
aggregation hierarchies 165–7
Ahner, A. 57
Alexandria Digital Library 222
American National Standards Institute (ANSI) 61, 74
  database architecture 129
  search inter-operability 132
analogue maps 203
analogue stereo-plotters 204
analytical plotters 204
Anders, K. H. 161
Anderson, Neil 1, 271
Annito, R. 67
application development standards 65–6, 73–4, 79
application layer 126–7
architectures 135–49
  database 129–30
  object management 64, 70
  systems 113–15
Arseneau, B. 247
Asia, global GDI 52
Association for Computing Machinery (ACM) 61
associations *see* object associations
Asynchronous Transfer Mode (ATM) 118
attribute accuracy, quality management 90
attribute data, standards 72–3
attributes, thematic classes 162–4
Australia, Public Sector Mapping Agencies (PSMA) 255–62, 273
Australian Surveying and Land Information Group (AUSLIG) 260
azimuthal projection 181

Bamberger, W. 59
Baranowski, K. 52
Berry, M. J. A. 223
Bessel's ellipsoid 177
Bijker, W. E. 98
Bishr, Yaser 170
  GDI architectures 135–49
Brandenberger, A. J. 52
Branscomb, Anne 1, 245, 271
Bregt, Arnold, Dutch clearinghouse for geo-information 264–8
Brenner, J. 137, 277
British Geological Survey 45
broadcast networks 118
Brodie, M. L. 161
Brown, M. D. 218
Brunt, R. 137
Burrough, P. A. 152

cadastral photogrammetry 212–13
cadastre 270
Cahan, Bruce, funding of geospatial data infrastructures 20–1
Campbell, H. 98, 99
Canada:
  Hydrographic Service 86
  national geospatial data infrastructure 2
  New Brunswick 245–51, 272–3
  and semantic accuracy 89

Canadian General Standards Board (CGSB) 60
Canadian Inter-Agency Committee on Geomatics 61
Cargill, C. F. 60
cartography 217–19
case studies:
 Australia: Public Sector Mapping Agencies (PSMA) 255–62, 273
 Dutch clearinghouse for geo-information 262–7, 273
 New Brunswick: modernization of land information service 245–51, 272–3
 North Carolina: geospatial data infrastructure 251–5, 273
Charged coupled device (CCD) 201
Chrisman, N. R. 152
Clarke's ellipsoid 177
classification schemes, standards 75
clearinghouses 131–3
 GDI architecture 142–8
client-server architecture 114–15, 138–9
coaxial cable 116
Coleman, David J. 71, 72, 245, 246, 247, 250, 271
 human resources 233–43
College diploma courses 239–40
Collins, M. 52
Cologne:
 buildings 209
 roof structures and vegetation 210
Comité Européen de Normalisation (CEN) 146
Comité Européen des Responsables de la Cartographie Officielle (CERCO) 45, 76
 QMS guidelines 86
commercial development:
 geospatial data infrastructure 19–22
 geospatial information products and services 14
commercial sector, expenditures 42
commercialization, public sector information 26–31
Commission of Economic Communities (CEC) 2
communication and network management standards 63–4, 66–8
community data warehouses 21
completeness, quality management 88–9
computer networks 115–28
confidentiality 25
conformal projection 178, 179
conical projection 181
consistency, quality management 88
consumers, and geospatial data infrastructure 18
Coopers and Lybrand, benefits of NGDI 52
copyright, and protection of investments 31, 32–3
costs:
 photogrammetry and remote sensing 213–15
 staff 43–4
 see also expenditure; funding
courses 237–42
Cox, B. J. 161
Craig, W. 71
Croswell, Peter L.:
 standards 57–80
Crystal Eyes 204
culture, factors concerning acceptance of technologies 97–110

Daedalus 200
Dale, P. F. 246
Dangermond, J. 196
Daniel, L. 71
Dassonville, L. 86
data co-ordination, technology 17
data coding, standards 75
data compilation and update, standards 80
data definition language (DDL) 130
data format standards 65, 80
 exchange and access 71–3
data layers 20
data manipulation language (DML) 130
data mining 223
data sharing, framework 135–7
database, definition 32
database architecture 129–30
database management system (DBMS) 128–9
 and technological advances 235, 236
database schemas, standards 74

database technology 128–31
Dataquest 235
datum 176–81
combining of sets 186–93
realization of 181–6
Dawe, P. 247
DECCA, positioning system 182
DeFanti, T. A. 218
Deignan, F. 137
Dent, B. D. 220
Didier, M. 21
Differential GPS 184
Digital Cartographic Model (DCM) 226
Digital Chart of the World (DCW) 221
digital elevation models (DEM), photogrammetry 207–11
Digital Geographic Exchange Standard (DIGEST) 72
digital image correlation 207–8
Digital Landscape Model (DLM) 226
digital maps 203
digital stereo workstations 204
distributed databases 131
distributed systems 137–8
Doig, J. F. 245
Domain Name Service (DNS) 126
Dood, Ruben, spatial referencing 175–94
Doucette, Mark, quality management 85–94
Doyle, A. 73
Dueker, K. 59
dynamic positioning 185

Earth, mathematical model 176–81
economics:
human resources requirements 234–5
see also costs; expenditure
edge matching 191
Egenhofer, M. J. 160, 161
electro-mechanical scanner 200
electro-optical scanner 200, 201
Electronics Industry Association (EIA) 62
ellipsoid, mathematical model of the Earth 176–81
Encarta Virtual Globe 229
Environmental Research Institute of Michigan (ERIM) 200
equal-area projection 180
equidistant projection 178, 180
equipotential surface 177
Estes, J. 52
Etzioni, A. 98
Euref project 76
European Commission:
commercialization of public sector information 30, 31
extraction right 32
European Committee for Standardization (CEN) 61
European Court of Human Rights, freedom of information 28
European Geostationary Navigation Overlay Service (EGNOS) 194
European Metadata Standard 146
European Union:
access and competition 33
economic benefits 2
expenditures 45–6
freedom of information 28
national geospatial data infrastructure 2
protection of personal data 33–6
see also under individual countries
Exite Travel website 221
expenditure:
distribution of 41–6
see also costs; funding
exploratory cartography 218
Extensible Markup Language (XML) 128
extraction, right of 31, 32

Fegeas, R. 72
Federal Geographic Data Committee (FGDC) (USA) 61, 75, 131
metadata 146
fibre optics 116–17
field approach 152–3, 154–6
File Transfer Protocol (FTP) 126
Finley, David, New Brunswick land information service 245–51
Floriani, L.de 160
foundation technologies 113–34
fragmentation 126

framework data, for NGDI 9
France:
commercialization of public sector information 29–30
global GDI 52
Frank, A. U. 161
free flow, and protection of personal data 35–6
Freisen, P. 71
Fritsch, D. 161
functionality, cultural factors 102–3, 104–6, 107
funding:
different models 48–51
and distribution of expenditure 41–6
geospatial data infrastructures 20–1
who should pay 46–8
see also costs; expenditure

Geobot 222
geodetic datum 175
Geodesey 176
Geographic Information Corporation (GIC) 246, 248, 249
Geographic Information Systems (GIS) 3, 14, 23
and culture 102–3
future 19
Geographic Information/Geomatics Committee 61
geographic phenomena, geospatial data modelling 152–4
geoid 177, 178
GEOID Network 235
Geolink 222
Geological Survey (USA) 77
Geomatics Canada 77, 86
Georgiadou, Y. 185
geospatial data:
accessing and use 217–30
definition 3
geospatial data exchange 71–2
geospatial data infrastructure (GDI) 3–5
first initiative on 2
Geospatial Data Service Centre (GDSC)
certification 86
definition 4–5
GDI architectures 148–9
and integrity of GDI 87
geospatial databases, protection of investments 31–3
geospatial descriptions, conceptual tools 151–71
geospatial information policy 15
geospatial information systems, photogrammetry and remote sensing 196–8
geostationary orbits 202
Germany:
map production 203–4
protection of personal data 33–4
Ghosh, S. K. 52
global geospatial data infrastructure (GGDI) 6, 52–3
GLObal NAvigation Satellite System (GLONASS) 185, 193–4
Global Positioning System (GPS) 92, 182–6, 193–4, 235
global server, GDI architecture 141
Glover, J. 73
Goodchild, M. F. 152
Gordon, L. 223
Grant, Don, Australia: Public Sector Mapping Agencies (PSMA) 255–62
graphical user interface (GUI) 70–1, 143–4, 145–6
Grelot, J.-P. 40, 48, 52
Groot, Richard 1–11, 149, 228
development of GDI concept 269–74
human resources 233–43
Grover, V. 101

Hanover, image of 205
hardware:
standards 62, 63, 66, 79
see also network hardware
Hayford ellipsoid 177
Hearle, F. R. 270
Hearnshaw, H. M. 218
heights, combining data sets 192–3
Helava, U. V.:
analytical plotters 204
Herring, J. R. 160
hierarchies:
aggregation 165–7
and object classes 161–5
high-level standards 62

Hofman, Marco:
spatial referencing 175–94
Hofstede, G. 97, 98
4D model of cultural differences 99–101
cultural indicators 99
social acceptance of technologies 103, 104
Htun, N. 52
Hughes, J. G. 161
Hydrographic Service (Canada) 86
Hypertext Markup Language (HTML) 128
Hypertext Transfer Protocol (HTTP) 127

image matching 207–8
image motion compensation 199
indexes, maps as 220–1
individualism versus collectivism (IDV), cultural differences 100, 102
information:
freedom of and commercialization 28–9
freedom of and protection of investments 32–3
Information and Communication Technology (ICT) 270
information highways 1
information infrastructure 271
information market place 1–2
information resources management 270
infrastructure 4
and geospatial information 20
Institute of Electrical and Electronic Engineers (IEEE) 61, 62
integrated mapping 270
Integrated Services Digital Network (ISDN) 118
integrity, quality management 87–90
inter-operability 137–9
inter-operability software interfaces 17
intermediaries, liability 25–6, 36
International Association of Assessing Officers (IAAO) 61
International Cartographic Association 72
International Geographical Congress (Fifth) 6 n.
international legal instruments 26
International Organization for Standardization (ISO) 20, 61
certification 85
Open Systems Interconnect 67, 122
and quality management 93–4
search inter-operability 132
SQL/MM Spatial 74
and TCP/IP reference model 124–6
International Telecommunications Union (ITU) 61
Internet Explorer 127
Internet layer 125–6
Internet Protocol (IP) 124–7
internetworks 120–1
investments:
geospatial data 4
protection of and freedom of information 32–3
protection of and geospatial databases 31–3
protection of 25

Janssen, L. L. F. 167
Java 128
Jiwani, Z. 71

Kabel, Jan 97
legal issues 25–37
Kainz, Wolfgang, foundation technologies 113–34
Kemp, Z. 161
Kindrachuk, E. 71
KINDS project 220–1
Kline, K. 52
Konecny, Gottfried, photogrammetry and remote sensing 195–216
Kottman, Dr Clifford A. 71, 72
market for geospatial data infrastructure 18–19
Kraak, Menno J., access to GDI and function of visualization tools 217–30

Labonne, B. 52
Landsat Thematic Mapper 200
landscape architect, as user of geospatial data infrastructure 16–17
languages, application standards 73–4

Laurini, R. 152
Law, J. 98
layer services, network software 121–2
Lazar, B. 72
Lee, G. 206
Lee, Y. C. 71
legal issues 25–37
  Guerra *v.*
  Italy, freedom of information 28
Leica scanner 206
levelling 182, 183
Levinsohn , A. 59
Li, R. 216
liability, intermediaries 25–6, 36
Libicki, M. C., standards 59–60, 61
lineage, quality management 88
local area networks (LANs) 118, 119
Local Government, expenditures 42
local server, GDI architecture 141, 143–4
location-awareness devices 13–14
Longley, P. 218
Loran positioning system 182
Loukes, D. K. 250
low-level standards 62
Lummaux, J. C. 22–3

McCormick, B. 218
McDonnell, R. A. 152
MacEachren, A. M. 218, 220
McKee, Lance 183, 235
  technology push and market pull 13–23
McLaughlin, John D. 1–11, 72, 245, 246, 250
  development of GDI concept 269–74
  human resources 233–43
MADTRAN 186
Majid, D. A. 52
Man, Erik de, cultural factors 97–110
management, human resources and technical changes 236–7
map digitizing, and photogrammetry 202–6
Map Library of Penn State University 221
map projections 178–81
Mapping Sciences Committee 1, 236, 270
maps:
  access to and visualization tools 217–30
  compilation and presentation standards 757
  digitization 202–6
market, geospatial data infrastructure 18–19
market pull, and technology push 13–23
masculinity *v.* femininity (MAS):
  cultural differences 100, 101–2
  functionality of GIS 105, 107
Masser, I. 41, 98
  technology diffusion 22
Mecator projection 178
Melody, W. H. 245
Messter, O., aerial survey camera 198
metadata:
  and GDI architectures 146–8
  standards 17, 75, 76
Meteorological Office (UK) 45
Microsoft Press 67
microwave remote sensing 201
Military Survey (UK) 45
Min, Erik de, spatial referencing 175–94
Mintzberg, H. 101
Molenaar, Martien, conceptual tools 151–71
Morrison, J. L. 218
Moyer, D. 57
multi-level server, GDI architecture 141
Multi-Satellite-based Augmentation Service (MSAS) 194

Nandlall, N. 250
Nanson, B. 46, 52
National Atlas of Canada 229
National Geospatial Data Clearinghouse 227
National Geospatial Data Framework 46
national geospatial data infrastructures (NGDI) 2
  definition 39–40
National Imagery and Mapping Agency (NIMA) 52
National Institute of Standards and Technology (NIST) 60

Netherlands:
clearinghouse for geo-information 264–8, 273
commercialization of public sector information 30–1
NGDI 41
Netscape Communicator 127
Network File System (NFS) 126
network hardware 118–21
network management, standards 63–4, 66–8, 69, 79
Network News Transfer Protocol (NNTP) 126–7
network software 121–2
networks 115–28
New Brunswick, modernization of land information service 245–51, 272–3
Newman, I. 75
Nichols, S. E. 245, 247
Niemann, B. J. Jr 57
non-governmental organizations, expenditures 42
North American Datum (NAD) 76, 177
North American Vertical Datum (NAVD) 76
North Atlantic Treaty Organization (NATO) 4
North Carolina, geospatial data infrastructure 251–5, 273
Nyerges, T. L. 161

Obermeyer, N. J. 98
object associations 167, 168
object dynamics 167–9
object hierarchies 160–7
object management architectures 64, 70
Object Management Group (OMG) 61
object representation 156–60
object-structured approach 153
Oddens Bookmarks 222
Ogilvie, M. 246, 247
on-screen monoplotting 204
Onsrud, H. J. 9, 75, 236, 271
Open Database Connectivity (ODBC) 60, 73
Open GIS Consortium™ (OGC) 7, 61, 73, 149, 272
Open GIS Specification 21
Open Group 61
Open Systems Interconnection (OSI) 67–8, 122–4
and TCP/IP reference model 124–6, 127
operating systems, standards 64–5, 68–70
Ordnance Survey (OS) (UK) 77, 227
funding 45, 46, 47–8, 50
Ormeling, F. J. 220
Ostensen, O. 61

Palmer, D. W. 236, 246, 271
Paresi, Chris:
quality management 85–94
Patterson, B. L. 67
Patton, B. 245
Peled, A. 71
Penck, Prof A.:
world map 6n.
personal data:
protection of 33–6
secondary use 35
Petersohn, C. 238
Peuquet, D. J. 152
photogrammetry:
and map digitizing 202–6
and remote sensing 195–216
physical connection standards 62, 63, 66
physical infrastructure, cost 43
Pinto, J. K. 98
platforms, and sensors 198–202
Plewe, B. 223, 247
point-to-point networks 118
Point-to-Point Protocol (PPP) 124
positional accuracy, quality management 90
power, cultural differences 101
power distance (PD):
cultural differences 100, 101
functionality of GIS 105, 107
Precise Positioning Service (PPS) 184
Price Waterhouse, benefits of NGDI 51–2
privacy 25, 26
protection of personal data 33–6

private sector:
  businesses and GDI 22
  expenditure 42
  funding of NGDI 46, 48–50
  and global GDI 52–3
product presentation, standards 80
programming standards 65–6, 73–4
public policy, geospatial data infrastructure 22–3
public sector:
  central government expenditure 42
  commercialization of information 26–31
  funding 46, 48–51
  legal issues 25
Public Sector Mapping Agencies (PSMA), Australia 255–62

qualifications, courses 237–42
quality management 85–94
Quality Management System (QMS) 85–6

radio positioning systems 182
Radwan, Mostafa, GDI architectures 135–49
raster data, photogrammetry 206–7
raster scanner, map digitizing 202
raster structure 154–6
  object representation 157–8, 159
raw data, capture and maintenance: cost 43
regional geospatial data infrastructure 6
remote sensing, and photogrammetry 195–216
retail, market for geospatial data infrastructure 19
Rhind, David W. 3, 97, 259, 270
  funding 39–54
Ridjanovic, D. 161
Roberts, W. 246
Robinson, A. H. 220
Rogers, E. M. 98
RS1, Rastermaster 206
Rushton, G. 75, 236, 271

Sandgren, U. 48
satellites, and imaging sensors 201–2
SCAI scanner 206
scanners 200–1
sea level 177
search engines 222–3
Seeber, G. 185
Selective Availability (SA) 184
semantic accuracy, quality management 89
sensors, and platforms 198–202
SeQueL (SQL) 73–4
Servicio Autonomo de Geografia y Cartografia Nacional 269
Shore, B. 106, 108
Siderelis, Karen, North Carolina 251–5
Simple Mail Transfer Protocol (SMTP) 126
Simpson, R. L. 246, 247
Slone, J. 67
Smith, D. C. P. 161
Smith Gunther Associates 235
Smith, J. M. 161
Smith, M. 52
Smith, N. S. 3, 40
social acceptability:
  GIS 98–9
  of technologies 103
society, effects of geospatial data infrastructure on 22–3
software, *see* application development; network software
Sondheim, M. 71
Spatial Data Transfer Standard (ASRA) 72
spatial referencing 175–94
SQL/MM Spatial 74
staff costs 43–4
stand-alone computers 114
Standard Positioning Service (SPS) 184
standards 57–60
  adoption and implementation 77–8
  information services 21
  organizations 60–1
  outline for manual 78–80
  and quality management 91
  types 62
  types of 62–77
standards organizations 60–1
Structured Query Language (SQL) 131
Strutz, S. 246
Synergy Guide Lines, access rights 33

synthetic aperture radar (SAR) 201
interferometry 208
system access, standards 80
system administration, standards 79
system architecture 113–15
Système probatoire de l'observation de la terre (SPOT) 201

tacheometry 182
Tapscott, D. 234
Taylor, D. R. F. 218
technology:
advances 235–6
influence of culture 101–2
technology diffusion 22
technology push, and market pull 13–23
TELNET 126
temporal accuracy, quality management 89
terrain feature-oriented approach 153
terrain objects 158
geometry of 156–60
thematic classes 162–5
thematic data, remote sensing 211–12
Thomas, E. 52
Thompson, D. 152
Thorpe, J. 206
Tiger map files 222
timeouts 126
Tissot's indicatrices 178
Tom, H. 58
Tomlin, C. D. 155
Toorn, Willem van den, cultural factors 97–110
Tosta, N. 40, 54
Total Quality 85
Total Station 182
transmission:
media 116–18
technology 118–21
Transmission Control Protocol (TCP) 124–7
transport layer 126
transversal cylindrical projection 181
Trudeau, M. 73
twisted pair 116

uncertainty avoidance (UA):
cultural differences 100, 101
functionality of GIS 105, 107
unfair competition:
commercialization of public sector information 30–1
and protection of investments 31
Uniform Resource Locator (URL) 128
United Kingdom:
expenditures 45–6
funding of NGDI 46–7
global GDI 52
United Nations:
and global GDI 52
surveys of world's maps 202–3, 206
United States of America (USA):
commercialization of public sector information 29
definition of NGDI 40
distribution of expenditures 41–5
funding 46–8
global GDI 52
investment 1
national geospatial data infrastructure 2
North Carolina: geospatial data infrastructure 251–5, 273
United States Geological Survey (USGS) 44, 47–8
Universal Transverse Mercator (UTM) projection 180
university courses 237–42
UNIX systems 68
Unshielded Twisted Pair (UTP) 116
Unwin, D. J. 218
Urban and Regional Information Systems Association (URISA), standards 58, 60, 61
User Datagram Protocol (UDP) 126
user design standards 65–6, 74–7
user interface, standards 64–5, 70–1

vector data, map digitizing and photogrammetry 202–6
vector structure, object representation 158–60
Venezuela 269
Venkatachalam, A. R. 106, 108
Ventura, S. 58
vertical highways 1

visualization tools 217–30
Vrana, R. 59

Weber, Max 101:
websites:
http://cgdi.gc.ca/frames.html (National Atlas of Canada) 229
http://fgdc.er.usgs.gov/framework/frameworkintroguide/ 40
http://gdc.er.usgs.gov 52
http://kartoserver.frw.ruu.nl/html/staff/oddens/mapsat14.htm (map projections and transformation software) 180
http://terraserver.microsoft.com (TerraServer) 48
http://www2.nas.edu/besr/223a.html 22
http://www.eurogi.org/gsdi 2
http://www.gov.state.nc.us/gGDI97 52
http://www.ngdf.org.uk 52
http://www.ngdf.org.uk/whitepapers/mass7.98htm 41
http://www.nima.mil/geospatial/geospatial.html ('Geodsey for the Layman') 176
http://www.nima.mil.geospatial/products/GandG/historic/hdatums.html 177
http://www.nima.mil.geospatial/products/GandG/madtran/madtran.html (MADTRAN) 186
http://www.ntis.gov/fcpc/cpn5198.htm (MADTRAN) 186
http://www.permcom.apgis.gov.au/index.html 52
http://www.w3.org (World Wide Web Consortium) 127
Wegener, M., technology diffusion 22
Wellar, B. 57, 58, 71
Wide Area Augmentation Service (WAAS) 194
Wide Area Networks (WANs) 118, 119–20
wireless communication, expansion of networks 13–14
wireless transmission 117–18
with-fault liability, intermediaries 36
World Wide Web 120–1, 127–8, 219

Zajac, G. 73
Zwart, P. R. 98

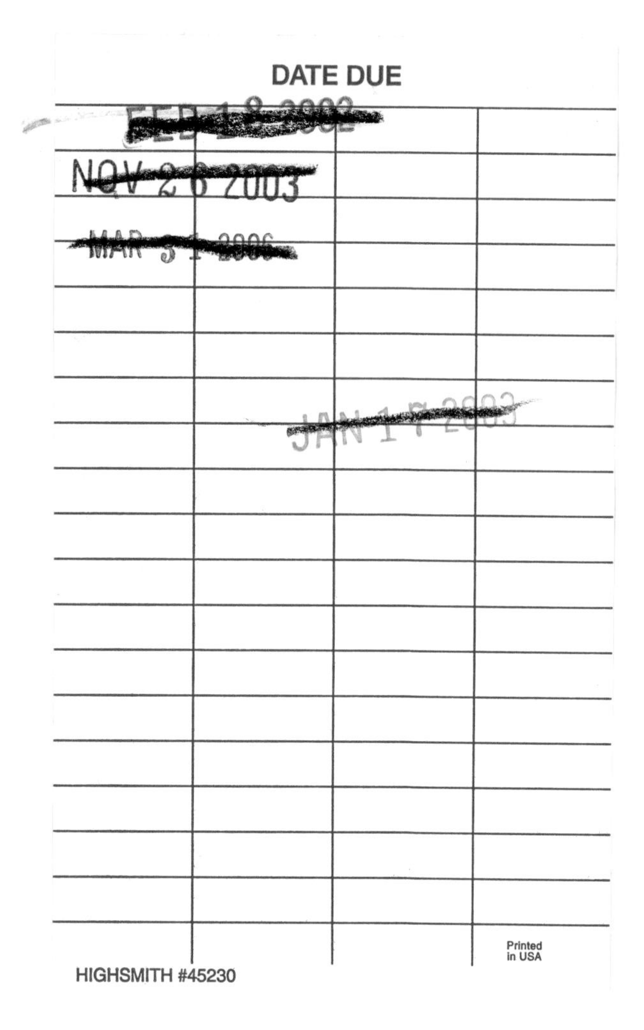
DATE DUE
HIGHSMITH #45230
Printed in USA